LEAD IN THE M
ENVIRONMEN

Proceedings of the International Experts Discussion on Lead Occurrence, Fate and Pollution in the Marine Environment, Rovinj, Yugoslavia, 18-22 October 1977

Edited by

MARKO BRANICA

and

ZDENKA KONRAD

Center for Marine Research, Ruđer Bošković Institute, Zagreb, Yugoslavia

Organizing Committee

M. BRANICA
B. ĆOSOVIĆ
M. JURKOVIĆ
Z. KONRAD
D. KRZNARIĆ
M. KRAJNOVIĆ-OZRETIĆ
D. MARTINČIĆ
B. OZRETIĆ

PERGAMON PRESS

OXFORD · NEW YORK · TORONTO · SYDNEY · PARIS · FRANKFURT

U.K.	Pergamon Press Ltd., Headington Hill Hall, Oxford OX3 0BW, England
U.S.A.	Pergamon Press Inc., Maxwell House, Fairview Park, Elmsford, New York 10523, U.S.A.
CANADA	Pergamon of Canada, Suite 104, 150 Consumers Road, Willowdale, Ontario M2J 1P9, Canada
AUSTRALIA	Pergamon Press (Aust.) Pty. Ltd., P.O. Box 544, Potts Point, N.S.W. 2011, Australia
FRANCE	Pergamon Press SARL, 24 rue des Ecoles, 75240 Paris, Cedex 05, France
FEDERAL REPUBLIC OF GERMANY	Pergamon Press GmbH, 6242 Kronberg-Taunus, Hammerweg 6, Federal Republic of Germany

First edition 1980

British Library Cataloguing in Publication Data

International Experts Discussion on Lead Occurrence, Fate and Pollution in the Marine Environment, *Rovinj, 1977*
Lead in the marine environment.
1. Marine pollution - Congresses
2. Lead - Environmental aspects - Congresses
I. Title II. Branica, Marko
III. Konrad, Zdenka
628.1'686'162 GC1083 80-40023

ISBN 0-08-022960-3

In order to make this volume available as economically and as rapidly as possible the authors' typescripts have been reproduced in their original forms. This method has its typographical limitations but it is hoped that they in no way distract the reader.

Printed in Great Britain by A. Wheaton & Co., Ltd., Exeter

Contents

Introduction

The general and the most essential problem in marine pollution is to recognize the natural biogeochemical cycle of the element and the real anthropogenic influence on it. In the case of trace heavy metals, very often a mistake is made in considering all heavy metals as belonging to a single group. Meanwhile, reactivity, the oxido-reduction and ionic forms in equilibria, i.e., chemical characteristics of each single metal, vary so much that only few of them can be treated as having similar properties. To one of such groups belong those metals that form stable organo-metallic compounds, the fate of which in the marine environment is much more similar to the fate of organic molecules than of other metallic forms. Precisely because organo-metallic compounds concentrate in organisms, their toxicity is multiplied by a certain order of magnitude. Although mercury in the sea is found in very low concentrations in its inorganic form, its organic compounds have very high concentrations in marine organisms. Mercury is the most typical representative of metals that, in nature, by a relatively fast process, change from inorganic into organo-metallic forms.

In the case of lead we are confronted with the opposite process - the release of lead by man in the environment is occurring partially in the form of organo-metals. Man is surrounded by relatively large amounts of lead in many of its forms, so that even objective determinations of actual lead concentrations in the natural-unpolluted sea have until recently presented insurpassable difficulties. As a consequence, there are very few laboratories in the world capable of carrying out reliable lead analyses in the marine environment. Presently we have found ourselves facing several unanswered questions, such as:

a) the transformation rate of organic lead into its inorganic form;
b) the existence of natural processes by which lead changes from its inorganic form into the organic one, and under what conditions;
c) the rate of incorporation of lead into the organisms, etc.

Lead occurs as divalent and tetravalent ion in nature, but Pb^{2+} predominates over Pb^{4+}. Most lead salts of Pb^{2+} with naturally common anions are only slightly soluble. Lead is a heavy metal which occurs in nature. It becomes of real concern as a contaminant because of its release into the air, water and soil in higher amounts. Industry and other technological uses of lead are the main sources of environmental pollution by man. Lead has no beneficial or nutritional effects. It is a toxic metal which tends to accumulate in the tissues of man and animals. In connection with the International Experts Discussion, of special importance is the ecological effect of lead in the marine environment.

The ability of many marine organisms to accumulate lead presents a potential hazard to humans through ingestion of contaminated marine produce. The toxic effects of lead on humans have been well documented, as lead has been used by man for thousands of years.

Lead enters the aquatic environment through precipitation, lead dust fallout, erosion and leaching of soil, municipal and industrial waste discharges and the runoff of fallout deposits from streets and surfaces. The toxicity of lead in water is affected by pH, hardness, organic material and the presence of other metals. Due to the poor solubility of most lead compounds in water, much of the lead entering marine systems via precipitation and runoff will be removed by sedimentation. Lead forms stable complexes with organic matter as well as with clay particles. Desorption of lead from sediments is facilitated by increased salinity.

In order to correlate the toxicity of dissolved lead to various marine organisms it is important to know its speciation and the possibility of bioconcentration, bioaccumulation and biomagnification in these species. Distinction of physico-chemical forms of lead in marine environment is a necessity when predicting its ecological effects.

The concentration of lead is altered in coastal waters as a result of the use of lead alkyls in the fuels for internal-combustion engines. Lead is classified as a metal, forming alkyls which are stable in aqueous systems and are reported to be biomethylated. Therefore, the shipwreck of the "Cavtat", which was loaded with many tones of tetraethyl lead and by an accident sunk in the South Adriatic, brought an urgent and important problem requiring advice and solution to the scientific community. Due to the fact that the residence time of lead is estimated to be 400 years (a rather low value), the interactions, involvement in biological reactions and the uptake by marine organisms could play an important role in the biochemical cycle of lead.

The entry of lead aerosols causes the concentrations of lead in surface seawaters to be always higher in comparison with the deep waters. The Mediterranean coastal surface waters are affected up to a depth of about 500 meters. It has to be mentioned that up to now the reported concentrations of lead might have been too high owing to contamination during sampling, sample preparation and analysis.

All these are closely related to the problem of methodology and techniques concerning the objective determination of the states and processes of various forms of lead in the marine environment. This is why we have tried during the meeting held on 18-20 October, 1977, at Rovinj, Yugoslavia, by an _ad hoc_ discussion of international experts to have in one place scientists whose views on the problems of marine environment pollution will give an overall picture of the present knowledge in this field.

The meeting has been organized in order to enable scientists from different disciplines to discuss topics in this particular field at length and detail. In contrast to many symposia and meetings, where the limitation of a rigid schedule reduces the lectures to little more than progress reports and "state of the art" statements, the organization of this workshop gave the lecturer an opportunity for ducumented and referenced presentation of the topic of his specialization.

The main topics of the International Experts Discussion on "Lead - Occurrence, Fate and Pollution in the Marine Environment" were:

(i) Physico-chemical state and processes of lead in the marine environment (ionic forms, stability and kinetics of "inorganic" and "organic" lead);

(ii) Bioaccumulation and toxicity (lethal and sublethal effects);

(iii) Biogeochemical cycle of lead in natural and polluted environment;

(iv) Fate of tetra and tri-alkyl lead with regard to different sources (rivers, atmospheric input and accidents in transport) in the marine environment.

Special attention has been paid to the problems concerning anthropogenic influence of the marine environment taking into account methodology and experimental techniques used in laboratory experiments and in field observations.

My appreciation is expressed to the scientists from many countries for their interest and participation in the program so that this meeting can be international in scope as well as interdisciplinary in nature.

It is hoped that the 26 papers presented will be useful to all scientists concerned with marine pollution and that this meeting has provided a forum for meaningful interdisciplinary discussions that will stimulate research towards elucidation and solving the complex problems concerned with lead in our environment and will eventually result in the improvement of the quality of our environment.

Marko Branica
Editor

Quality Assurance Measures for Environmental Data

J. K. Taylor
National Bureau of Standards, Washington, D.C., U.S.A.

ABSTRACT

An understanding of the occurrence, fate, effects, and pollution potential of suspected toxic substances requires a wide variety and large quantity of analytical data, obtained over extended time intervals, in diverse locations, and by independent investigators. Unfortunately, much of the data reported in the literature is often difficult to interpret and correlate because of lack of information on the reliability of sampling and measurement. This paper discusses general principles of quality control of both measurement and sampling and recommends minimum requirements for reporting environmental data. The use of generally accepted reference materials to intercalibrate measurement laboratories is emphasized.

INTRODUCTION

An understanding of the occurrence, fate, effects, and pollution potential of a suspected toxic substance requires a wide variety and large quantity of analytical data. In the case of lead as an example, information is needed on sources, on its distribution in air, water, food, and biota, and the concentrations may vary from major constituent down to trace levels too low for measurement in pristine environmental samples. Such measurements must necessarily be made over long intervals of time, in diverse locations, and by a number of independent investigators. Accordingly, stringent requirements are placed upon its reliability if data are to be intercompared and correlated. The present paper presents guidelines which should be useful to those engaged in the design of data acquisition networks, and to laboratories making environmental measurements. It should also be useful in decisions as to whether published data may be accepted in compilations for environmental purposes.

GENERAL PRINCIPLES

Reliable analytical data require close attention to all steps of the measurement process. These include:

- Sampling
- Measurement methodology
- Calibration

- Quality control
- Quality assurance
- Data reduction
- Data reporting

Variability of data due to both measurement variance and heterogeneity of samples is to be expected. Furthermore, the differences between normal and pollution levels are typically small and may be masked by such variability. Accordingly, the advice and collaboration of a statistician experienced in measurement and sampling problems may be necessary to design an experimental plan that will provide interpretable data.

Sampling

The actual sample measured is a critical part of environmental measurement. It should be obtained according to a strategy (see Table 1) carefully developed by the environmental scientist in close cooperation with the analyst. Once developed, a protocol should be established and adhered to unless it is conclusively demonstrated to be defective. In such a case, a revised protocol may be required which may necessitate discarding data obtained under the initial protocol. Variation in sampling, including short cuts and omissions of samples, can lead to data gaps and complicate statistical analysis. Technical deviations such as improper cleaning, handling, and storage of containers may introduce both random and systematic errors into the data, or even make important samples useless (Ref. 1).

TABLE 1 Sample Strategy

- Locations sampled
- Time of sampling
- Frequency of sampling
- Type of sample
  - spot, integrated, continuous
- Sample equipment
- Sample container
- Sample pre-treatment
- Sample storage
- Retention
- Documentation

The procedure for pre-treatment of the sample prior to analysis is a major consideration. Whether the total sample is to be analyzed or some specific portion such as the particulate fraction, the dissolved portion, or that chemically or physically bound needs to be decided: The analysis of sediments will depend upon whether total or "available" constituent needs to be measured. Tissue samples again present special problems.

The question of moisture often is perplexing. Should the data be reported "as is", on a "wet" basis, or "dry"? The neglect of such matters can lead to incompatibility of data and the incomplete specification of sample conditions can limit the usefulness of otherwise excellent data.

Measurement

Selection of the measurement methodology can be a complex decision. The essential requirements for a reliable method are:

Sensitivity. The sensitivity must be adequate to detect the substance at the concentration level of interest or to measure the desired characteristic above instrumental noise levels.

Specificity. The response must be specific for the substance of interest or the cross-sensitivity to other substances, likely to be present, shall be insignificant with respect to that being measured.

Precision. The measurement must be sufficiently repeatible over a short period of time to insure acceptable confidence limits for a reasonable number of replicates. The methodology must be sufficiently well defined to permit reproducibility by the same laboratory or others over an extended period of time.

In addition, the following characteristics are desirable: large dynamic range; ease of operation; low cost of equipment; low unit-cost for each measurement; low cost of maintenance; portability of equipment.

Some of the considerations in the choice of methodology for environmental analysis have been reviewed in an earlier publication (Ref. 2).

Reliable measurements result from trustworthy methodology, diligently applied and as free from error as possible. Measurement errors are of two kinds, random and systematic. Some examples of each are discussed below.

Random Error. Contamination of samples prior to or during the course of measurement can result from introduction of foreign substances, incomplete removal of a previous sample from the analytical system, or pick-up or loss of constituent from sampling lines.

Maladjustments of critical operational parameters are usually random in nature. Departures from established procedures, either accidental or intentional, can cause serious problems which may be unsuspected unless the deviations are documented.

Personal errors are often the chief cause of problems in all but the most automated measurements. Operator fatigue, boredom, carelessness, and lack of motivation are underlying causes of error. Experience in routine clinical laboratories (Ref. 3) has shown that a minor percentage of operators often account for the major percentage of "mistakes". Machines are less prone to such "mistakes" and the elimination of human error may be the most significant argument for automation.

Systematic Error. Systematic errors may be independent of measurement level, i.e., constant, or correlated with it, i.e., multiplicative. They can arise from improper physical calibration or standardization, i.e., chemical calibration. Physical calibrations may exhibit gradual changes with time, so that periodic checks may be needed to detect them. Chemical calibrations may show abrupt changes as a new supply of standard material is introduced into a system.

Blanks introduced by reagents, carrier gases, etc. ordinarily cause long term constant errors which can be detected and eliminated by the proper procedure. However, such corrections often involve subtleties so that improper blank corrections are a frequent source of analytical biases.

Quality Control

Quality control measures are those procedures, preventive in nature, which anticipate problems and seek to avoid or minimize them. Reagent control requires high quality materials from reliable sources with additional checks to confirm quality. Cleanliness of the laboratory environment may require state-of-the-art practices and even the use of glove-boxes or clean rooms for all or at least critical steps of a measurement process.

The maintenance of good records, both of samples and measurements is essential, especially if regulatory data is to be obtained. A chain of custody of samples may mean that they are kept under special custody and that witnesses may need to verify that a particular sample was indeed measured.

Eternal vigilance is the price that must be paid for good measurements. Periodic and aperiodic inspections and checks should be made. More use should be made of a logical interpretive procedure to screen data for credibility (Ref. 4). The reasonablness of a measurement is evaluated on the basis of expected limits and departures are verified before they are reported. Such a procedure can eliminate errors, detect redundancies, detect or validate anomalies, and improve both sampling and measurement techniques.

Quality control procedures should also be applied to sampling procedures. Sources of error in each step need to be anticipated and appropriate control measures introduced into the sampling protocol. Sample takers must be fully aware of the consequences of poor technique.

It is axiomatic that no laboratory should report data until it has demonstrated that all aspects of the measurement process are in a state of quality control. Every item of data reported should be accompanied by a statement of the limits of confidence claimed for it, and the evidence for this should be retained and be available for review as it may be necessary.

Quality Assessment

Quality assessment refers to those procedures used to evaluate the effectiveness of quality control procedures and the data output. Internal standards should be measured at intervals depending upon expected performance deviations. Large departures should be self evident while control charts can evaluate operational variance and detect departures from the norm before they become serious. Such internal procedures can attest to the consistency of the analytical measurements.

Compatibility of data with that from other sources can only be evaluated by external checks (Ref. 5). Participation in collaborative tests is a time-honored method of peer evaluation. Audit samples distributed from an independent source may be used to evaluate laboratory proficiency. Reference samples of unknown composition can show relative performance of laboratories by departures from consensus values but "true values" are needed to evaluate biases.

A number of Standard Reference Materials are now available from the National Bureau of Standards for the evaluation of environmental measurements. While not intended for routine use, they can be invaluable for intercalibration of stations in a measurement network, which are otherwise demonstrated to be in a state of quality control (Refs. 6 and 7).

Data Reporting

Data should be reported in sufficient detail, so that its significance can be evaluated and that it can be correlated with respect to other data taken by the same laboratory or elsewhere. This requires documentation of all aspects of the sample, including the number of samples examined and their accurate description both in time and space. The measurement methodology must also be documented with respect to: methodology used; calibration; quality control; intercalibration; precision; accuracy. Above all, each piece of data should include the confidence limits claimed by the measurement laboratory. Such confidence limits should include those related to measurement error and to sample variability.

CONCLUSION

When examined in the light of the guidelines presented in this paper, serious dificiencies will be found in much of the environmental data reported in the literature. As a result, much potentially useful data cannot be unequivocally interpretated or correlated with other data. By adequate planning and often with little extra effort, such situations can often be avoided.

Each step in the analysis, including sampling, measurement, and data reduction should be carried out according to planned procedures-protocols - so that any dificiencies or sources of error can be identified. Such identification would be difficult or impossible unless the sampling and measurement are done systematically. A well designed quality control program must be in operation to reduce the magnitude of both random and systematic errors. Furthermore, the laboratory should strive to insure that its measurements are intercalibrated with those of other laboratories making the same kind of measurements. The reporting of estimated confidence limits for the data and full experimental details will provide a basis for its independent evaluation and correlation that from other sources.

REFERENCES

1. E. J. Maienthal and D. A. Becker, A survey of current literature on sampling, sample handling, and long term storage for environmental samples, NBS Technical Note 929 (1976), National Bureau of Standards, Washington, D. C. 20234, U. S. A.

2. B. Greifer and J. K. Taylor, Pollutant Analysis Cost Survey, EPA-650/2-74-125 December 1974. Office of Research and Development, U. S. Environmental Protection Agency, Washington, D. C. 20460.

3. J. D. Lott, Laboratory personnel: the most important aspect of quality control, Medical Instrumentation 8, 22-25 (1974).

4. J. M. Rogers, I. D. Macaulay, B. Sundby, and D. E. Buckley, Data are for looking at, or quality control through interpretation, Water Quality Parameters, Special Publication STP 573, American Society for Testing and Materials, Philadelphia, Pa. 19103, U. S. A. (1975).

5. J. K. Taylor, The importance of intercalibration in marine analysis, Proceedings, IV Inter. Symp. on Chemistry of the Mediterranian, Thalassia Jugosl., 13 (1977) (in press).

6. NBS Special Publication 260, "Catalogue of Standard Reference Materials", National Bureau of Standards, Washington, D. C. 20234, U. S. A.

7. J. K. Taylor, The status of Standard Reference Materials for environmental analysis, in "Methods and Standards for Environmental Measurements", NBS Special Publication 464 (1977), National Bureau of Standards, Washington, D. C. 20234, U. S. A., 503-7.

Impact of Man on Coastal Marine Ecosystems

M. Burnett, A. Ng, D. Settle and C. C. Patterson
Division of Geological and Planetary Sciences, California Institute of Technology, Pasadena, California 91125, U.S.A.

ABSTRACT

Dissolved Pb, ranging from 5 - 15 ng/kg in open-ocean surface waters up to 50 ng/kg in highly polluted coastal waters, enters marine food chains at the 10^{-8} g Pb/g fw level by adsorption onto surfaces of algae which are consumed by herbivores. Dissolved Pb and biological productivity tend to be inversely correlated in surface ocean waters. Benthic filter-feeders consume particles containing 10^{-6} g Pb/g wet wt, yet the 10^{-9} g Pb/g fw concentrations in their tissues are the same as those in benthic grazers whose food contains 10^{-8} g Pb/g fw. The isotopic composition of dated coastal basin sediment cores conclusively documents the increasing pollution of the marine biosphere by industrial Pb, derived primarily from gasoline Pb-alkyls.

It is now believed by some of us that lead concentrations in coastal surface waters that are involved in natural circulation processes do not increase to levels above 50 ng/kg unless the waters are within sewage plumes or within heavily contaminated plumes from harbors and industrial bays (Ref. 1 - 4). Contrary to widely accepted beliefs, we and our co-workers do not believe that lead concentrations in most coastal surface waters commonly approximate 1000 - 2000 ng/kg. Among the largest sewage outfalls in the world, those emanating from the Southern California metropolitan area do not elevate lead concentrations in surface seawaters above 50 ng/kg, when mixed over a relatively small area of 10000 km^2 to an average depth of 300 m (Ref. 1). Total lead concentrations within the direct plumes from these sewage outfalls do range from 100 - 1500 ng/kg in coastal regions, however. Where industrial outputs are smaller, lead concentrations in surface waters range from 5 - 20 ng/kg off the coast of western United States. It has been found that surface concentrations of lead in the oceans parallel the concentrations of ^{210}Pb whereby both common lead and ^{210}Pb surface concentrations are low in biologically productive regions near the continental shelves, and are high in biologically barren regions of the open oceans (Ref. 2 and 5). In the north Pacific this means that the concentrations of common lead in surface waters increase from levels of about 5 ng Pb/kg seawater over the shelf to 15 ng Pb/kg seawater in the open waters of the central northeast Pacific (Ref. 2). The relationship of concentrations of ^{210}Pb in these same surface waters have been found to be parallel (Ref. 2 and 5). Sparse measurements indicate that the vertical distribution of common lead does not change significantly in going from deep waters near the continental shelf of western North America to deep waters in the open NE Pacific (Ref. 2). Surface concentrations of common lead near the coast of western North America vary from 5 - 15 ng Pb/kg seawater mainly as a result of either temporal variations in biological productivity or temporal variations in the amount of industrial lead introduced to the coastal waters. These two processes have been distinguished by parallel variations, or lack of them, in ^{210}Pb concentrations (Refs. 2 and 5).

The above data for lead in surface waters refer to total lead which is the sum of both highly reactive, or dissolved lead, presumable existing in the form of molecluar $Pb(CO_3)$ (Refs. 6,7,8) and chemically inactive, or particle lead. In surface waters containing 15 ng of total Pb/kg seawater or less, the fraction of particle phase lead is found to be about 10%, or less. This fraction is maintained at depth, except for the possible existence of a layer extremely rich in lead particles at 3000 to 3500 m in the NE Pacific (5 ng particle Pb/kg coexisting with 2 ng dissolved Pb/kg) (Ref. 2). It is not yet known if this layer is an artifact related to collection procedures, although it has been confirmed in the samples that were collected by replicate analyses of them. In shallow waters more highly polluted with industrial lead which are very near the coasts, the fraction of particle lead is generally higher, increasing to as much as 90% of the total (Ref. 1). Within a few km of the coasts, lead concentrations decrease in progressing seaward from the coast, and most of this decrease is associated with a decrease in the concentration of particle lead, rather than dissolved lead (Ref. 1).

The concentration of industrial lead in seawater particles is usually extremely high, averaging about 200 ppm Pb in wet sewage particles and 1 ppm Pb in wet coastal seawater particles for example(Refs. 1, 2 and 9). When anthropogenic particles are introduced to seawater, a fraction of their original lead is released in soluble form. For sewage particles this immediately released soluble fraction is only about 10% (Ref. 1), while for atmospheric particles enriched with lead from auto exhausts and introduced as wet and dry fallout the immediately released soluble fraction is about 50% (Ref. 10). The soluble fraction from sewage does not increase with time in seawater, but that from aerosols containing auto exhaust lead does. The soluble form of industrial lead pollution in coastal waters enters marine ecosystems by sorption on algal surfaces, while industrial lead pollution in the form of particles does not, for the most part, enter marine food chains by surface adhesion to algal surfaces. Instead, particle lead pollution suspended in the water column enters marine food chains through filter feeders. Even though soluble lead is enriched with respect to calcium by sorption on algal surfaces in going from seawater to algae, herbivore grazers feed on material which contains far smaller polluted Pb/Ca ratios than do filter feeders, because lead concentrations in detritus particles are so extremely high. That is, concentrations of pollutant lead in the detritus particle food of filter feeders are in the 10^{-6} g/g range, but only in the 10^{-8} g/g range for the algae food of grazers, while concentrations of calcium in both detritus and algae are similar (Ref. 11). Surprisingly, the tissues of both herbivore grazers and filter feeders show about the same Pb/Ca ratios (Ref. 11). This is illustrated in Table 1. It is believed that high pHs in the stomachs of the filter feeders which depresses the availability of lead relative to calcium, may account for this. It must be emphasized, however, that for a given species of animal the effects of increasing the concentration of pollutant lead in the environment are reflected by correspondingly elevated Pb/Ca ratios in both soft tissues and shell of that species. This is shown in Table 1 where increases of lead in the food of the animals produces increase in Pb/Ca ratios in muscle and shell (Ref. 11). Despite enormous (200-fold to 4000-fold) biodepletions of lead relative to calcium in going from food to the main calcium reservoirs in bone tissues of these animals, the amounts of lead in these reservoirs are still delicately regulated by quite small (4-fold) changes in the Pb/Ca ratio in food caused in turn by small variations in levels of lead pollution. We use the term "small" for 400% increases in lead pollution because lead pollution increases in city and town environments are 10000-fold to 100000-fold above natural levels. Lead is bioenriched relative to calcium in going from food to the small and highly dynamic calcium reservoir in

TABLE 1 Effects of Different Levels of Lead Pollution on Lead Concentrations in Tissues of a Filter Feeder and a Grazer (Ref. 11). Seawater Values are Estimated from only a Few Measurements in the Face of Large Temporal Variations of these Values (Refs. 1 and 2)

Animal	Substance	Cortes Banks (diss. Pb poll. ~ 4-fold above natural levels) Pb/Ca (wt)	Abalone Cove (diss. Pb poll. ~ 20-fold above natural levels) Pb/Ca (wt)
-	seawater detritus	10000 x 10^{-7}	200000 x 10^{-7}
Scallop (Hm)	gut contents	17000 "	83000 "
"	muscle	250 "	360 "
"	shell	4 "	12 "
-	seawater diss. metals	0.5 "	1 "
Gastropod (Nn)	food	170 "	680 "
"	muscle	290 "	1200 "
"	shell	1 "	2 "

the muscle tissue of N. norrisii. This contrast between bioenrichment and biodepletion of lead relative to calcium in different tissues of the same animal is not unusual in marine animals, because it has also been observed in abalone and in tuna (Ref. 11 and 12). This contrast between muscle and bone has not been observed within 7 different kinds of terrestrial animals (Ref. 13) although reports in the literature (which may be erroneous because muscle is easily contaminated with lead) indicate that the same contrasting phenomenon does exist in humans and other terrestrial animals. If, as is probable, the Pb/Ca ratio in plasma supplying nutrients to the muscle cells is less than that within the cells, then the residence time of lead within those cells must be longer than that for calcium. Despite the close similarity of Pb/Ca ratios in shell and dissolved in seawater, studies with radioactive ^{45}Ca, show that most of the Ca in the shell enters these animals through the stomach (Ref. 14). It should also be explained that the 5-fold lead pollution effect on animal food illustrated in Table 1 is super-imposed on an overall lead pollution of the ambient environment, which on a world-wide scale, seems to amount to about a 4-fold increase above natural prehistoric levels, producing an overall lead pollution factor of about 20-fold in the Abalone Cove region cited in Table 1. This is justified in the southern California region by the data in Table 2, where it can be seen that the industrial lead input to coastal waters near Los Angeles is an order of magnitude greater than the prehistoric natural input of lead to the same region. These same data show that the advection of ambient seawater lead (which is about 75% industrial) in and out of the region is the same order of magnitude as the industrial input at this locality.

The increase of industrial lead in coastal marine ecosystems is recorded in basin sediments. Goldberg and Chow pioneered these observations in the southern California region by a combination of reliable ^{210}Pb dating of the sediments with identification through stable lead isotopic tracers of the industrial source of the increased amount of lead in the more recent sediments (Ref. 15). We have confirmed these lead isotopic findings by more detailed studies of the Santa Barbara Basin sediments that are shown in Fig. 1 and 2 (Ref. 16). Increasing concentrations of lead are associated with changes in the isotopic composition of lead which match changes in both the amounts and isotopic composition of atmospheric industrial lead aerosols in Southern California. Atmospheric industrial lead aerosols generated mainly from auto exhausts in California possess $^{206}Pb/^{207}Pb$ ratios which changed

TABLE 2 Mass Balance of Lead in the Southern California Bight
This is a 40 km wide coastal water strip between San Diego and Pt. Conception bounded by chain of channel islands on seaward side

Type of Input	(tons Pb/yr/12000 km^2)
Ambient seawater lead entering Bight by advection	280
Dry aerosol deposition lead	280
Sewage lead	200
Storm runoff lead plus dry season flow from storm drains	170
Prehistoric natural river clay lead	40
Rain lead	20
Industrial river dissolved lead	20
Prehistoric natural river dissolved lead	1
Type of Output	
Ambient seawater lead leaving Bight by advection	280
Addition of lead pollution to ambient seawater leaving Bight	530
Industrial lead accumulated in sediments within Bight	160
Natural clay lead accumulated in sediments within Bight	30
Natural clay lead leaving Bight by advection	10

Data revised from: C. C. Patterson and D. Settle "Contribution of lead via aerosol impact to the So. California bight", Journal de Recherches Atmospheriques, Numero Special 8, 957-960 (1974); and J. J. Huntzicker, S. K. Friedlander and C. I. Davidson, "Material balance for automobile emitted lead in the Los Angeles basin", Environmental Science and Technology 9, 448 (1975).

from 1.145 in 1964 to 1.20 in 1972 to 1.25 in 1977. The calculated lead isotopic compositions of mixtures of industrial lead coming from such aerosols with lead in the sediments at 1940 agree within experimental error of measured isotopic values. This is conclusive proof that the lead is anthropogenic, that it originates mainly from gasoline lead, and that the lead concentration changes are not derived from natural climatic, biologic, or diagenetic processes. It must be emphasized that the changes in the concentrations of lead in the sediments do not necessarily reflect changes on a 1 to 1 basis of lead concentrations in overlying waters.

The anthropogenic lead input flux we observed in this inner coastal basin is the same as that reported earlier by Goldberg and co-workers, that is, about 2 μg Pb/cm^2/yr in excess of a prehistoric flux of about 1 μg Pb/cm^2/yr. As might be expected, coastal basins further out on the continental shelves show smaller but still observable anthropogenic lead additions, about 1/20th in excess of natural fluxes in some cases (Ref. 17). Other measurements have been made of increased lead deposition fluxes in recent decades in sediments at other coastal locations in the world, but these measurements have not used lead isotopic tracers to identify the sources of the extra lead. It is hoped that the isotopic tracer experiment for leaded gasoline now being carried out in Italy will provide additional helpful information on this subject within a few years. Most studies relate to bays that have been heavily industrialized and highly perturbed by lead effects that cannot be generalized on a world scale. On the east coast of North America, for example, Narragansett Bay shows an anthropogenic lead input of 170 μg Pb/cm^2/yr, compared to a prehistoric flux of 3 μg Pb/cm^2/yr (Ref. 18). However, much more modest lead pollution effects are also observed in bays near heavily industrialized regions.

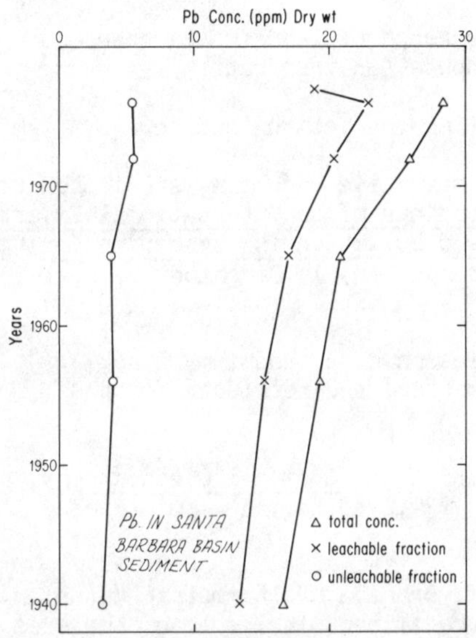

Fig. 1. The occurrence of lead in sediments from the anoxic region of the Santa Barbara Basin, southern California. Samples collected and dated by A. Soutar and co-workers.

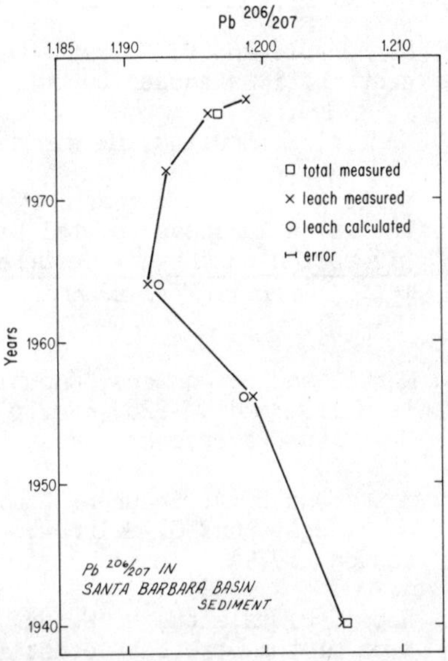

Fig. 2. Secular changes in the isotopic compositions of lead in Santa Barbara Basin sediments.

West of Kiel, in the Baltic Sea, for example, Exkernförder Bucht shows an anthropogenic lead input of 2 μg Pb/cm^2/yr in excess of natural background (Ref. 19). These two examples represent the range of local anthropogenic effects reported in coastal regions of the world. It has been claimed, on the basis of the observed presence of coal residues in the two examples given above, that the source of the anthropogenic lead is coal, but this is not necessarily true, because the isotopic compositions of the leads have not been determined. It has been estimated on the basis of temporal changes in coal residue concentrations that at least 1/3rd of the excess anthropogenic lead flux in each of the above examples probably originates from auto exhaust lead. The latter source may account for much of the recent industrial lead, while losses from lead smelters and widespread use of lead account for most of the rest in earlier times.

This work was supported by the IDOE Office of the National Science Foundation through grant OCE74-24362 A02 and by the Bureau of Land Management, US Department of Interior contract 696-16. This article is a summary of an oral presentation which includes discussion of some unpublished work by Schaule and Patterson (Ref.2), Hirao, Elias, and Patterson (Ref. 13), as well as by the listed authors. Manuscripts of these studies are in preparation and will be published in full technical detail. Contribution No. 3050: Publications of the Division of Geological and Planetary Sciences, California Institute of Technology, Pasadena, California 91125, USA.

REFERENCES

(1) C. Patterson, D. Settle and B. Glover, Analysis of lead in polluted coastal water, Marine Chemistry 4, 305-319 (1976).

(2) B. K. Schaule and C. C. Patterson, The occurrence of lead in seawater and perturbations caused by anthropogenic inputs (in preparation).

(3) J. Martin, K. Bruland, H. Windom, R. Duce, Personal communication.

(4) L. Mart, H. W. Nürnberg, and P. Valenta, Comparative base line studies on Pb-levels in European coastal waters, Proceedings of an International Expert Discussion on "Lead - Occurrence, Fate and Pollution in the Marine Environment", Rovinj 1977, M. Branica and Z. Konrad (eds.), Pergamon Press, 1979, 155-179.

(5) J. Martin and co-workers, Unpublished studies on zinc, cadmium, lead, lead-210, radium-226, etc. particulate matter, and nutrients for the Hawaiian Traverse of August, 1977.

(6) W. Stumm and P. A. Brauner, Chemical Speciation, in Chemical Oceanography, J. P. Riley and G. Skirrow, eds., 1, 173-239, 2nd Ed., Academic Press, London (1975).

(7) L. Sipos, P. Valenta, H. W. Nürnberg, and M. Branica, Voltammetric determination of the stability constants of the predominant labile lead complexes in sea-water, in Proceedings of the International Experts Discussion on "Lead - Occurrenece, Fate and Pollution in the Marine Environment", Rovinj, 1977, M. Branica and Z. Konrad (eds.), Pergamon Press, 1979, 61-76.

(8) M. Whitfield and D. R. Turner, Theoretical studies of the chemical speciation of lead in sea-water, in Proceedings of the International Experts Discussion on "Lead - Occurrence, Fate and Pollution in the Marine Environment", Rovinj, 1977, M. Branica and Z. Konrad (eds.), Pergamon Press, 1979, 109-148.

(9) K. Bruland, The occurrence of metals in filtered solids from seawater, in Baseline studies of outer continental shelf, Southern California, Vol. 3, Study report prepared under contract for US Bureau of Land Management, Department of Interior (1976)

(10) M. Koide, J. J. Griffin and E. D. Goldberg, Records of plutonium fallout in marine and terrestrial samples, Journal of Geophysical Research 80, 4153-4162 (1975).

(11) M. W. Burnett and C. C. Patterson, Distribution and occurrence of alkaline earths and lead in marine ecosystems, (in preparation).

(12) C. C. Patterson and D. Settle, Comparative distributions of alkalies, alkaline earths and lead among major tissues of a tuna, Marine Biology 39 289-295 (1977).

(13) Y. Hirao, R. W. Elias and C. C. Patterson, Anthropogenic elevation of lead levels in some food chains in a subalpine ecosystem (in preparation).

(14) J. Nicol and A. Colin, in The Biology of Marine Animals, Wiley, New York, 638 (1967).

(15) T. J. Chow, K. Bruland, K. Bertine, A. Soutar, M. Koide and E. D. Goldberg, Lead pollution: records in Southern California coastal sediments, Science 181, 551-552 (1973).

(16) Amy C. Ng and C. C. Patterson, Profiles of lead, barium, and lead isotopic composition in the Santa Barbara basin sediments as indicators of anthropogenic lead and barium pollution, (in preparation).

(17) K. K. Bertine and E. D. Goldberg, History of heavy metal pollution in Southern California coastal zone - reprise, Environmental Science and Technology 11, 297-299 (1977).

(18) E. D. Goldberg, E. Gamble, J. Griffin and M. Koide, Pollution history of Narragansett Bay as recorded in its sediments, Estuarine and Coastal Marine Science 5, 549-561 (1977).

(19) H. Erlenkeuser, E. Suess and H. Willkomm, Industrialization affects heavy metal and carbon isotope concentrations in recent Baltic Sea sediments, Geochemica et Cosmochimica Acta 38, 832-842 (1974).

Analysis of Natural and Industrial Lead in Marine Ecosystems

M. Burnett and C. C. Patterson

Division of Geological and Planetary Sciences, California Institute of Technology, Pasadena, California 91125, U.S.A.

ABSTRACT

Sequential food chain relationships in benthic and pelagic systems, obscured previously by gross analytical errors, are in fact characterized by large initial bioenrichments of the Pb/Ca ratio in going from seawater to plant because of selective passive adsorption processes. This ratio is subsequently biodepleted in successive transfers to higher trophic levels because of inefficient transport of lead across membranes in animals. Concentrations of Pb in organism tissues are quite sensitive to changes in the ambient concentrations of dissolved rather than particulate Pb in seawater. The preparation and standardization of water containing 1 ng Pb/kg is an essential prerequisite to be met before embarking on a program to reliably determine Pb levels in any component of a marine ecosystem. Gross analytical errors for Pb in US and UK regulatory agency laboratories have been responsible, in part, for the exposure of entire populations to deleterious chronic Pb insult.

In order to illustrate the effect that analytical errors have on interpretations of lead in marine ecosystems, the general relationships of lead in marine food chains and organisms must be described concurrently with a discussion of analytical errors. It has been found that lead is passively sorbed by chelating agents on the surfaces of algae, and this produces high concentrations of lead in bulk algal material (Ref. 1). This characterizes the first step in the flow of lead through all marine food chains. In subsequent stages in progressing from primary producer to primary consumer and subsequently to higher carnivorous levels, lead is not readily absorbed into cellular constituents of organisms; instead only a fraction of the ingested lead is retained within organisms. This means that in sequential food chain relationships, higher organisms tend to contain less lead than do their precursors because of the biodepletion of lead. The general aspect of lead in marine food chains is therefore an initial bioenrichment followed by stages of successive biodepletion (Ref. 1). When comparing variations in lead concentrations among different organisms to show the effects of biodepletion, data scatter may result from variations in ash or water contents, and in order to eliminate them it is best to consider the concentrations of lead in only the metal fraction of organisms. Furthermore, since lead is such a close relative of calcium that the morphological distributions of both metals in organisms are nearly identical, and since calcium is a nutrient metal which flows easily along food chains, it is best to consider ratios of lead to calcium in organisms at the different trophic levels in food chains. When this is done it can be seen, as shown in Fig. 1, that the Pb/Ca ratio increases dramatically at the primary producer level compared to seawater but then declines.

This is the general aspect for most animal food chains, even though specific examples are given here because the main reservoir of both calcium and lead in most animals is the skeleton, and the calcium in this reservoir is highly purified of lead relative to the less pure calcium in food because of selective biochemical processes. Even though proteins in some tissues may tend to concentrate lead relative to calcium the masses of such reservoirs of lead and calcium are insignificant compared to the skeletal reservoir, and the net effect for the entire organism is biodepletion of lead compared to food. Note that the variations cover 3 orders of magnitude. Trustworthy data for phytoplankton are lacking. Valonia is a good model, not only for phytoplankton, but for macro algae as well, because the Pb/Ca bioenrichment ratio is relatively insensitive to the type of complexing agent on algal surfaces. Trustworthy data for zooplankton are lacking and our spiny lobster data provide only an approximate model for these organisms. The anchovy sample analysed was taken from the stomach of an albacore.

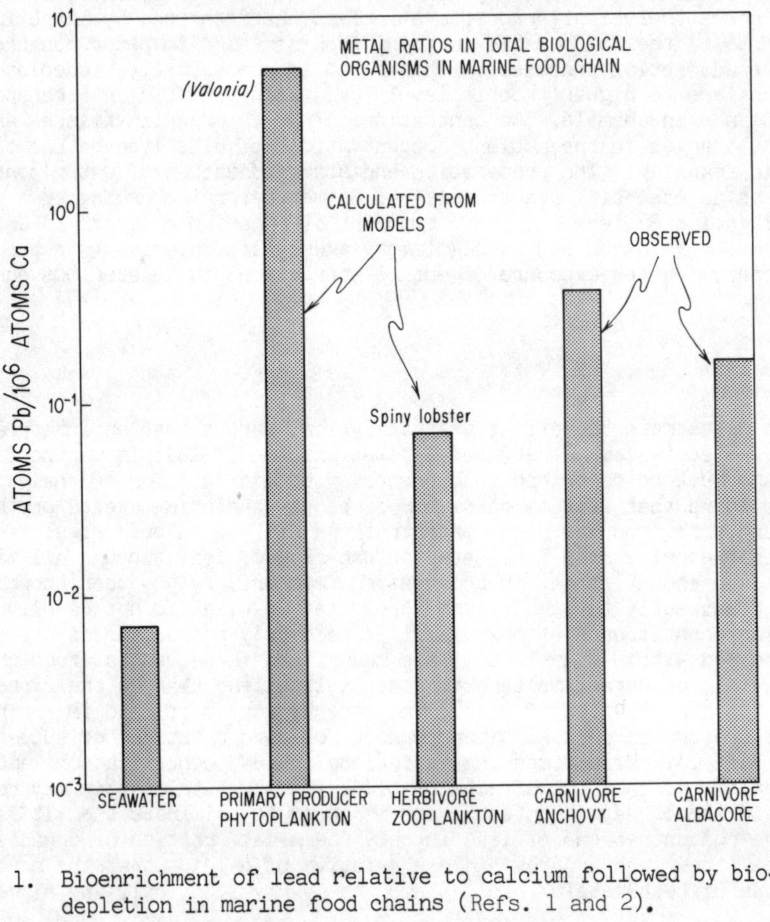

Fig. 1. Bioenrichment of lead relative to calcium followed by biodepletion in marine food chains (Refs. 1 and 2).

Figure 1 shows that the biodepletion of lead relative to calcium should be generally applicable to most components of most food chains because calcareous tissues comprise the major reservoir for both calcium and lead in most organisms, and lead is far less efficiently biotransported to that bony reservoir than is calcium. Biodepletion effects continue to operate more or less independently of any particular lead concentration level that may exist because the decimal fraction of lead absorbed into animals from the gut remains relatively fixed, whatever this value may be for a given species, despite large fluctuations of lead concentrations in nutrient mediums. This happens because lead is only inadvertently and inefficiently absorbed as a trace constituent of calcium, and the factors which actually control the decimal fraction of lead absorption, as long as lead remains a trace constituent, are those which control the decimal fraction of calcium absorption (Ref. 3).

In past work by other investigators, these relationships have been obscured, as shown in Fig. 2, by lead contamination during collection, handling, and analysis. The incorrect classical view that all marine organisms concentrate lead from seawater, is based upon erroneous data which suggest uniformly high concentrations of lead at all trophic levels in marine food chains.

The concentrations of strontium and barium, which are trace metal relatives of calcium, have been studied in marine food chain components as well as lead and calcium (Ref. 1). It has been found that the observed concentrations and enrichments of these four metals in total Macrocystis pyrifera relative to seawater are directly proportional to the stabilities of the metal alginates shown in Table 1, and that a brief 0.1 N HCl leach removes these metals in inverse proportion to their respective binding affinities to alginate.

That is, lead contamination of marine algae is related to the amount of lead in the dissolved phase of lead in seawater, and not to the amount of lead in the particle phase. This ionic sorption process on algal surfaces for ionic lead is quite different from that for radionuclides of polonium, plutonium, and americium, where, as has been shown in work pioneered by Folsom and coworkers, particle sorption appears to be the predominant process (Ref. 6). The reason for this difference is not yet clear. There may be a similar difference between common lead and ^{210}Pb (Ref. 7), but the actuality of such a difference is still in the process of being established. Near sewage outfalls where the concentration of lead in particles is extremely high, the adherence of relatively few particles on plant surfaces may, in some instances, give anomolously high lead values. In general, however, it is only the dissolved phase that determines the concentrations of lead in marine algae (Ref. 1) as is shown in Fig. 3.

An increase in dissolved lead, caused by pollution, elevates lead concentrations in animals, as the data in Fig. 3 show. The concentration of lead in herbivores is related to the concentration of dissolved lead in seawater through their precursor algae.

When other investigators analyse lead in samples of seawater collected and standardized by the CIT laboratory only a few investigators can obtain correct values. This is shown in Table 2.

Outside the Caltech laboratory the following investigators are known by us to have correctly analysed for lead at concentrations of 3 ng Pb/kg seawater or in distilled water: H. W. Nürnberg, Director of Institute for Chemistry, KFA Jülich, ICH-4, Postfach 1913, D-517 Jülich, F. R.; Y. Hirao, College of Science and Engineering, Aoyama Gakuin University, 6-16-1 Chitosedai Setagaya-Ku, Tokyo 157, Japan; and M. Murozumi, Muroran Institute of Technology, 27-1 Mizumoto - cho, Muroran, Japan 050. For correct analyses for lead at 10 ng Pb/kg seawater: J. Martin, Moss Landing Marine Laboratories, Moss Landing, California 95039, USA; and

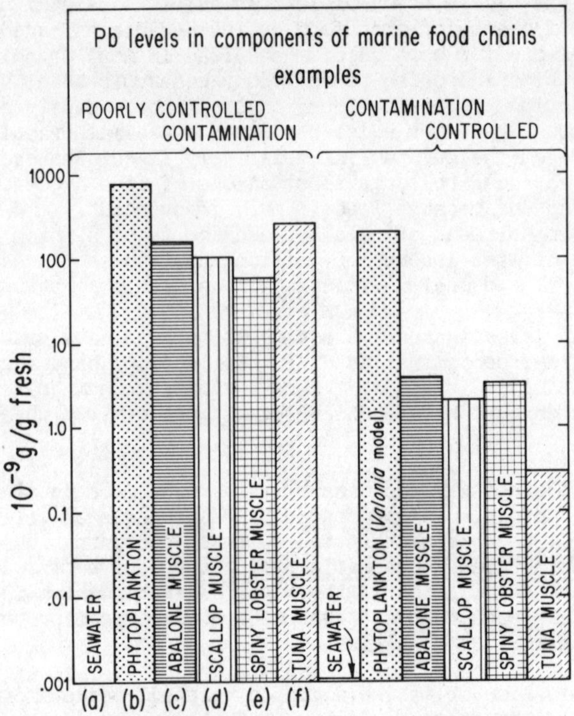

Fig. 2. The obscuring of true relations of lead in marine food chains by improper control of lead contamination, which yields erroneous data. Note that the plotted variations of lead in tissues cover 4 orders of magnitude. Values on the left are from recent literature commonly regarded as the best available: (a) M. Tatsumoto and C. Patterson, Common Pb in seawater, Nature, Lond. 199, 350 (1963) T. J. Chow and C. Patterson, Pb and Ba in seawater, Earth and Planetary Science Letters 1, 397 (1966); (b) J. H. Martin and G. A. Knauer, The Elemental Composition of Plankton, Geochimica et Cosmochimica Acta 37, 1639-1653 (1973); (c) J. Stewart, M. Schulz-Baldes, Long term Pb accumulation in abalone fed on Pb treated brown algae, Marine Biology 36, 19-24 (1976); (d&e) T. J. Chow, C. Snyder, H. Snyder, and J. Earl, Pb content in some marine organisms, J. Environmental Science Health - Environmental Science Engineering A-11, 33-44 (1976); (f) See note (+) in Table 6. Values on the right are from data obtained at the CIT biogeochemical laboratory (see Refs. 1, 2 and 4).

K. Bruland, University of California, Santa Cruz, California 95064, USA. For correct analyses for lead at 25 ng Pb/kg seawater: C. S. Wong, Ocean Chemistry Division, Marine Sciences Directorate, Pacific Region, Department of the Environment, 211 Harbour Road, Victoria, B. C., Canada. For correct analyses for lead at 100 ng Pb/kg seawater: B. J. Presley, Department of Oceanography, Texas A & M University, College Station, Texas 77843, USA. For correct analyses for lead at 150 ng Pb/kg seawater: J. M. Bewers, Chemical Oceanography Division, Marine Sciences Directorate, Atlantic Oceanographic Laboratory, Bedford Institute of Oceanography, Dartmouth, Nova Scotia B2Y 4A2; R. Duce, Graduate School of

TABLE 1 Numbers in second column (Ref. 5) are mequiv. H^+ liberated per ml by addition of metal salt (0.02 mequiv./ml) to solutions containing 0.027 mequiv. half-neutralized polyanions per ml. The alginate titrated is 90 % guluronic acid. *M. pyrifera* contains both guluronic and mannuronic alginic acids. The latter forms stronger Pb complexes but less strong alkaline earth complexes than the former. These second column numbers show that the stabilities of the metal alginates increase in a series Ca $<$Sr $<$ Ba$\ll$ Pb. The equimolar conditions for these data are different from actual field conditions. Numbers in the third column (Ref. 1) are normalized to calcium because in the field, alginates in *M. pyrifera* are predominately calcium alginates. The relation among these numbers, which show an increasing enrichment in the series Sr $<$ Ba $<<$ Pb relative to Ca in going from seawater to *M. pyrifera*, parallels exactly the relation of numbers in the second column.

Metal	mequiv. H^+ liberated from guluronic acid	$\left[\frac{\text{metal}}{\text{Ca}}\right]_{Mp} \Big/ \left[\frac{\text{metal}}{\text{Ca}}\right]_{sw}$
Pb	.0024	470
Ba	.0010	56
Sr	.0008	38
Ca	.0004	1

Oceanography, University of Rhode Island, Kingston, Rhode Island 02881, USA; and W. Fitzgerald, Marine Sciences Institute, University of Connecticut, Avery Point, Groton, Connecticut 06340, USA. Some of these investigators may be able to analyse lead correctly at lower concentrations, but this has not been confirmed by interlaboratory calibration with IDMS in an ultraclean laboratory. Concentrations of lead in seawater above 15 ng Pb/kg are not significant on a world-wide scale. A number of other investigators, some in prominent oceanographic laboratories, are known by us to have made serious errors in analyses for lead in standardized samples of seawater in recent years, where concentrations ranged from 1 to 40 ng Pb/kg, and whom we have not been able to recheck in subsequent work. On the basis of a comparison of reasonable estimates of the probable extent of lead contamination and its probable control in most chemical oceanographic laboratories in the world with those that are required for reliable lead analysis of seawater, it is believed that very few, if any, out of the hundreds of such laboratories in the world, that we have not carried out interlaboratory calibrations with, can obtain correct values for lead in most of the seawaters of the world at the present time. The widely recognized lead-in-open-and-deep-oceans results of Tatsumoto, Chow, and Patterson (Refs.8 and 9) reported before 1974 are seriously wrong because of contamination during collection, although the analytical methods they developed in their pioneering work were accurate and serve as a foundation for the recent studies that have provided more reliable data.

The discovery that the shallow waters of the open oceans contain only 10 ng Pb/kg and that the deep waters of the oceans contain only one or two ng Pb/kg. instead of

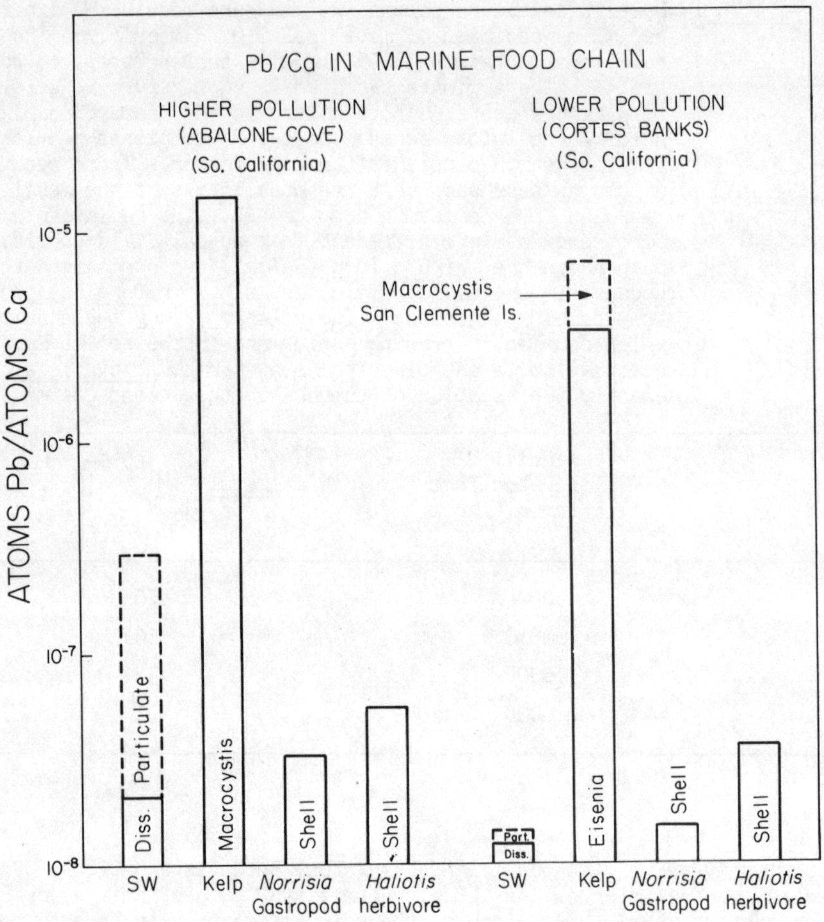

Fig. 3. Pollution effects caused by simultaneous small increases of dissolved Pb/Ca and large increases of particle Pb/Ca in seawater. The 2-fold increase of dissolved Pb/Ca that occurs in going from the less polluted Cortes Banks region to the heavily polluted Abalone Cobe region is paralleled by a 2-fold increase of Pb/Ca in kelp and in the shells of herbivores. These latter changes are far smaller than the 20-fold difference between particle Pb/Ca at the Cortes Banks region and the Abalone Cove region. As noted in Fig. 1, the biodepletion of lead relative to calcium operates independently of lead levels in nutrient mediums, whatever they may be, as long as they are not more than 100-fold above natural levels.

many hundreds or thousands of ng Pb/kg, as is still commonly believed, is crucially important, because the correct very small concentrations are an index of ocean reservoirs of lead that are small with respect to the annual input of industrial lead to the oceans, which correctly signifies a large lead pollution effect, while erroneous very large concentrations of lead in seawater incorrectly suggest that ocean reservoirs of lead are hundreds of times larger than annual lead inputs and that lead pollution effects in the oceans are insignificant. In this same sense,

Table 2 Comparison of actual and observed values of lead in seawaters known by us to have been obtained by different investigators within the last 4 years

Method	ng/kg Reported Values	ng/kg True Value	No. of Invest.
Waters Standardized at CIT and then Analysed by Other Investigators			
AA	120 & 150	140	3
AA	100	100	1
DPP	20	40	1
AA	22 & 100	40	2
AA	80 & 1200	25	2
IDMS, AA	30	25	1
ASV	180 to 1300	14	3
AA	50 to 120	14	3
AA	10	10	2
ASV	70 & 500	3	2
DPP	3	3	1
IDMS	2	2	1
AA	1	1	1
Waters Known to have been Contaminated During Collection and Reported by Other Investigators			
AA	100 to 5000	5 - 30 (est)	7
ASV	500 to 5000	5 - 30 "	4
IDMS	700	1 - 2 "	1
IDMS	30	1 - 2 "	(pre 1974 by CIT)

true low baseline levels of lead in marine plants and animals correctly show that these substances are much more susceptible to lead pollution effects than when baseline lead concentrations in marine plants and animals are incorrectly reported as being high.

Although there are many thousands of lead analyses reported each year for marine animals, it is doubtful that much of these data are scientifically significant. Most analyses of invertebrates for example, are reported at levels of 10^{-6} g/g fresh wt. lead in total soft parts. Since the concentrations of lead in the soft tissues of invertebrates are actually on the order of 10^{-9} g/g fresh wt., reports of 10^{-6} g/g fresh wt. levels of lead in whole soft parts can refer only to concentrations of lead in detritus contained in the stomachs of the animals, which can be correct only in those cases of extreme lead pollution where lead is > 10 ppm in wet detritus (>100 ppm dry detritus). The true distribution of lead among the various organs of a scallop are listed in Table 3. The region from which this scallop was taken is 200 km from the coast west of San Diego on the seaward side of the channel islands, and it is polluted with industrial lead on an estimated world-scale of about 4-fold above prehistoric natural values. Analyses show that 10 ppm Pb (dry) and 1 ppm Pb (wet) in filtered seawater detritus near the surface in these same waters (Ref. 10), and about 1000 ppm Pb (dry) and 200 ppm (wet) in sewage solids (Ref. 11) discharged at the coast of this region. In this outer shelf region, where there is an absence of particles highly polluted with lead from local sources, 80 % of the lead in this total animal is in the bone reservoir. Most of the remainder is in food residues in the gut, which lead is not biochemically incorporated into the tissues of the animal. If lead in the "total soft parts" which includes the gut contents but excludes bone, were to be reported in this animal, about the same lead concentration would be observed as for the total animal, as shown in the last column in Table 3, but 90 % of the lead in this mixture would not be in tissues at all, but in food residues. Actually, 97 % of the total lead biochemically incorporated in tissues of the animal resides in bone, and the concentration of lead in "total soft parts" minus food residues in the gut would be only 1/10th of that in soft parts which include gut contents, shown in the next to the last column of Table 3.

When this same species grows within several km of sewage outfalls, the concentration of lead in the gut contents is observed to increase 10-fold as a consequence of the addition of sewage solids, shifting the bulk of lead in the total organism away from bone (30 %) to the gut contents (70 %) (Ref. 1). If lead in the "total soft parts" of this animal were to be reported, the observed lead concentration would jump 10-fold to more than 1 ppm, even though the concentration of lead incorporated in total tissues (~ 95 % in bone) had increased only 3-fold, and the concentration of lead in muscle tissue was only 0.01 ppm (Ref. 1). Thousands of Pb determinations in *Mytilus* have been made by investigators monitoring global heavy metal pollution. Although we have only begun studies of *Mytilus*, our new data, listed in Table 4 and Table 5, make it highly improbable that many, if any, of the analyses previously reported by other investigators are scientifically significant. The 0.2 to 1 ppm Pb concentrations reported earlier in the muscle tissue of this filter-feeder (Ref. 12) exceed, in many cases, the Pb concentrations determined for digestive organs or shell, erroneously suggesting that Pb metabolism in *Mytilus* is significantly different than for similar bivalves, such as scallop. However, as may be seen by comparison of Table 3 and Table 5, the morphological distribution of Pb and its significance are essentially identical in both scallop and mussel. Common lead is not highly concentrated in gut walls of invertebrates, as has been suggested (Ref. 13), nor in kidney (Ref. 14), nor even in fish caecum (Refs. 13 and 15).

The true distributions of lead among the various organs of a tuna fish are listed in Table 6.

We have not yet found any gross differences between the distributions of lead in bony fish and in invertebrates. Table 7 lists the concentrations of lead in tuna muscle, ranging from prehistoric times, to present day ocean environments, to that found in non-soldered, and in lead soldered cans. The 20,000-fold lead pollution factor observed in canned tuna, a marine food commonly eaten by Americans provides stark, indisputable, and chilling proof of the extremity of the lead pollution situation today in the human environment (Ref. 17). The average person in the United States and the United Kingdom today contains a body burden of lead which is more than 100 times above natural levels prevailing in prehistoric times (Refs. 16, 17, and 18). Present day exposures for the average person in the US and UK are about 1/3 of those that will cause symptoms of classical lead poisoning upon prolonged exposure. The 100-fold lead pollution effect in humans constitutes severe chronic lead insult whose effects probably lead to significant reductions in mental acuity, increases in mental irrationality, reduction in resistance to malignancies and infectious diseases, and damages to germ plasm (Ref. 17). This

TABLE 3 Distribution of Lead in Organs of Scallop (Hinnites multirugosa), 433 g Total Wt from Cortes Banks, So. California (Ref. 1)

Organ	% Body	Pb Conc. (µg/g fresh wt)	% Pb Body Burden (incl. gut cont.) (& shell)	% Pb Body Burden (in tissues only)	% Pb Body Burden ("total soft parts") (excl. gut cont.)	% Pb Body Burden ("total soft parts") (incl. gut cont.)
Shell	75	0.14	77	97	-	-
Muscle	11	0.004	0.3	0.4	11	1.4
Gonad	5.1	0.023	0.9	1.1	31	3.8
Mantle	4.6	0.04*	1.3	1.7	48	5.9
Gut contents	2.3	1.2	20	-	-	88
Gills	1.6	0.01*	0.2	0.2	4	0.2
Gut wall	0.2	0.06	0.1	0.1	3	0.4
Kidney	0.2	0.05**	0.1	0.1	3	0.4
[Total Pb conc. (µg/g fresh wt)]			(0.14)	(0.11)	(0.017)	(0.13)

* est. from bony fish data (Ref. 2); mantle ≈ epidermal mucus; gills ≈ epithelial gill tissue.

** est. from abalone from same location (Ref. 1).

TABLE 4 Distribution of Ca, Sr, Ba, and Pb in the Mussel Mytilus Californianus, 54 g Total Wt from Punta Banda, Baja California, Mexico

	Pb	Ba	Sr	Ca
Muscle	.025	.006	2.89	179
Gonad	.068	.007	3.56	228
Gut wall	.309	.081	1.78	132
Periostracum	.949	3.29	115.9	29700
Shell	.320	-	-	360000
Gut Conts.	.635	2.39	9.45	960
Mantle	.042	.078	3.73	240

Note that some tissues are easily cross-contaminated by epidermal mucus containing high concentrations of metals relative to underlying tissues. This is illustrated by the following analyses of Pb and Ba in a sample of muscle tissue from Mytilus from which outer layers were not shaved away and discarded, although it was dissected in a clean room.

Cross-contaminated muscle	.056	.008	2.18	-

situation which applies to entire populations has been ignored by responsible authorities in the US and the UK, while they have devoted their attention instead to deleterious effects on small segments of populations caused by lead pollution exposures in excess of typical ones. This statement is so extreme it is difficult to take seriously. Unfortunately, the cold facts are that the entire earth's biosphere is polluted by an annual production of industrial lead that is 100-fold larger than the total amount of lead circulating each year through the earth's biomass, that cities and towns throughout the world are vast sewers of industrial lead, that the average person in the industrialized regions of the earth does indeed contain body burdens of lead that are 100-fold larger than natural ones existing in prehistoric times, that such persons are hovering on the brink of classical lead poisoning, and that there are no thresholds of exposure for deleterious effects from lead (Ref. 17). One of the main reasons for the existence of this situation is that for decades a vast profusion of erroneous lead analyses engendered by engineering technology have obscured the true magnitudes of lead pollution in foods.

As shown in Table 7, the shocking and certainly poisonous contamination of fish muscle by lead in lead-soldered cans was overlooked through analytical error on the part of a prominent US Federal surveillance laboratory that was carrying out a study program on behalf of the US Food and Drug Administration. The report issued by that laboratory in effect reinforced the erroneous belief by the US FDA that Pb-soldered cans elevated lead levels in foods only a few-fold and that natural levels of lead in one of the purest of foods available with respect to lead were an erroneous 10^{-7} g/g instead of the true 10^{-11} g/g fresh wt.

Table 5 Distribution of Pb in Mussel Mytilus Californianus, 54 g Total Wt from Punta Banda, Baja California, Mexico

Organ	% Body	Pb conc. µg/g fw	% Pb Body Burden Incl. Gut Conts. & Shell	% Pb Body Burden In Tissues Only	% Pb Body Burden "Tot. Soft Pts." Excl. Gut Conts.	% Pb Body Burden "Tot. Soft Pts." Incl. Gut Conts.
Shell (calcitic)	63	.032	74.6	85.5	-	-
Periostracum	1	.95	3.5	4.0	-	-
Muscle	.4	.025	0.4	0.4	3.8	1.8
Mantle	12	.042	1.9	2.0	21.5	8.5
Gills	2	.01*	0.1	0.4	1.3	0.6
Gonad	5.5	.068	1.4	1.6	15.2	6.1
Kidney	3	.05**	0.6	0.8	6.3	2.4
Gut walls	4	.31***	4.6	5.2	51.9	20.7
Gut conts.	5.5	.64	12.9	-	-	59.8
[Total Pb conc. (µg/g fw)]			(0.270)	(0.249)	(0.079)	(0.164)

* teleost fish gills ≃ mussel gills (Ref. 2)

** abalone kidney ≃ mussel kidney (Ref. 1)

*** value believed elevated as a result of cross-contamination by gut contents

TABLE 6 Distribution of Lead in Organs of Tuna (*Thunnus alalunga*) (Ref. 2)

Tissue	wt% H_2O	Organ % of body	Pb (μg/g fresh wt)	% of body burden Pb
Bone (vertebra)	29	4.6	0.074	56
Scales	25	1.5	0.059	15
Teeth	3	0.005	0.24	0.2
Total skeleton	28	6.1	0.070	70
Muscle	71	73	0.0003	3.6
Liver	64	1.4	0.009	2.1
Spleen	75	0.06	0.019*	0.2
Kidney	75	0.55	0.006	0.5
Caecum wall	65	1.2	0.004	0.8
Gills (epithelium & lamellae)	73	0.5	0.014	1.1
Dermis	44	1.6	0.013	3.4
Epidermal mucus	80	0.03	0.035	0.2
Stomach contents (anchovy)	58	2.1	0.021	7.2
Remainder	-	13	0.005**	10
Whole fish	-	100	0.006	100

* now believed too high - reinvestigation incomplete

** assumed

Lead has the same distribution as Ca, Sr, and Ba. Furthermore, this distribution is the same as that in a wild terrestrial carnivore (pine marten - *Martes americana*) (Ref. 16), except for lead in the fur of the marten.

Tuna muscle that had been dissected from a whole fresh fish at CIT was sent to the Federal surveillance laboratory. The lead concentration reported for this sample was 20-fold less than their previously reported average for lead in muscle of 27 fresh fish of the same species from the same region. This value is listed in Table 7. Part of this reduction for the later sample may be attributed to a different and more careful analytical technique (flameless AA earlier, ASV later), but part is probably also due to reduction of lead contamination during dissection, because this process was carried out in the CIT laboratory. The later analysis was still 50-fold (5000 % if such a term exists) too high. Variations of lead concentrations in muscle among tuna are small (Ref. 2).

This is not an isolated incident. It is rather certain that only a handful of ultra-clean, IDMS laboratories in the world have the potential ability to correctly analyze lead in tuna muscle at this time, and each of these, now devoted to studies

TABLE 7 Failure of US Federal surveillance laboratory (Fsl) to discover gross Pb contamination in tuna muscle (albacore - Thunnus alalunga) in its report to the US Food and Drug Administration

Type of Muscle		Reported Pb (μg/g fresh wt)
Prehistoric in sea (natural)		0.00007 (est. from 4 x Pb poll. of oceans)(Ref.4)
Present day in sea (contaminated)		0.00030 (dissect. & anal. CIT) (Ref. 2)
Non Pb-soldered can*		0.00700 (indus. cut, canned; anal. CIT) (Ref.17)
Dried powder in plastic bag	"	0.40000 (indus.cut,fried,ground;anal.CIT)(Ref.17)
Pb-soldered can	"	1.40000 (indus. cut, canned; anal. CIT)(Ref.17)
Present day in sea	"	0.40000 (dissect. & anal. Fsl) (error)++
Pb-soldered can	"	0.70000 (indus. cut, canned; anal. Fsl)++
Present day in sea**	"	0.02000 (dissect. CIT; anal. Fsl) (error)++

* "Chicken of the Sea" chunk albacore in small die-punched cans which lack both a soldered side seam and a non-soldered bottom crimp seal. "Chicken of the Sea", prepared by Van Camp Co., is also packed in small Pb-soldered cans which show the above high lead content, therefore examine can for this soldered side seam and non-soldered bottom crimp seal.

** CIT standardized Pb conc. = 0.00040

++ In our cooperative laboratory intercalibration program, laboratories reporting erroneous results are not identified. This is done to ensure continued participation by other laboratories in the intercalibration program and the improvement of their analytical techniques for lead.

of lead in igneous minerals, would require a year or more of preparatory work before they could report publishable data on fish muscle. This means that the US Federal surveillance laboratory mentioned here does not stand alone, but is one in a population of thousands at this time in the world. This is clear evidence that man does not have the ability to control the environment that he, through engineering technology, has created. In nearly every case where we have been able to check analyses of plant and animal materials for lead by other laboratories that use conventional rapid methods of analyses, serious errors have been revealed. One instance involves the participation of the CIT laboratory in the US Bureau of Land Management Baseline Study of Outer Continental Shelf Regions (Ref. 10). In another instance it has been found that all of the many reported analyses of lead in old tree stemwood are, without exception, erroneously high by about 1000-fold (Ref. 17). Entire human populations have been unnecessarily exposed to harmful concentrations of lead in their foods for decades as a direct consequence of the mistaken belief that sophistication of instrumentation by itself is the better modern successor of plain human integrity and perserverance.

Lead analyses in the CIT laboratories are carried out under ultra clean conditions using thermal ionization mass spectrometric isotope dilution. This is an absolute, yield independent method, which does not use standards or working curves, and is the method used to prepare standardized references for DPP, AA, SRF, AE, NA, etc. analytical methods (Refs. 2, 10, and 19).

A weighed amount of one lead isotope tracer is added to the sample and mixed with the unknown weights of different sample lead isotopes of the lead in that sample by chemical homogenization. The purified lead isotope mixture is chemically isolated by solvent extraction, or by ion column chelation, or by electrolysis, and then analysed in a mass spectrometer to find the ratio of the tracer lead isotope to the sample lead isotope. The measurements can easily be made with less than 1 % error using nanogram amounts of isotopes. Lead concentrations determined in the CIT laboratory are obtained using contamination control procedures that are exceedingly thorough and they are not low because solvent extraction steps fail to extract all the lead, or because some of the lead adsorbs on the walls of collectors, or because the analytical method overlooks lead in particles. Total leads in seawater are measured in samples boiled to dryness with isotope tracer in aqua regia in ultra-pure quartz, for example. The amounts of lead adsorbed in various surfaces under different conditions have been studied carefully in the CIT laboratory.

Most of the reports of lead adsorptions and of lead contamination from containers (Ref. 20) are of little significance in marine lead chemistry because the actual amounts of lead involved in these two processes under the optimum conditions required for correct lead analysis are much smaller than typical amounts of lead contamination encountered in most laboratories, and are far below the trustworthy analytical working limits for lead in most laboratories. Most of the lead in seawater that becomes adsorbed on conventional polyethylene or teflon surfaces at pH 8 is associated with particles, and is not of the readily exchangeable, or active type of lead. That is, it is charged particles containing lead that migrate to and adhere to the plastic surfaces. This process occurs within hours, and it is enormously modified by the pretreatment history of the plastic surface, being less effective for recently acidified surfaces, and most effective after > 6 weeks exposure to water at pH 8. It generally involves only a fraction of the particle form of lead, and amounts, in most cases of open seawater, to < 0.1 ng Pb/kg unacidified seawater. This adsorbed lead can be released or its formation prevented with 0.01 N acid. Contamination by container walls is far more serious. After washing conventional polyethylene with acetone briefly, followed by pure 4 N HNO_3 at 55 ^{o}C for 3 days, followed by ultra-pure 0.01 N HNO_3 at 55 ^{o}C for 1 day, followed by ultra-pure 0.01 N HNO_3 at 55 ^{o}C for 5 days, followed by storage with cold ultra-pure 0.01 N HNO_3 until just before use, that plastic, if virgin and not made of recycled plastic, will generally yield 1 picogram Pb/cm^2/day at 55 ^{o}C and 0.1 picogram Pb/cm^2/day at 20 ^{o}C to 0.01 N acid solutions every day for many months. Pb contamination from FEP teflon cleaned in a similar manner is generally about twice as high. It must be emphasized that floods of lead contamination, amounting to tens and sometimes hundreds of nanograms will be released from plastic containers upon exposure to dilute acids if they are previously cleaned only by heating with acid and then rinsed and allowed to dry before use. Proper measurement of lead in reagents and in contamination from containers is an absolute requirement for proper control of lead contamination in the laboratory during analyses. These measurements cannot be made by simply omitting the sample from the total procedure. Each separate phase of the procedure must be broken down into its many separate parts and each part then analysed for lead as though it were a sample. This is extremely difficult, but analysts must perservere and devote far more time and effort to this work than to the actual analysis of samples, because it is the degree of success at this level which determines their integrity and the degree of reliability of their lead analyses in samples at higher levels of lead concentrations. For relative methods such as AA, ASV, XRF, etc., this success cannot be achieved with absolute assurance without cross checking by the absolute method of IDMS. The single most crucial indicator of laboratory success is the preparation of distilled water containing less than 1 ng Pb/kg together with proof of that value. This, of course, is an absolute prerequisite for reliable monitoring of lead contamination of natural water samples during collection, storage, and analysis.

This work was supported by the IDOE Office of the National Science Foundation through grant OCE74-24363 A02 and by the Bureau of Land Management, US Department of Interior, Contract 696-16. This article is a summary of an oral presentation which includes discussion of some unpublished work by Elias and Patterson (Ref. 3), by by Schaule and Patterson (Ref. 4), by Hirao, Elias, and Patterson (Ref. 16), and by Settle and Patterson (Ref. 17), as well as by the listed authors. Manuscripts of these studies are in preparation and will be published in full technical detail. Contribution No. 3049: Publications of the Division of Geological and Planetary Sciences, California Institute of Technology, Pasadena, California 91125, USA.

REFERENCES

(1) M. W. Burnett and C. C. Patterson, Distribution and occurrence of alkaline earths and lead in marine ecosystems (in preparation). Although the sorption of radioactive lead on algae has been studied and reported, it is not yet clear whether these observations are applicable to common lead in seawater. Published observations of common lead in algae are in doubt because of analytical error.

(2) C. Patterson and D. Settle, Comparative distributions of alkalies, alkaline earths and lead among major tissues of a tuna, Marine Biology 39, 289-295 (1977).

(3) R. W. Elias and C. C. Patterson, Theory of biodepletion of lead relative to calcium in food chains (in preparation).

(4) B. K. Schaule and C. C. Patterson, The occurrence of lead in seawater and perturbations caused by anthropogenic inputs (in preparation).

(5) A. Haug, The affinity of some divalent metals to different types of alginates, Acta Chemica Scandinavica 15, 1794 (1961).

(6) K. M. Wong, V. F. Hodge, and T. R. Folsom, Plutonium and polonium inside giant brown algae, Nature 27, 460-462 (1972); V. F. Hodge, F. L. Hoffman, and T. R. Folsom, Rapid accumulation of plutonium and polonium on giant brown algae, Health Physics 27, 29-35 (1974); V. F. Hodge, M. Koide, and E. Goldberg, Particulate uranium in the biogeochemistry of the marine coastal zone (in preparation).

(7) E. Goldberg and K. Turekian, By analogy with ^{210}Pb Turekian believes that only a small fraction of soluble lead from rivers escapes sedimentation on the shelves and passes into the open oceans. Goldberg and coworkers, on the other hand, by anology with ^{210}Pb and from their studies of common lead deposition in coastal basins believe that a substantial fraction of soluble common lead from rivers escapes depositon on shelves, personal communication.

(8) M. Tatsumoto and C. C. Patterson, Concentration of common lead in some Atlantic and Mediterranean waters and in snow, Nature 199, 350-352 (1963); T. J. Chow and C. C. Patterson, Concentration profiles of barium and lead in Atlantic waters off Bermuda, Earth and Planetary Science Letters 1, 397-400 (1966).

(9) M. Tatsumoto and C. C. Patterson, The concentration of common lead in seawater, in Earth Science and Meteoritics, (Geiss and E. Goldberg, eds.) North-Holland Publishing Co., Amsterdam, 1963, 74-89.

(10) C. C. Patterson, D. M. Settle, B. K. Schaule, and M. W. Burnett, Standardization of reference samples for certain trace metals: the first step in a multistage program designed to insure the reliability and significance of baseline data for the trace metals, Baseline Studies of Outer Continental Shelf, Southern California, California Institute of Technology, Pasadena, Ca., USA., 1976, Vol. 3, Report 4.4.

(11) C. Patterson, D. Settle, and B. Glover, Analysis of lead in polluted coastal water, Marine Chemistry 4, 305-319 (1976).

(12) T. J. Chow, H. G. Snyder, and C. B. Snyder, Mussels as an indicator of lead pollution, The Science of the Total Environment 6, 55-63 (1976); V. Talbot, R. J. Magee, and M. Hussain, Lead in Port Phillip Bay Mussels, Marine Pollution Bulletin 7, 234-237 (1976); M. Schulz-Baldes, Lead uptake from seawater and food and lead loss in the common muscle Mytilus edulis, Marine Biology 25, 177-193 (1974).

(13) E. Goldberg, personal communication.

(14) G. L. Hoffman, personal communication.

(15) T. Folsom, personal communication.

(16) Y. Hirao, R. W. Elias, and C. C. Patterson, Anthorpogenic elevation of lead levels in some food chains in a subalpine ecosystem (in preparation).

(17) D. M. Settle and C. C. Patterson, Lead in albacore: guide to pollution in Americans (unpublished manuscript).

(18) C. C.Patterson, Contaminated and natural lead environments of man, Archives of Environmental Health 11, 344-360 (1965).

(19) C. C. Patterson and D. M. Settle, The reduction of order of magnitude errors in lead analyses of biological materials and natural waters by evaluating and controlling the extent and sources of industrial lead contamination introduced during sample collecting and analysis, in Accuracy in Trace Analysis: Sampling, Sample Handling, Analysis (P. LaFleur, ed.), US National Bureau of Standards Special Publication 422, 1976, 321-351.

(20) D. E. Robertson, The adsorption of trace elements in sea water on various container surfaces, Analytica Chimica Acta 42, 533-536 (1968); A. W. Struempler, Adsorption characteristics of silver, lead, cadmium, zinc, and nickel on borosilicate glass, polyethylene, and polypropylene container surfaces, Analytical Chemistry 45, 2251-2254 (1973).

The Occurrence of Lead in the Northeast Pacific and the Effects of Anthropogenic Inputs

B. Schaule and C. C. Patterson
Division of Geological and Planetary Sciences, California Institute of Technology, Pasadena, California 91125, U.S.A.

ABSTRACT

After employing a special seawater sampler which strictly excludes artifact Pb-contamination, Pb profiles have been determined in the northeast Pacific revealing deep water concentrations as low as 1 ng/kg and surface concentrations between 5 and 14 ng/kg. The surface enrichment appears to be due to atmospheric input of anthropogenic lead. Mass balance flows of lead through North Pacific open ocean waters for both prehistoric and present times are depicted.

Unfortunately at present no investigator at any oceanographic laboratory in the world, other than that at Caltech, can correctly collect samples of seawater below a meter depth without some probability of serious lead contamination (Ref. 1). Samples obtained from various depths of the surface mixed layer down to depths of 100 m by means of plastic hoses and pumps that have been washed in the field with acid, and that have the hose supported from a low-contamination raft away from the research vessel (Ref. 2) have not yet been subjected to unambiguous interlaboratory calibration at CIT, but tentative indications are, excluding surface adsorption effects, that lead contamination may be less than +20% at the 10 ng Pb/kg level for this type of collector. Surface samples collected by hand dipping bottles from the bow of a small raft or craft moving upwind and upcurrent from the support vessel, using properly cleaned containers, proper handling methods in the field to exclude contamination, and proper seals during transport, are the only type that can be collected by other laboratories at present to provide non contaminated seawater samples. Table 1 illustrates the degree of contamination contributed by seawater samplers in common use today. Only the Caltech seawater sampler shown in Fig. 1 will at present provide equivocation samples of seawater uncontaminated with artifact lead (Ref. 1). Four of the samplers listed in the lower half of Table 1 were used in parallel with the Caltech sampler, and in each of those cases the seawater samples were analyzed in the Caltech laboratory. The Caltech sampler rigorously excludes local artifact lead contamination present as an aura in waters surrounding the research vessel from contacting any surface of the sampler that also contacts the sampled water. Its internal parts, which are made of conventional polyethylene, are cleaned, assembled, and enclosed in cleaned bags in an ultra-clean laboratory. Although the outside seals are broken when the sample container is inserted into the sampler mechanism aboard ship just before use, the internal seals, including the protecting cap which is individual with each sample container, are never displaced except at the time the sample is collected at the appropriate depth. After its contents have been drained into container bottles, each sample collector unit, including its individual protecting cap, is brought back to the laboratory and re-cleaned and resealed for subsequent use.

TABLE 1 Contamination from Deep Water Samplers

Sampler Type	Volume (L)	Features	Measured Pb Conc. ng/kg Last 5 values Anal. in CIT Laboratory	True Pb Conc. ng/kg In same waters but collected in CIT Sampler
Niskin	30	PVC, internal rubber band	3000 - 5000	1 - 15 (est.)
Gerard-Ewing	270	All stainless	90 - 200	1 - 5 (est.)
Modified Niskin	100	All stainless external closures	30 - 200	1 - 15 (est.)
Swedish lucite	5	All plexiglass	20	4
Top Drop	30	PVC, teflon coated	10	1.7
Go Flo	30	PVC, Teflon coated, on plastic wire	40	4
Modified Go Flo	30	Teflon drain plugs	5.7	2.9
Caltech Sampler	10	CPE, FEP teflon, stainless housing	1.0	1.0

Although sensitive instruments have been developed for measuring small concentrations of lead in seawater, it has been found through intercalibration studies that the accuracy of measurements of lead concentrations provided by these instruments in most laboratories today is so poor that most of these measurements are without scientific significance (Ref. 3).

The only trustworthy data concerning the occurrence of lead in the open oceans which are available at this time are those obtained at two field stations off the California coast and obtained during a cruise from Honolulu, Hawaii, to Monterey Bay, California in a collaborative research program with J. Martin and coworkers and K. Bruland and coworkers. Lead concentrations in waters obtained from these experiments are shown in Fig. 2. Surface samples were collected by hand dipping bottles from the bow of a small raft moving upwind and upcurrent away from the research vessel. Depth profile samples were collected with the sampler illustrated in Fig. 1. Analyses were performed by isotope dilution mass spectrometry under ultra-clean laboratory conditions. Dissolved lead was determined by dithizone extraction into chloroform, while total lead was determined by boiling the sample to salts with aqua regia. These procedures have been shown to properly distinguish between dissolved and particle lead (Ref. 4).

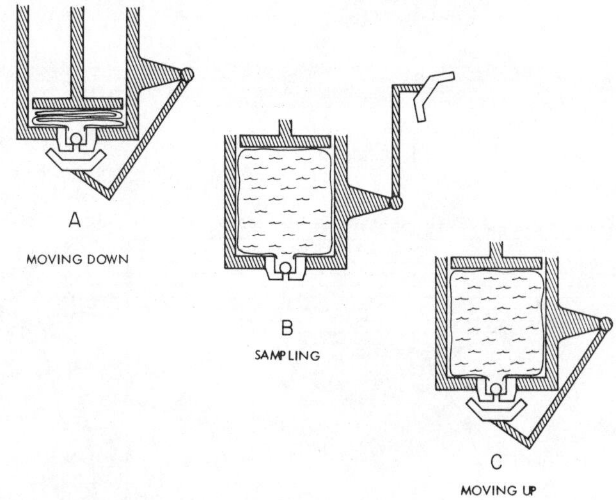

Fig. 1. Schematic drawing of CIT deep water sampler.

A Entry port is protected by retractable cup. Sample bag is collapsed. Sampler is lowered at end of hydrowire, trailing contamination.

B Protection cup is retracted and sample bag is filled by withdrawing piston. Water is retained by check valve. Sampler moves down through uncontaminated waters while filling.

C Protection cup is replaced. Sampler is rising to ship.

It should be recognized that the present occurrence of lead with depth in the North Pacific ocean is distinctly different from the occurrences of some other trace heavy metals such as Ba, Zn and Cd. As discovered only very recently by Edmond, Bender and coworkers, and by Martin, Bruland and coworkers, the concentration profiles of the latter metals adhere closely to those for nutrients, with the distributions of Ba and Zn paralleling the distributions of SiO_2 and the distributions of Cd paralleling those of PO_4^{-3} and NO_3^- (Ref. 5). The general trend for these constituents is for their concentrations to increase rather sharply with increasing depth of about 700 - 1200 m with considerably smaller concentration changes at greater depths. The bulk of these elements are found in the deep water. For common lead, however, the bulk is located in the surface mixed layer and in the thermocline. The Pb profile essentially parallels those of unsupported ^{210}Pb and tritium as shown in Fig. 3. The concentrations do not change significantly to depths of 250 to 400 m, then concentrations decrease sharply within small depth intervals to very small values, after which concentration changes become more gradual. Mass balance considerations indicate that today a large proportion of the lead introduced to the oceans enters from the atmosphere, originating mostly from combustion of leaded automobile fuels (Ref. 8). Since both tritium, as debris from nuclear bomb tests, and unsupported ^{210}Pb, derived from continental ^{222}Rn emanations, enter the oceans basically by this same atmospheric route, it is not surprising to observe that their distributions are similar to that for common industrial lead. This depth profile, however, provides reliable experimental evidence for these estimates and for the large scale perturbations caused by man.

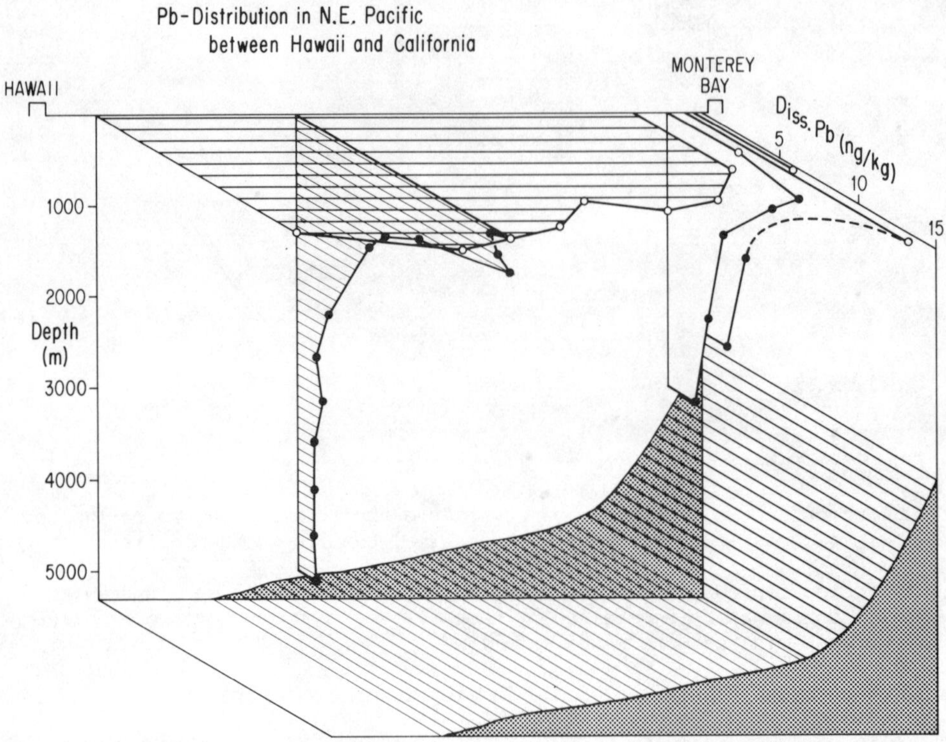

Fig. 2. Occurrence of dissolved lead in surface and deep waters of NE Pacific between Hawaii and California (Ref. 8).

In a series of surface water samples taken between Hawaii and California, concentrations of common lead also parallel unsupported ^{210}Pb concentrations, which is explained mainly by similar input distributions and parallel removal mechanisms which are very efficient near the ocean boundaries (depending on biological productivity and suspended particle load for example). In barren waters in the open North Pacific, the concentrations of both species of lead are relatively high, while in highly productive shelf waters the concentrations are relatively low, as shown in Fig. 4.

The concentration of lead in deep ocean waters is about 1 ng Pb/kg seawater, and if we generalize from the single mid-ocean profile shown in Fig. 2 and Fig. 3, we have some indication of both the sizes of lead reservoirs in the North Pacific, and the magnitudes of the prehistoric and present day fluxes of lead via the atmosphere and rivers to this ocean reservoir.

It is not generally recognized that the isotopic compositions of authigenic lead in Pleistocene pelagic sediments (lead that was originally soluble and transformed

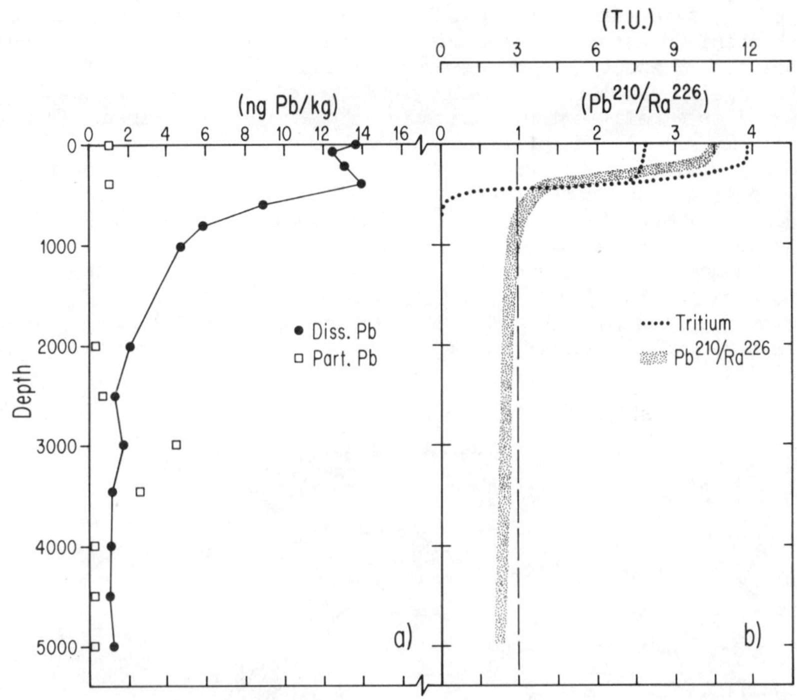

Fig. 3. a) Depth profile of common lead concentration in the central northeast Pacific (Ref. 8).

b) General structure of Tritium and ^{210}Pb distribution in the central northeast Pacific (based on Refs. 6 and 7).

after biological sedimentation) are distinctly different from region to region on the ocean floor and that they reflect the various isotopic compositions of lead in the surface rocks of the terrestrial drainage basins adjacent to their corresponding pelagic regions (Ref. 10). This imposes the restriction that soluble lead must have a short residence time in shallow surface waters that move rapidly in horizontal directions, otherwise the various isotopic compositions would become too well homogenized, and the observed isotopic patterns in the pelagic sediments could not exist. The residence time of common lead in surface waters can be estimated for the upper 100 m from ^{210}Pb data (Ref. 9), assuming ^{210}Pb is a suitable analogue for common lead and that below this depth removal mechanisms are different. Calculations using measured ^{210}Pb distributions in surface waters and estimates for the atmospheric input flux of ^{210}Pb give a residence time of about 2 years for ^{210}Pb in open ocean surface waters, which can also be attributed to common lead (Ref. 9).

In thermocline waters, common lead probably has a much longer residence time and the distributions of lead in these waters of the northeast Pacific are affected by vertical transport of sinking particles which both release and scavenge lead, and by the lateral movement of tongues of water containing different concentrations of lead.

The residence time of common lead in waters deeper than 100 m probably lies in the range of 50 to 300 yrs with some indication that the longer time is more likely. One way to estimate this is to assume that the shape of the depth profile of common lead shown in Fig. 3 has not been anthropogenically disturbed seriously below 2000 m. If, for prehistoric times, this profile is extrapolated upward to 100 m such that the prehistoric reservoir of soluble lead in the north Pacific was determined by a concentration of about 1 ng Pb/kg over the depth interval of 100 to 4300 m, a deep water standing crop of about 420 ng/cm^2 of lead would be obtained. The deposition flux of originally soluble seawater lead would be obtained. The deposition flux of originally soluble seawater lead through the water column to pelagic sediments during the Pleistocene is known rather well from measured rates of accumulations of sediments and the concentrations of authigenic lead in the ferro-manganese phases of those sediments (Ref. 10). That flux, combined from both the biologically active surface layer and the deeper waters amounted to ~ 3 ng Pb/cm^{2}yr during that time period. The residence time of the deep water lead during the prehistoric era would therefore range from 300 to 1500 yrs, according to whether 50 % or 10 % of this total measured flux came from the deep water reservoir, while the remainder came from the biologically active surface layer.

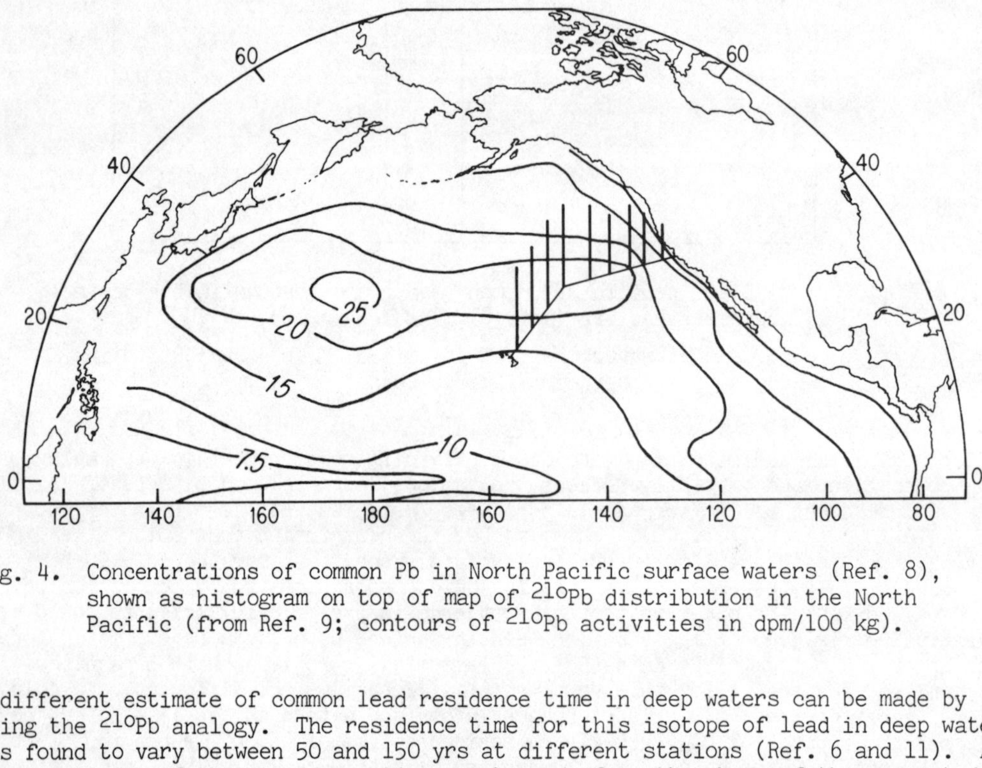

Fig. 4. Concentrations of common Pb in North Pacific surface waters (Ref. 8), shown as histogram on top of map of ^{210}Pb distribution in the North Pacific (from Ref. 9; contours of ^{210}Pb activities in dpm/100 kg).

A different estimate of common lead residence time in deep waters can be made by using the ^{210}Pb analogy. The residence time for this isotope of lead in deep waters was found to vary between 50 and 150 yrs at different stations (Ref. 6 and 11). A third estimate of this residence time can be made from the shape of the present day observed profile between 4500 m and 1500 m (shown in Fig. 3a) together with calculations based on vertical diffusion-advection model (Ref. 8). Residence times (standing crop/scavenging rate) of about 1000 yrs are obtained. This time would drop to 200 yrs if the lead profile were altered by only 0.8 ng/kg, which is close to our analytical resolution and blank control capabilities.

The distribution of common lead in the North Pacific open ocean at present can be compared with that during prehistoric times by comparing Figs. 5 and 6 which depict tentative models of input and output fluxes and standing crops in shallow and deep portions of a water column located in the central North Pacific for those two times. The water column is divided into a shallow surface layer in which the residence time of lead is about 2 years, and a deep compartment in which the residence time of lead in on the order of a century. This deep water compartment is subdivided into upper and lower portions in which, at present, the upper section contains a large standing crop of lead which we believe has accumulated during the past several centuries.

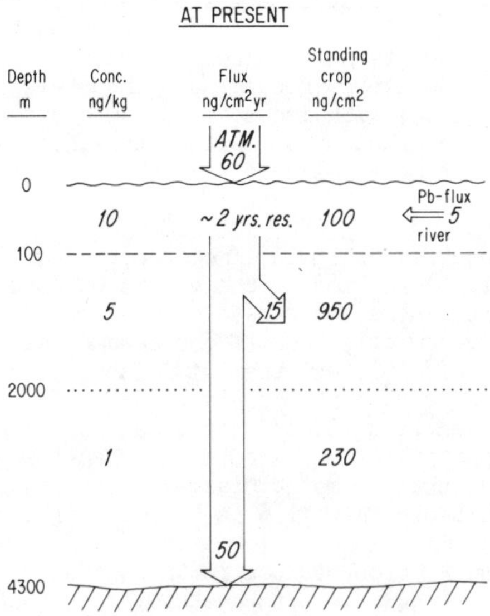

Fig. 5. Oceanic outputs, river and atmospheric inputs and reservoirs of soluble lead at present times in the open North Pacific (Ref. 8).

In the North Pacific today, the atmospheric input of 60 ng Pb/cm^{2}yr shown in Fig. 5 is estimated by two methods. One method is based on measurements of lead in rain at the NOAA Observatory in Samoa. This rain contains lead > 100 times in excess of natural dust or salt concentrations (Ref. 12). This wet deposition flux of lead was multiplied by 10 to account for the greater atmospheric concentration of lead in the North Pacific to yield a flux of 80 ng Pb/cm^{2}yr in the North Pacific. By far the largest proportion of industrial lead aerosols is produced in the northern troposphere, and since the residence time of lead aerosols in the atmosphere (~ 1 to 2 wks) is short compared to the half life of the interhemispheric exchange (~ 1 to 2 yrs), the atmospheric concentration of lead in the northern hemisphere should be much larger than that in the southern hemisphere.

The high concentrations of lead in marine wet and dry deposition fluxes reported by other investigators in northern latitudes (Ref. 13) cannot be accepted as relevant to the world flux calculations because those samples probably have been seriously contaminated with lead both during collection and by highly local sources of industrial lead that cannot be generalized on a world scale. In American Samoa, for example, it was proved by lead isotopic tracers that lead concentrations (50 ng Pb/kg) would be elevated 5-fold by local auto exhausts, despite the remoteness of this island and the extremely modest amount of leaded gasoline burned, if care was not exercised to exclude the collection of rain that had first formed over the island and then moved laterally to the collection site located on a remote promontory of the island (Ref. 12).

A second method of estimating the atmospheric input of common lead to the North Pacific is based on measurements of common lead /^{210}Pb ratios in surface waters and on atmospheric ^{210}Pb input fluxes to the sea. The atmospheric input of common lead of ~40 ng/cm^{2}yr is estimated by assuming that the ratio of concentrations of these two species of lead in surface waters reflects the ratio of their inputs (Refs. 6 and 8). The mean of the above two estimates of the atmospheric input was used in Fig. 5. World production of far spreading lead aerosols derived from the burning of ~300 000 tons of gasoline lead annually is estimated to account for about half of this calculated flux. Additional sources of industrial lead aerosols are both lead and other base metal smelters, resuspended industrial lead dusts, and fires burning plant foliage polluted with industrial lead dry deposition. This agreement within a factor of 2 is satisfactory and within the uncertainties of the tentative model. This picture is far more reliable than the earlier one offered in 1963 (Ref. 14).

The present day river input is estimated from reported dissolved lead concentrations in the Mississippi (Ref. 15) applied to the volume of river water discharged annually into the North Pacific, increased two-fold to account for the additional input of industrial lead from storm drain runoff and sewage that is discharged directly into the North Pacific (Ref. 16), and then diminished to one third to account for loss by biologic uptake on continental shelves (Ref. 17). This is probably a lower limit because 99 % of river lead is attached to particles and it is probable that a small fraction of this particle lead is released when the particles enter seawater. If 10 % is released, then the estimated river input is increased by an order of magnitude, and the input shown in Fig. 5 is the mean of those two values.

The present average concentration of lead in the surface layer of the North Pacific is ~10 ng Pb/kg. This is equivalent to a standing crop of 100 ng/cm^2, which, divided by the annual input flux, yields a residence time of about 2 years for lead in the surface layer. This time is feasible, because it has been shown that this mass of lead can be removed from the surface layer within about a year by means of sorption and phytoplankton surfaces, followed by zooplankton uptake and excretion (Ref. 18). The residence time of ^{210}Pb in surface waters of the open North Pacific has been estimated from observed atmospheric input fluxes and water concentrations and also found to be ~ 2 years (Ref. 9).

Studies of ^{210}Pb indicate that the bulk of this nuclide removed from the surface layer is transported to the sediment with not more than about one tenth being released into subsurface waters during transit (Ref. 6). If it is assumed that one fifth of the output flux of the common lead in the North Pacific is released during transit from the surface layer to the sediments, then this input of common lead to the upper part of the deep waters can account for the common lead now found there. The increased concentration of anthropogenic lead in the atmosphere during the last centuries, as recorded in Arctic ice strata (Ref. 19), would provide a sufficient amount of lead for this process.

The estimated prehistoric reservoirs and fluxes of common lead in the North Pacific open ocean are shown in Fig. 6. Methods for estimating the output flux and standing crop of common lead and the residence time of this lead in the deep water portion of this model have been mentioned above.

The estimated prehistoric atmospheric input of soluble lead to the oceans was < 0.5 ng/cm^{2}yr. This prehistoric atmospheric input is obtained by multiplying the crustal Pb/Al ratio with the concentration of aluminum observed in rain at the NOAA Observatory at American Samoa (Ref. 12), adjusted to estimated Al fluxes in the North Pacific.

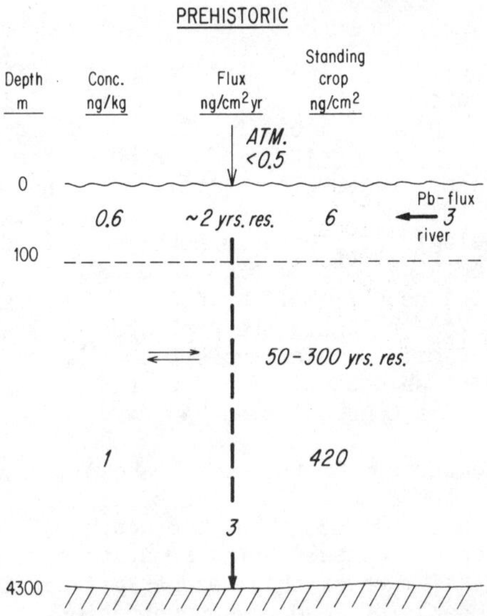

Fig. 6. Oceanic outputs (Ref. 10), river and atmospheric inputs, and reservoirs (estimated) of soluble lead during prehistoric times in the open North Pacific. Soluble lead entering the shallow reservoir is removed with particles to the sediments where it is chemically transformed to authigenic minerals. Fluxes of lead in terrigenous silicates which do not contribute significant amounts of soluble lead are not shown.

It is rather certain that Pb/Al ratios in atmospheric aerosols were not > 100-fold higher than crustal values during prehistoric times, as they are now in all remote regions of the earth, including American Samoa (Ref. 12), and that volcanic emissions did not contribute significantly to atmospheric inputs of lead to the oceans during prehistoric times. First, direct measuremtns of Pb/S ratios in volcanic gases ($\sim 7 \times 10^{-8}$) (Ref. 12), multiplied by total world volcanic emission of sulfur today (3×10^{6} tons S/yr) (Ref. 20), yield inputs of lead to the atmosphere from volcanoes of less than 1 ton Pb/yr. It is indeed observed that Pb/Al ratios are highly enriched in volcanic gases above crustal values (Ref. 12 and 21), but it must be emphasized that the Pb/S ratio is so enormously depleted relative to crustal values (30 000-fold) that the lead flux to the atmosphere from this source was inconsequential.

Second, direct measurement of the Pb/Si ratio in Arctic snow layers 3000 yrs old (Ref. 19) show that it was not enriched above crustal values at that early date. Investigators who have recently discounted this single observation as inconsequential in the face of their numerous recent measurements of a 50-fold enrichment of Pb/Al ratios above crustal values in 1000 yr old Arctic show (Ref. 22) used samples known to be seriously contaminated with lead, and in addition used analytical sensitivities and controls for lead contamination that were inadequate and deficient by about the same 50-fold factor needed to discover whether the enrichment was absent. It is very expensive to collect 3000 yr old snow samples under conditions where lead contamination is rigorously controlled (Ref. 19), and we are now carrying out cooperative studies with these investigators to see whether drill core ice, which can be obtained at much less cost, can be properly cleaned of its large lead contamination before analysis.

Third, the sedimentary record in the oceans makes it difficult for a > 100-fold enrichment of Pb/Si above crustal values to have existed in the atmospheric input to the oceans during prehistoric times, for if it did, it would have been inconsequential with respect to the magnitude of the Pb in the output flux from the oceans to pelagic sediments, because the total Pb/Si ratio recorded in prehistoric sediments is only ~3-fold, and not > 100-fold higher than crustal value. Two thirds of the lead in pelagic sediments is authigenic lead contained in ferro-manganese minerals, which originated from dissolved lead in river water that escaped biological removal in continental shelf waters. Therefore the atmospheric input flux of lead was <.5 ng/cm^2yr, while the prehistoric river input of dissolved lead reaching the open North Pacific was equivalent to 3 ng/cm^2yr. Assuming that the residence time was 2 years as determined today, the standing crop calculations to be 6 ng/cm^2, equivalent to a concentration of ~.6 ng/kg in surface waters.

It can be seen from a comparison of the models shown in Figs. 5 and 6 that man has probably substantially altered the occurrence of lead in the open oceans of the northern hemisphere through the industrial production of lead products. Although industrial lead pollution of coastal marine ecosystems is substantial on a worldwide scale and the input of much of this lead originates from rivers and sewers, the effect of river discharge of lead on concentrations of lead in surface waters of the open oceans is far less than the influence of the input of atmospheric industrial lead. The effect of river input on the natural concentration of the surface waters of the open ocean is uncertain at this time, but it would appear that it has resulted in about a 2-fold elevation of lead concentrations. Atmospheric inputs of industrial lead to the surface waters of the open oceans in the northern hemisphere on the other hand have probably elevated lead concentrations about 10-fold.

Half the marine biomass is produced in continental shelf regions, where the effects of industrial lead pollution are much greater than in the open oceans, because of the concentration of inputs over limited areas. The input of industrial lead has elevated the lead concentration in this shelf region fraction of the marine biomass 3-fold to 10-fold according to mass balance considerations and studies of temporal changes of lead concentrations in coastal marine sediments (Ref. 16). The average effect of the anthropogenic increase of lead concentrations in the open oceans on the remaining half of the marine biomass is less clear at this time. The increase of lead concentrations in these surface waters, brought about since prehistoric times as shown by comparison of Figs. 5 and 6, is clearly about 10-fold, and has elevated lead concentrations in the biomass there by this factor. However, these data apply to nonproductive waters and the magnitude of the elevation of lead concentration in the marine biomass in productive areas of the open ocean remains to be determined by further studies.

Temporal changes in the atmosphere input flux into the oceans as estimated from (Ref. 19) are characterized by a ~ 3-fold increase above natural river inputs about two centuries ago, followed by an additional gradual two-fold increase until ~ 1945. After that time the atmospheric input tripled by 1965, and then remained approximately constant to the present. The residence time of lead in surface waters is so short that lead concentrations closely followed the above probable changes in atmospheric input fluxes. However the residence time of lead in deep ocean waters is on the order of centuries, so that the influence of the large increase of atmospheric flux since 1945 in the deep ocean waters has not yet reached steady state. It is probable that the standing crop of lead in this reservoir is increasing with time and is propagating downward. The mass of anthropogenic lead in this reservoir which probably amounts to ~ 1 million tons in the northern hemisphere is estimated to be 10 to 20 % of the mass of anthropogenic lead transported to the sediments during the last two centuries. This 15 million tons originated from the 200 million tons of industrial lead produced during that period (Ref. 8).

A tentative measurement of the downward flux of lead attached to particles at 1000 m depth has been made at the northeast Pacific profile station and a lead flux lower than that shown in Fig. 5 was found. However the particle collection efficiency for the prototype device used is not yet known nor are the influences of temporal variations in productivity and heterogeneities in particle concentrations, and further studies of particle interceptor traps are needed to provide reliable data. The model shown in Fig. 5 is suggested for the region of the Pacific north of 10°N. It is possible that higher lead input fluxes in the North Atlantic and significantly lower lead input fluxes in the South Pacific and in the South Atlantic have resulted in quite different lead distributions in the water column. Further modifications are likely in different areas of the world oceans due to differences in river inputs, productivity, suspended particle loads, and ages of deep water masses. These findings initiate a new era in studies of the chemical oceanography of lead, and to know the general occurrence of lead in the world oceans, to understand the influence of the various factors which determine that occurrence, and to recognize properly the extent of anthropogenic lead perturbations in the oceans, more studies should be carried out.

Lead in seawater has been seriously studied for more than 45 years. We now know with certainty that most of this work, including a quite large amount of recent effort, is without scientific significance because of both inaccurate analyses and contaminated samples. Sixteen years ago Chow and Patterson outlined principles for lead transport from land to marine surface waters and from marine surface waters to sediments, and they measured the Pleistocene output flux of soluble lead from the oceans, which indicated that the present input flux of lead greatly exceeds that of the past (Ref. 10). Later, Tatsumoto, Chow and Patterson developed the IDMS-clean laboratory analytical methods for lead, which laid the foundation for the studies reported here (Refs. 14 and 23). Although those earlier IDMS analyses for lead in seawater were accurate, the samples had been badly contaminated during collection by auras of dirty water surrounding the research vessels, and their reported results were incorrect. It should be understood, however, that the studies reported here are an extension of those earlier contributions by our colleagues, Chow and Tatsumoto.

This work was supported by the IDOE Office of the National Science Foundation through grand OCE74-24362 A02, and by the Bureau of Land Management, US Department of Interior, Contract 696-16. This article is a summary of an oral presentation which includes discussion of some unpublished and ongoing work by Martin and coworkers and Bruland and coworkers (Ref. 6), by Duce and coworkers, Fitzgerald and coworkers, and Patterson and Settle (Ref. 12), by Chesselet (Ref. 21), by Burnett and Patterson (Ref. 18), as well as by the listed authors (Refs. 1 and 8). Results of these studies will be published in full technical detail. Contribution No. 3051: Publications of the Division of Geological and Planetary Sciences, California Institute of Technology, Pasadena, California 91125, USA.

REFERENCES

(1) B. K. Schaule and C. C. Patterson, A deep water collection device which provides samples free of lead contamination (in preparation).

(2) J. Martin and K. Bruland, Collection devices used for Zn, Pb, and Cd designed and used in research (not yet published).

(3) C. C. Patterson, Lead in seawater workshop meeting report. Science 183, 553 (1974); Participants of the lead in sewater workshop, Interlaboratory lead lead analyses of standardized samples of sewater, Marine Chemistry 2, 66-84 (1974);12 Authors, Comparison determinations of lead by investigators analyzing individual samples of seawater in both their home laboratory and in an isotope dilution standardization laboratory, Marine Chemistry 4, 389-392 (1976).

(4) C. Patterson, D. Settle, and B. Glover, Analysis of lead in polluted coastal seawater, Marine Chemistry 4, 305-319 (1976).

(5) M. L. Bender and C. Gagner, Dissolved copper, nickel, and cadmium in the Sargasso Sea, Journal of Marine Research 34, 327 (1976); E. A. Boyle, F. Sclater, and J. M. Edmond, On the marine geochemistry of cadmium, Nature 263, 42 (1976); J. Martin, K. Bruland, and W. Broenkow, Cadmium transport in the California current, in Marine Pollutant Transfer, 1976, pp 159-184 (Windom and R. Duce, eds.); K. Bruland, G. Knauer, and J. Martin, Cadmium in northeast Pacific waters, J. Limnology and Oceanography (in press); K. W. Bruland, G. A. Knauer, and J. H. Martin, Zinc in north-east Pacific water, Nature 271, 741 (1978).

(6) M. P. Bacon, D. W. Spencer, and P. G. Brewer, Pb^{21o}/Ra^{226} and $^{21o}Po/^{21o}Pb$ disequilibria in seawater and suspended particulate matter, Earth and Planetary Science Letters 32, 277-296 (1976); K. Bruland and coworkers, and J. Martin and coworkers, Unpublished studies in zinc, cadmium, lead, lead-21o, radium-226, etc., particulate matter, and nutrients for the Hawaiian Traverse of August, 1977.

(7) B. L. Michel and H. E. Sues, Bomb tritium in the Pacific Ocean, Journal of Geophysical Research 80, 4139-4152 (1975).

(8) B. K. Schaule and C. C. Patterson, The occurrence of lead in seawater and perturbations caused by anthropogenic inputs (in preparation).

(9) Y. Nozaki, J. Thomson, and K. K. Turekian, The distribution of 21o-lead and 21o-polonium in the surface waters of the Pacific Ocean, Earth and Planetary Science Letters 32, 304-312 (1976).

(10) T. Chow and C. C. Patterson, The occurrence and significance of lead isotopes in pelagic sediments, Geochimica et Cosmochimica Acta 26, 263-308 (1962).

(11) B. L. K. Somayajulu and H. Craig, Particulate and soluble ^{21o}Pb activities in the deep sea, Earth and Planetary Science Letters 32, 268 (1976); Y. Nozaki and S. Tsunogai, 226-radium, 210-lead, and 210-polonium disequilibria in the western North Pacific, Earth and Planetary Science Letters 32, 313-321 (1976); J. Thomson and K. K. Turekian, 21o-Po and 21o-Pb distributions in ocean water profiles from the eastern South Pacific, Earth and Planetary Science Letters 32, 297 (1976).

(12) Data obtained under NSF Contract OCE74-24362 A02, collaborative work by R. Duce and coworkers, by W. Fitzgerald and coworkers, and by D. Settle and C. Patterson, Wet and dry deposition fluxes of heavy metals in the southern Pacific. Emissions of heavy metals from Hawaiian volcanoes (in preparation).

(13) R. S. Cambray, D. F. Jeffries, and G. Topping, An estimate of the input of atmospheric trace elements into the North Sea and the Clyde Sea (1972-73), Atomic Energy Authority Harwell Report AERE-R7733 (1975) pp 30. It should be realized that these data cannot be extrapolated to the open oceans without major corrections for effects of local pollution and analytical contamination.

(14) M. Tatsumoto and C. C. Patterson, The concentration of common lead in sea-water, in _Earth Science and Meteoritics_, (Geiss and E. Goldberg, eds.), North-Holland, Amsterdam , 1963, 74-89.

(15) J. H. Trefey and B. J. Presley, Heavy Metal Transport from the Mississippi River to the Gulf of Mexico, in _Marine Pollutant Transfer_, (Windom and R. Duce, eds.), Lexington Books, 1976, pp 39-76.

(16) M. W. Burnett, A. Ng, D. Settle, and C. Patterson, Impact of man on coastal marine ecosystems, in _Proceedings of the International Experts Discussion on "Lead - Occurrence, Fate and Pollution in the Marine Environment"_, Rovinj, October 1977, M. Branica and Z. Konrad (eds.), Pergamon Press, 1979, 7-13.

(17) E. Goldberg, and K. Turekian, By analogy with ^{210}Pb Turekian believes that only a small fraction of soluble lead from rivers escapes sedimentation on the shelves and passes into the open oceans. Goldberg and coworkers, on the other hand, by analogy with ^{210}Pb and from their studies of common lead deposition in coastal basins believe that a substantial fraction of soluble common lead from rivers escapes deposition on shelves, _personal communication_.

(18) M. W. Burnett and C. C. Patterson, Distribution and occurrence of alkaline earths and lead in marine ecosystems (in preparation).

(19) M. Murozumi, T. J. Chow, and C. C. Patterson, Chemical concentrations of pollutant lead aerosols, terrestrial dusts, and seasalts in Greenland and Antarctic snow strata, _Geochimica et Cosmochimica Acta_ 33, 1247-1294 (1969).

(20) R. E. Stoiber and A. Jepsen, Sulfur dioxide contributions to the atmosphere by volcanoes, _Science_ 182, 577-578 (1973).

(21) R. Chesselet, Metals in volcanic gases of Mt. Aetna, unpublished manuscript.

(22) M. Herron, C. Langway, H. Weiss, and J. Cragin, Atmospheric trace metals and sulfate in the Greenland ice sheet, _Geochimica et Cosmochimica Acta_ 41, 915-920 (1977).

(23) T. J. Chow, Lead isotopes in sea water and marine sediments, J. of Marine Research 17, 120-127 (1958); T. J. Chow and C. Patterson, Concentration profiles of barium and lead in Atlantic waters off Bermuda, _Earth and Planetary Science Letters_ 1, 397-400 (1966).

Investigations into the Formation and Behaviour of Aqueous Solutions of Lead Alkyls

J. R. Grove

Associated Octel Company Limited, Ellesmere Port, U.K.

ABSTRACT

The behaviour of tetramethyllead (TML) and tetraethyllead (TEL) in water has been investigated. Saturation solubilities are low: about 15 mgPb/l for TML and not greater than 100 g Pb/l for TEL in fresh water. Concentrations taken up into flowing water are lower than the saturation values. In the case of water flowing over TEL a degradation product, the trialkyllead cation, is found to be present in the water to a higher concentration than TEL itself. Breakdown of dissolved TEL and TML in solution takes place at rates giving half lives measurable in days.

INTRODUCTION

Lead tetraethyl (TEL) and tetramethyl (TML) are manufactured as antiknock additives for gasoline. TML is manufactured in the presence of toluene to give a mixture containing about 80% TML by weight. Before transport the lead alkyls are blended with alkyl halides. The materials referred to in this paper are therefore: TEL and TML/toluene (the unblended materials), TEL-B (an equimolar blend of TEL and 1,2-dibromoethane) and TML-CB (a blend of TML plus one half mole of 1,2-dibromoethane, one mole of 1,2-dichloroethane, and toluene).

If released into water, lead alkyls or antiknock additives will sink since their specific gravities are all greater than 1.5. A series of experiments has been carried out over a number of years to determine the fate of these materials under water. Throughout this period, techniques particularly for analysis have been continuously improved.

The objective of the first experiments was to determine the solubilities of the tetraalkyl species TEL and TML in water. Feldhake and Stevens (Ref. 1) reported values for TEL of between 0.13 and 0.19 mgPb/l between 0 and 34°C. (N.B. Concentrations of organolead species are expressed throughout this paper on the basis of the lead content only). Problems in determining the solubilities of tetraalkyllead species arise from the fact that ionic breakdown products rapidly accumulate in any water in contact with the lead alkyls. Before each experiment the lead alkyl used was washed with a dilute ammonia solution to remove as much decomposed material as possible. It never proved possible to obtain lead alkyl totally free from breakdown products. The major ionic species found is the trialkyllead ion but the dialkyllead ion and inorganic lead ion are also found in small amounts. The solubility determination thus requires the tetraalkyl species to be differentiated from the breakdown products; this is most easily carried out if the concentration of the breakdown products is minimised. In addition, it is desirable to minimise the possibility of emulsion formation. Use was therefore made of the equilibrator described by Jones and Kaplan (Ref. 2).

EQUILIBRATOR EXPERIMENTS

The equilibrator consists of two glass cylinders each of which contains a liquid phase. The cylinders are interconnected at each end through the gas phase. When the equilibrator is set at a specified angle to the vertical and rotated, the movement of the two liquid phases causes vapour to be pumped from one side of the equilibrator to the other. The whole is immersed in a water bath at constant temperature and kept in the dark. If rotation is continued for a sufficient time enough material is transported from one liquid phase to the other through the vapour, so that each liquid becomes saturated with the vapour of the other without the two liquid phases coming into direct contact. The temperature in all experiments was 20°C.

Water was placed in one cylinder and lead alkyl in the other. At the time of the experiments described, the analytical procedure was as follows: an aliquot of the water layer was treated with iodine monochloride solution to bring all the organic lead species to the dialkyllead state which was then determined by the use of PAR reagent. The tetraalkyl species was then extracted from a further aliquot using hexane and the residual organolead ionic species determined using ICl/PAR; the tetraalkyllead concentration was determined as the difference between the total organic lead and the ionic organic lead. Dialkyllead was determined directly using PAR and inorganic lead ions determined as the difference between the total lead content and the total organic lead. The results obtained are illustrated in Tables 1 and 2 which show results for TEL and TML/toluene in sea-water. In these experiments the results for each period of equilibration represent a separate experiment since the whole of the water phase from the equilibrator was used for the analysis. It is seen that breakdown products are not entirely eliminated from the water phase, but are sufficiently low in amount to give confidence in the concentrations of the tetraalkyllead species. The experiments illustrated were carried out under a nitrogen atmosphere; the presence of air in the reactor did not materially alter the apparent solubilities of the tetraalkyllead species.

TABLE 1 Jones-Kaplan Equilibrator

Liquid phases: TEL and sea-water
Lead compounds in sea-water (mgPb/l)

Equilibration Time	R_4Pb	R_3Pb^+	R_2Pb^{2+}	Pb^{2+}	Total Pb
1 hour	0.3	0.2	0.4	(nil)	0.7
2 hours	0.5	0.3	y.3	(nil)	1.0
4 hours	1.2	0.2	0.3	(nil)	1.6
6 hours	1.4	0.2	0.3	(nil)	1.9
18 hours	1.9	0.3	0.2	0.2	2.6
24 hours	2.0	0.3	0.4	(nil)	2.7

TABLE 2 Jones-Kaplan Equilibrator

Liquid phases: TML/toluene and sea-water
Liquid compounds in sea-water (mgPb/l)

Equilibration Time	R_4Pb	R_3Pb^+	R_2Pb^{2+}	Pb^{2+}	Total Pb
1 hour	5.4	0.2	0.1	0.7	6.3
2 hours	8.4	0.3	0.2	1.6	10.4
4 hours	8.5	0.4	0.2	1.6	10.7
6 hours	9.2	0.7	0.1	2.6	12.6
18 hours	9.5	1.5	0.2	2.5	13.7
24 hours	9.2	1.6	0.1	1.9	12.8

The maximum water phase concentrations of the tetraalkyllead species obtained in this were as follows:

	(mgPb/l)	
	TEL	TML (toluene)
Distilled water	0.8	15.0
Sea-water	2.0	9.0

Some of these runs have been repeated more recently using more modern analytical techniques (Ref. 3). These gave concentrations in sea-water of 1.5 mg Pb/l for TEL and 22 mg Pb/l for TML (toluene). This suggests that the older values are reliable as to order of magnitude, but that either the whole procedure is not accurately reproducible or else the differences reflect improvements in the analytical methods.

After these experiments had been brought to a conclusion, intimations were received that the saturation solubility of TEL was lower than the value observed in the equilibrator (Ref. 4). The report was confirmed that, if solutions of TEL prepared in the equilibrator were centrifuged in an analytical centrifuge at 3,000 rpm, separation of TEL could be observed. The separation appeared to have reached equilibrium in less than 90 minutes. The following illustrates the results obtained. A solution of TEL prepared in the equilibrator by rotation for one and a half hours showed a TEL concentration of 510 μg/l. Each contrifuge tube holds 37 ml. After centrifuging, the top 10 ml of sample contained 54 μgPb/l and the bottom 10 ml, 16 μgPb = 1.6 mgPb/l as TEL. The TEL which did not move in the field of the centrifuge is presumably in true solution. The material which moved is presumably held in suspension. This latter material had been transferred to the water through the vapour phase. Presumably the driving force to transfer this material would only arise if the suspension were stabilised in some way.

In a number of different experiments inconsistent residual concentrations of TEL were found after centrifuging, although a sufficient number of values were found in the range 60 to 70 μg/l to suggest that this is the order of the true solubility of TEL in water. The solubility is thus less than the value found in the equilibrator by at least an order of magnitude.

In the case of TML, a much smaller proportion of the TML in solution moves during centrifugation; the solubility of TML recorded in the equilibrator is probably high but of the correct order of magnitude.

OPEN TANK EXPERIMENTS

Attention was now turned to simulating a more realistic situation. A glass tank containing 20 litres of water and sufficient lead alkyl (as blended Antiknock) to cover the bottom of the tank (about 80 ml) was gently stirred, and aliquots of water removed for analysis at intervals. The temperature was approximately constant at 20° C, and the tank shielded from daylight. The results are illustrated in Tables 3 and 4. The trialkyllead species are seen to increase with time. The TML concentration builds up to the value found in the equilibrator experiments, but the TEL concentration remains significantly lower. The observed tetraalkyllead concentrations are therefore compatible with the solubilities of these materials revised in the light of the centrifuging experiments.

TABLE 3 Open Tank

TEL-B Distilled Water (mgPb/l)

Time	Et_4Pb	Et_3Pb^+	Et_2Pb	Pb^{2+}	Total Pb
1 hour	0.06	1.50	0.05	0.04	1.65
2 hours	0.07	1.75	0.04	0.02	1.88
4 hours	0.07	1.82	0.02	0.05	1.96
6 hours	0.09	1.88	0.02	0.04	2.03
25 hours	0.18	2.15	0.03	0.02	2.38
2 days	0.18	2.31	0.01	0.02	2.52
3 days	0.24	2.34	0.01	Nil	2.59
7 days	0.20	2.69	0.01	Nil	2.90
10 days	0.14	2.66	Nil	Nil	2.80
15 days	0.16	2.79	0.01	Nil	2.96

TABLE 4 Open Tank

TML-CB Distilled Water (mgPb/l)

Time	Me_4Pb	Me_3Pb^+	Me_2Pb^{2+}	Pb^{2+}	Total Pb
1 hour	0.58	0.20	0.04	0.04	0.86
2 hours	1.16	0.36	0.02	0.03	1.57
4 hours	2.66	0.40	Nil	0.03	3.09
6 hours	3.68	0.84	Nil	0.01	4.53
1 day	5.32	0.92	Nil	0.02	6.26
2 days	11.97	1.25	0.02	0.01	13.25
4 days	11.98	3.65	Nil	0.01	15.64
7 days	16.98	6.52	Nil	0.03	23.53
14 days	18.50	10.00	Nil	0.06	28.56
21 days	18.98	11.32	Nil	0.02	30.32
30 days	19.96	11.72	0.02	0.08	31.78

TROUGH EXPERIMENTS

The rates of uptake of organolead species into flowing water were measured in a trough made of "Perspex", 4.8m long, 2.5cm wide and 10cm deep. It was found necessary to prevent loss of lead alkyl to the air by closing the trough with a lid. The trough was kept completely filled with water at each rate of flow by adjusting the height of a weir gate at the trough exit. The lead alkyl was held in a sump 60cm long, 2.5cm wide and 2.5cm deep attached beneath the trough. When the sump was filled with lead alkyl, the surface of the lead alkyl was level with the bottom of the trough. At water velocities up to about 4m/min flow was laminar. At greater flow rates the water flow started to become turbulent, and some rippling of the surface of the lead alkyl was seen. At flows of 10m/min or more, turbulent mixing in the trough was shown by dye injection to be complete and rapid. As the water flow increased the waves on the surface of the lead alkyl increased in frequency but increased little in amplitude, and no liquid droplets appeared to be taken up into the flowing water.

All experiments were carried out using fresh (filtered river) water. The results are illustrated by the values obtained at 15m/min flow rate (Table 5). With TML (toluene), TML was the lead species present in solution in greatest amount; ionic species constituted less than 10% of the total organic lead. With TEL in the trough, however, the major species found in solution was triethyllead ion; TEL was detectable by mass spectrometry but was at top low a concentration to be determined and therefore could not have been present to greater than about 1 μgPb/l.

TABLE 5 Trough Experiments

Water Flowrate: 15m/min

Experimental		
Material in Sump	TML (toluene)	TEL
Concentration in Water	0.020 mg Pb/l (TML)	0.004 mg Pb/l (Et_3Pb^+)
Rate of Uptake into Water	2.1 g Pb/m^2h	0.4 g Pb/m^2h (Et_3Pb^+)
Calculation (W.J. McManamey, 1976)		
Saturation Solubility Used	15 mg Pb/l	0.8 mg Pb/l
Rate of Uptake with Water	0.5 to 1.8 mg Pb/m^2h	0.02 mg to 0.08 mg Pb/m^2h
Concentration in Water Corresponding to Higher Rate of Uptake	0.017 mg Pb/l	0.8 μg Pb/l

Dr. W. J. McManamey (Ref. 5) has estimated mass transfer coefficients and calculated rates of uptake of lead alkyls into flowing water. The rates of uptake obtained are directly proportional to the values adopted for the saturation solubilities of each material. These are shown in Table 5. The lower calculated rates are for flat liquid surfaces. The higher rates include the maximum allowance expected for the effect of surface waves - a factor 3.5 over the rate for a flat surface. For TML the higher calculated value for the rate of uptake is seen to agree closely with the experimental value.

For TEL the measured concentration of triethyllead ion is five times the maximum calculated concentrations of TEL expected, even though the saturation solubility is assumed to be the concentration observed in the equilibrator. It is therefore likely that the triethyllead ions were extracted from the bulk lead alkyl and did not arise from the solution and subsequent breakdown of TEL. The calculated maximum TEL concentration depends on the value adopted for the saturation solubility of TEL; taking the value of 0.8 mg Pb/l observed in the equilibrator leads to a predicted concentration of 0.8 μgPb/l TEL in the trough. The lower solubility value indicated by the centrifuging experiments would lead to concentrations at least an order of magnitude lower than this. At the present time, the experimental techniques cannot distinguish between these predicted concentration levels. It is therefore not clear whether the mechanism of TEL uptake involves only true solution, or whether the suspended forms also participate.

BREAKDOWN OF LEAD ALKYLS IN SOLUTION

Liquid lead alkyls accumulate breakdown products, trialkyllead ions, dialkyllead ions and inorganic lead ions. The carbonates and hydroxides of the tri- and dialkyllead species are water soluble and so these are readily leached from the lead alkyl by water. In the tank experiments, relatively stationary concentrations of dissolved tetraalkyl lead species were obtained. By observing the steady build-up of trialkyllead ions which follows the frequently observed initial rapid accumulation of these ions at the start of the experiment, rates of breakdown of the dissolved tetraalkyllead species can be calculated. Because of the possibility of continuous leaching of some decomposition products from the pool of liquid lead alkyl, these apparent rates will be maximum rates of breakdown. In the case of TEL the half-life of the dissolved material calculated in this way ranged from about half a day in sea water to about two days in fresh water. On holding TEL solutions in syringes completely filled with fresh water which had not been de-aerated, the half-life of the TEL appeared to be longer and nearer to seven days. It does not appear that diffused daylight accelerates decomposition; the presence of air produced some increase in breakdown in the equilibrator experiments as compared with experiments carried out after ordinary nitrogen purging.

The decomposition rate of TML similarly determined is less than that of TEL.

The triethyllead ion is remarkably stable. In the tank experiments at about the 4 mg Pb/l level, decay amounted to about 25% over 60 days, diffused daylight making little difference. Trimethyllead ions at the level of 12 mg Pb/l showed virtually no decay over 60 days. The dialkyllead species were rarely observed in other than very small amounts in any of our experiments.

DISCUSSION

In attempting to explain the high values of TEL in water found in the equilibrator, and why these values, whether before or after centrifuging, are not particularly reproducible, it is possible to postulate the existence of droplets of tetraalkyllead stabilised by a layer of organolead ions at the interface between the water and the TEL. Alternatively, or additionally, it is possible to speculate about the role played by particulate matter in adsorbing dissolved organolead species; thus variable amounts of particulate matter could account for inconsistent values of apparent solubilities both before and after centrifuging the "solutions". Particulate matter will be present in sea-water, and centrifuging of the sea-water sample before equilibration was shown to affect the results. The nature of possible particulates in distilled water is less clear, but may be derived from the glass of the vessels used. The part played by particulate matter is known to be important in explaining the distribution of inorganic lead in sea-water (Ref. 6).

Theoretically the rate of uptake into flowing water depends on the saturation solubility of the material; it is not established whether the existence of a suspended phase will influence the rate of uptake in flowing water. The presence of solid particles will presumably affect the concentrations of lead alkyl observed in flowing water. Suspended particles may increase the rate of uptake; on the other hand layers of sediment on the surface of the pool will presumably have the reverse effect.

CONCLUSIONS

In spite of the experimental difficulties encountered with these systems, certain semi-quantitative generalisations are possible.

Tetraalkyllead species are sparingly soluble in water. The saturation solubilities in fresh water are probably about 15 mg Pb/l for TML and not greater than 100 μg Pb/l for TEL. Apparent solubilities greater than these values can be observed due to material held in suspension in some unidentified form. There is evidence that the solubility of TML is somewhat lower in sea-water than the value in fresh water.

Concentrations of tetraalkyllead taken up into flowing water are much lower than the above saturation values. For TML observed values accord reasonably well with theory. For TEL they are below the limit of measurement under the experimental conditions used.

Breakdown of TEL and TML in solution takes place at rates which give half lives measurable in days. The trialkyllead species are much more persistent.

ACKNOWLEDGMENTS

The author is indebted to the following colleagues who carried out the work described: P. Ashworth, D. J. Hodges, A. Slater, A. J. Stringer, W. Towers and A. Whittingham.

REFERENCES

(1) C. J. Feldhake and C. R. Stevens, J. Chem. Eng. 8, 196 (1963).

(2) G. Jones and B. B. Kaplan, J A C S, 50, 1600 (1928).

(3) F. G. Noden, The determination of tetraalkyllead compounds and their degradation products in natural waters, Proceedings of the International Experts Discussion on the "Lead - Occurrence, Fate and Pollution in the Marine Environment", Rovinj, October 1977, M. Branica and Z. Konrad (eds.), Pergamon Press, 1979, 83-91.

(4) Ethyl Corporation, Private Communication, (1977).

(5) W. J. McManamey, Private Communication (1976).

(6) L. Mart, H. W. Nürnberg, P. Valenta, Comparative baseline studies on Pb levels in european coastal waters, Proceedings of the International Experts Discussion on the "Lead - Occurrence, Fate and Pollution in the Marine Environment", Rovinj, October 1977, M. Branica and Z. Konrad (eds.), Pergamon Press, 1979, 155-179.

A Dehydration Method to Prevent Loss of Trace Elements in Biological Samples

E. D. Mor and A. M. Beccaria

Consiglio Nazionale delle Ricerche, Laboratorio per la Corrosione Marina dei Metalli, Genova, Italy

ABSTRACT

A dehydration method to avoid loss of trace elements in biological samples is described.

The samples are dehydrated in an apparatus connected to a vacuum pump until a vacuum of 10^{-6} Torr is obtained and complete dessication is effected.

The volatile lead compounds and the tissue water are collected by means of a trap cooled with liquid nitrogen.

The condensate, containing both organic and inorganic lead compounds, is extracted with benzene in order to separate the organic compounds. The organic lead compounds are mineralized with 6N HCl. The lead is determined by a polarographic technique.

INTRODUCTION

Several methods, employing polarographic techniques or flame or flameless atomic-absorption spectrometry for the determination of the total amount of lead in specimen of different origin (minerals, biological samples, etc.) have been described.

Especially the polarographic methods give fairly accurate and precise results (Ref. 1).

The main problem in the determination of lead in seaweeds and in some marine species, especially filter feeders (mussels, oysters) having a high water content, is the drying and decomposition of the biological samples.

A widely used dehydration system consists in heating the samples in an air-oven at 105 °C for a period ranging from one to several hours (Ref. 2-6).

Fourie and Peisach (Ref. 7) dried the samples at 50 °C in air. All these drying methods cause loss of elements which can be present in the shape of organic or inorganic volatile constituents. Drying methods in vacuum at 75 °C (Ref. 8) or at low temperatures (ranging between -10 °C and -24 °C) has therefore been suggested (Ref. 7).

Serious losses of some metals, including lead, have been described (Ref. 7) in both cases, since the volatile constituents were not retained.

Severe lead losses may also occur during calcining at 400 - 600 °C (Ref. 7, 9 and 10). Therefore, in order to obtain more accurate results methods of acid digestion at low temperatures have to be used to decompose organic specimens.

Recently, some organo-metallic lead compounds have been analyzed by Sirota and Ute (Ref. 11) and Chau et al.(Ref. 12 and 13). Sirota and Ute (Ref. 11) homogenized the specimens without drying them, by stirring them in a shaker for 10 minutes with benzene. They separated the benzene phase by centrifugation, and determined the inorganic Pb content in the homogenized textures submitted to centrifugation.

Chau et al.(Ref. 12 and 13) have developed a methodology for the measurement of the alkyl constituents of Pb in the atmosphere. However, this method is not easily applicable to the analysis of Pb in animal tissues having a high moisture content.

Since the technique developed by us (Ref. 14) provides basis for the determination of both organic and inorganic Pb constituents in marine sediments, it was deemed advisable to use it also for the determination of Pb in inorganic tissues.

EXPERIMENTAL

Instruments

A Tacussel polarograph PRT 30-01 for alternating superimposed sinusoidal current was used to determine the quantity of lead. The qualitative features of the organometallic constituents of lead were analyzed with the aid of a Varian-Aerograph series 2800 gaschromatograph fitted with a flame-ionizing detector.

Reagents

Only ARISTAR-BDH reagents were used.

Solutions were obtained with bidistilled water in a quartz still. Pb $(Et)_4$ and Pb $(Met)_4$ for the titration of the solution were supplied by S.I.A.C.

Procedure

The biological matter was dried in the apparatus illustrated in Fig. 1. The sample container C was connected to the vacuum pump through a capillary tube R, until a vacuum of 10^{-3} torr. was obtained. At this point, the sample container was directly connected to the vacuum by means of the tube L, up to 10^{-6} torr and to complete desiccation.

The volatile lead constituents and tissue water were collected by means of the liquid nitrogen cooled trap T.

Two dual samples of 10 μl each were taken from the condensate at room temperature, for gas-chromatographic analysis, under the following operating conditions:

- injection temperature: 120 °C
- temperature of the detector: 170 °C
- carrier gas (N_2)-hydrogen-air at 30 - 30 - 300 ml/minute;
- column: 1/8", l = 3m, T=80 °C packed with 10 % SE - 30 on chromosorb W 80 -100 mesh;
- electrometer: sensitivity 1 x 10^{-12} Amps/MV.

The condensate was then thoroughly mixed and stirred together with 10 ml of benzene to separate the organic from the inorganic Pb constituents. The organic constituents were extracted from the benzene, whereas the inorganic constituents remained in the aqueous phase.

The benzene phase was heated in a Hopkins still to 90 ^{o}C, together with 10 ml of concentrated HCl for 30 minutes, to decompose the organic Pb constituents (Ref. 15).

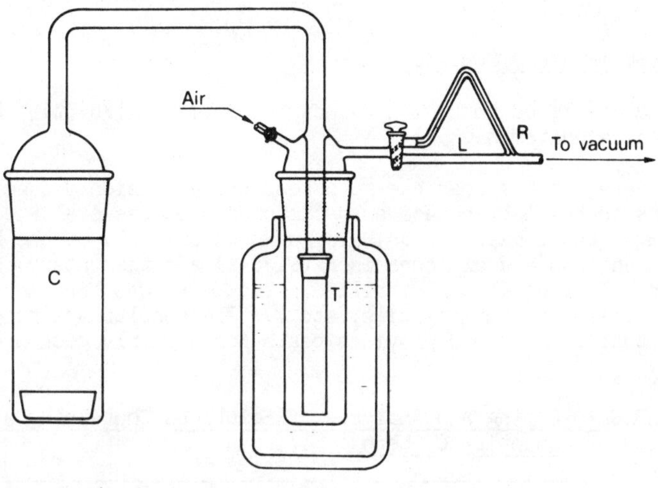

Fig. 1. Apparatus for specimens dehydration.

After this period, the benzene was evaporated and the same method was adopted to determine the inorganic Pb content in the aqueous phase, i.e., the Pb containing solution was desiccated while keeping the temperature below 50 ^{o}C and then eluted with HCl in order to obtain a 0.12 M final HCl solution. Furthermore, a small quantity of KCl was added to bring the solution to a 0.26 M value with respect to this salt. The solutions containing Pb had a final vclume of 25 ml. The half-wave potential reading was -325 mV (S.C.E.).

The end scale value of the Pb^{++} to Pb^{o} reduction current intensity was 50 nA with a recorded peak height of 25 cm.

The concentration (X) of Pb based upon the i_r value was derived by means of the following equation obtained from the calibration curve plotted with a $PbNO_3$ standard solution:

$$y = 0.4367 \pm 0.06485 + (14.5 \pm 0.2133)\ X$$

in which:

y = height of the i_r peak expressed in cm.

The dehydrated samples were decomposed by acid digestion with concentrated HNO_3 (65 %) and a few drops of $HClO_4$ (70 %) were added at the end of the dissolution. The solution was repeatedly evaporated to dryness by steam bath, in order to remove all N_2O_4 and $HClO_4$. The residue was dissolved with cold HCl (1:1), Fe^{3+} was reduced with hydroxylamine chlorhydrate and sodium acetate and the polarographic determination of the lead was accomplished after adjustment of the molarity of the solution at HCl 0.6 M and KCl 2.8 M.

The half-wave potential reading was -325 mV (S.C.E.). The concentration rate (X) of Pb, based upon the i_r (y) value was derived from the equation:

$$y = (13.0 \pm 0.198)\ X$$

in which Y = the height of the i_r peak, expressed in cm.

The precision of the polarographic measurements was $\pm$ 1 %.

RESULTS AND DISCUSSION

This method was developed both with standard solutions [$Pb(Et)_4$; $Pb(Met)_4$; $PbNO_3$] and with biological material.

In the former case, it was possible to assess the precision and accuracy of the method, whereas in the latter case only the precision was evaluated. This method was used throughout all experiments, i.e., a known quantity of the Pb standard solution was poured into a small container (Fig. 1) and the watery phase was transferred into trap T to be vacuum distilled. The lead was then measured in its various forms (volatile, non-volatile, etc.). The results are reported in Table 1 and they show that this method provides basis for a fairly good precision and accuracy.

TABLE 1 Drying and Analyses of Solutions Containing a Known Quantity of Lead

Compound	Pb theor.	Pb ex.				acc. %
		Pb vol.org.	Pb vol. inorg.	Pb non vol.	Pb tot.	
$Pb(Et)_4$	19.4	19.2 $\pm$ 0.2	---	---	19.21$\pm$0.2	| 1 %|
$Pb(Met)_4$	21.6	20.9 $\pm$ 0.3	---	---	20.0 $\pm$0.3	|3.2 %|
$Pb(NO_3)$	20.0	---	2.6 $\pm$ 0.4	17.3$\pm$0.5	19.9 $\pm$0.6	|0.5 %|

The values are expressed in μg and are the average of 5 determinations.

Full development tests were carried out in biological specimen, using filter feeders (mussels) collected on the island of Palmaria (Bay of La Spezia).

In some cases only the soft part of the mussels was analized and the Pb content of their shell was neglected. The precision of the findings was fairly good, though not as good as of the standard solution. This may be due either to greater operating difficulties or to the nonhomogeneity of the samples.

Furthermore, to check whether their residing in deionized water would possibly solubilize any Pb adhering to the shells, thus making them more homogeneous, the samples were twice washed in deionized water, stirred for two minutes.

The Pb values obtained after this treatment were compared with the values that had been obtained from the analysis of the specimen just as they were. The values reported in Table 2 show that the treatment did neither affected the volatile nor the non-volatile Pb content, and that it had no influence on the grade of precision of these measurements.

Table 2 also features the Pb values obtained after drying of the specimens and destruction of the organic matter (Ref. 7, 9 and 10) by the usual methods. A comparison of these values reveals that the Pb content determined according to these techniques is much lower than the values obtained by vacuum desiccation.

TABLE 2 Comparison between Different Methods for the Dehydration and Decomposition of Mussel Flesh Samples

Trial	Volatile Pb		Non volatile Pb	Total Pb
	inorg. Pb	org. Pb		
	μg/g	μg/g	μg/g	μg/g
1. Drying under vacuum with collection of the evaporated fraction in liquid N_2 trap; acid digestion	10.84±0.8	13.32±1.6	57.82±5.7	81.98±5.97
2. Washing in deionized water and drying under vacuum as in 1.	9.87±0.9	7.63±0.3	71.15±6.3	88.65±6.37
3. Oven-drying at 105 °C; acid digestion	---	---	---	45.94±5.70
4. Oven-drying at 105 °C and dry-ashing at 600 °C	---	---	---	33.44±3.50

These values are the mean of 7 analyses.

The difference is not due only to a loss of the organometallic compounds of the lead but also to the volatilization of a small part of inorganic Pb.

The presence of organometallic Pb compounds ($Pb(Et)_4$) in the mussels was further confirmed, qualitatively by gas chromatography and it was demonstrated that these Pb compounds were also present in the mussels collected in other Italian seas (Table 3).

The results of analyses of mussels collected from various Italian waters, assumed to be more or less polluted, are reported in Table 3. These analyses were carried out on mussels taken from buoy located above a ship (Cavtat) wrecked near Otranto, on mussels in the Genoa harbour, off-shore Crotone, in the Bay of La Spezia, and near the island of Palmaria.

The first two sites were assumed to be strongly polluted; the first because of the presence of the alkyl lead cargo of the ship found in this area (Cavtat) and the second because harbour waters are very polluted.

On the other hand, the sea near Crotone was assumed to be little polluted and the Bay of La Spezia, near the island of Palmaria was even thought to be completely unpolluted.

TABLE 3 Comparison between the Pb Content in Mussels Sampled in Different Environments

Sample	Volatile Pb		Non volatile Pb	Total Pb
	inorg. Pb µg/g	organic Pb µg/g	µg/g	µg/g
Palmaria	10.84±0.8	13.32±1.6	57.82±5.7	81.98±5.97
S/S CAVTAT Buoy	29.34±0.9	18.52±1.2	103.86±7.5	151.72±7.66
Genoa Harbour	142.44±1.3	9.69±0.5	56.40±5.3	208.53±48
Crotone	45.06±0.7	47.87±2.0	152.45±8.9	245.38±9.15

The values are the mean of 7 analyses.

All the mussels examined, except for the specimens collected in the last mentioned area, were of the same size and had been collected in winter. It could, therefore, be reasonably assumed that their Pb content would be comparable. The mussels taken from the Bay of La Spezia were, however, of a much larger size (4-5 cm instead of approx. 2 cm at the other sites) and had been collected during summer. Therefore, their comparison may be misleading.

The results of these analyses are reported in Table 3. This Table shows the presence of organic Pb compounds in all mussels taken into consideration.

Furthermore, the presence of $Pb(Et)_4$ was confirmed in all mussels.

While the presence of alkyl lead compounds could be expected in the specimens collected from the buoy of s/s "CAVTAT" (Otranto) and in the Genoa harbour, it is interesting to note that the specimens of other origin also had a high organic and inorganic lead content.

The presence of $Pb(Et)_4$ in the analyzed mussels may indicate that the alkyl lead compounds are not immediately metabolized by living organisms and may remain in the shape of alkyl compounds in living tissues for a long time. This is corroborated by the research work done by Stevens *et al.*(Ref. 10) on the metabolic kinetics of $Pb(Et)_4$ in rats. The presence of alkyl lead might also mean that inorganic lead is metabolized by the mussels themselves which transform this inorganic Pb into alkyl compounds.

Studies are now under way to determine its possible origin.

CONCLUSIONS

From the foregoing, it can be concluded that the determination of volatile metals, such as lead, in specimens having a very high water content, must be accomplished by using a proper technique, by collecting and analyzing the removed water as well as all other volatile constituents.

Moreover, the decomposition of biological samples must be accomplished by acid digestion rather than by ashing.

It has also been found that the use of conventional dehydration techniques and the decomposition of these samples, will cause a severe loss of lead and give results that are lower than the true values.

ACKNOWLEDGEMENT

This work was supported by the C. N. R. under Research "Promozione Qualità Ambiente" Series n. AC/2/27. The authors express their thanks to Mr. Gildo Poggi, to Mrs. Laura Scotti and RosaMaria Cannoni for their collaboration in this work.

REFERENCES

(1) T. M. Florence and G. E. Batley, Talanta 24, 152-157 (1977).

(2) R. Geldiay and H. Uysal, Origin and fate of chemical residues in food, Agriculture and Fisheries STI/PUB/399 TAEA, 1975, p. 69.

(3) B. Patel, P. G. Valanju, C. D. Mulay, M. C. Nalani and S. Patel, Radioactive contamination of the marine environment STI/PUB/313 IAEA, 1973, 307.

(4) R. R. Brooks and M. G. Rumsley, Limnol. Oceanogr. 10, 521 (1965).

(5) D. A. Segar, J. D. Collins and J. P. Riley, J. Mar. Biol. Ass. U. K. 51, 131 (1971).

(6) R. J. Pentreach, J. Mar. Biol. Ass. U. K. 53, 127 (1973).

(7) H. O. Fourie, M. Peisach, Analyst 102, 193-200 (1977).

(8) L. Favretto, Atom. Absorpt. Newsl. 15, 98-101 (1976).

(9) R. Machiroux and J. C. Dupont, Analytica Chim. Acta 85, 231-239 (1976).

(10) Tatsuo Hayakawa, Tgaku to Seibutsugaku 66, 207-209 (1963).

(11) G. R. Sirota and J. F. Ute, Analyt. Chem. 40, 823-826 (1977).

(12) Y. K. Chau, P. T. S. Wong and P. D. Goulden, Analyt. Chim. Acta 85, 421-424 (1976).

(13) Y. K. Chau, P. T. S. Wong and P. D. Goulden, Proc. Int. Conf. Heavy Metals in the Environment, Toronto, 1975.

(14) A. M. Beccaria, E. D. Mor, G. Poggi, Comunicazione al 2° Congresso Nazionale di Chimica Analitica, Catania 6-10 settembre 1977.

(15) G. W. C. Milner, The principles and applications of polarography, Longmans, 1957, 424.

(16) C. D. Stevens, C. J. Feldhake, and R. A. Kehoc, J. Pharmacol Exptl. Therap. 128, 90-4 (1960).

Voltammetric Determination of the Stability Constants of the Predominant Labile Lead Complexes in Sea Water

L. Sipos*,**, P. Valenta*, H. W. Nürnberg* and M. Branica**

*Institute 4 Applied Physical Chemistry, Nuclear Research Centre (KFA) Juelich, F. R. Germany

**Center for Marine Research, "Rudjer Bosković" Institute, Zagreb, Croatia, Yugoslavia

ABSTRACT

Traces of heavy metals and other toxic trace metals occur dissolved in the sea mainly in form of labile inorganic complex species with the anionic constituents of sea water as ligands. Therefore, the identification of the preferentially by the respective metal formed complex species, their physicochemical characterization and the determination of their stability constants are important actual topics in marine trace metal chemistry. Suitable polarographic and voltammetric methods, provide a rather unique access to this kind of speciation problems. Experimental data obtained in aqueous model solutions with I of 0.7 and in artificial sea water by single sweep voltammetry and by a particular semiautomated ASV-procedure have established that the two labile complexes of Pb(II) prevailing in the sea are $[PbCO_3]$ and $[Pb(CO_3)_2]^{2-}$.

INTRODUCTION

The tendency and rate by which a metal participates in chemical, geochemical and biological processes in the marine environment, depends on the physicochemical form in which the respective metal occurs in the dissolved state. Thus, the chemical speciation is one of the most important parameters to understand the mechanisms by which pollutants act in the marine environment. Toxic trace metals, as Pb, belong to the very hazardous pollutants and their speciation is of special importance. They occur in the sea usually in form of labile complexes with inorganic anionic constituents of sea water as ligands and possibly to some extent also as rather strong chelates with certain components of dissolved organic matter. The dissolved state is an important stage in the biogeochemical cycle of toxic metals as from there they interact with sediments and enter the aquatic food chains and webs as consequence of toxic metal accumulation by aquatic organisms from various trophic levels. Due to their particular general potentialities with respect to sensitivity, accuracy and precision and their selective ability to determine chemical species instead of just elements suitable voltammetric methods offer a very efficient approach to study trace metal speciation in natural waters, even at the actual trace level of the dissolved substances (Ref. 1-4).

The usual way to approach hitherto the problem of speciation of trace metals in natural waters was to design, mainly based on theoretical considerations of known data from fundamental coordination chemistry in aqueous media, a chemical model for the equilibrium distribution of the species present (Ref. 5-13). However, large disagreement regarding the conclusions of the authors is noticed due to large variations between the values of equilibrium constants available and used for the calculations. The experimental verification of the chemical model of a trace metal in sea water was not possible so far, due to the very low concentrations at which trace metals occur dissolved in the sea and in inland waters, even if moderate pollution prevails. However, it is possible in principle by measurements of polarographic half-wave potential shifts $\Delta E_{1/2}$ or shifts of corresponding potential values, as the half peak potential $E_{p/2}$ in single sweep voltammetry (Ref. 15 and 36), as function of the ligand concentration to determine the stability constants and composition of labile complexes, if they undergo a reversible electrode process (Ref. 14 and 15). These experimental results can be then compared with the theoretical predictions from the respective models. The only difficulty to be overcome with respect to realistic conditions operative in sea water was to measure half-wave potentials at the natural very low concentration level of the trace metals.

The aim of the present communication is to describe the voltammetric determination of those labile Pb(II)-complexes, which are predominant in sea water due to the two intercorrelated parameters, stability constant and available ligand concentration. At the same time these data furnish the under realistic conditions experimentally established foundation for the equilibrium distribution of inorganic Pb(II)-species in sea water, a topic to be treated in detail elsewhere (Ref. 16).

The investigation of chelates with dissolved organic matter, which requires an alternative voltammetric approach is treated for Pb(II) in a further contribution (Ref. 34) and has been described for Cd(II) previously (Ref. 35).

METHODOLOGICAL APPROACH

Two voltammetric methods have been applied depending on the level of the Pb(II)-concentration.

In model solutions, where a rather high Pb(II)-concentration of 4×10^{-6}M had been adjusted, single sweep voltammetry (Ref. 15 and 36) at the hanging mercury drop (HMDE) as working electrode has been applied to follow directly the function of the half peak potential $E_{p/2}$ of Pb(II)-reduction on the ligand concentration. The HMDE was polarized under potentiostatic control with a voltage sweep using the Polarographic Analyzer, PAR 174A.

For the measurement in the 10^{-9}M range of Pb(II), a concentration level encountered in reality in coastal waters (Ref. 25), the subsequently outlined anodic stripping procedure was applied.

The required half-wave potential data needed for the evaluation of the stability constants of labile Pb(II)-complexes can be determined even at the natural ultra trace level of Pb(II) in a reliable but also more elaborate manner from the anodic stripping peak current-deposition potential relationship (ASV-PCDPR). This method has been recently introduced to elucidate the speciation of Cd-chlorocomplexes in sea water (Ref. 17 and 18). The method enables the determination of pseudo half-wave potentials ($E^{*}_{1/2}$), which are equivalent to the classical polarographic half-wave potential ($E_{1/2}$) (Ref. 15) or the half-peak potentials $E_{p/2}$ obtained from single sweep voltammograms. The ASV-approach is described in Fig. 1 for the case of the determination of the stability constants of the carbonato complexes of Pb(II). As working electrode served the mercury film electrode (MFE) (Ref. 19). The MFE

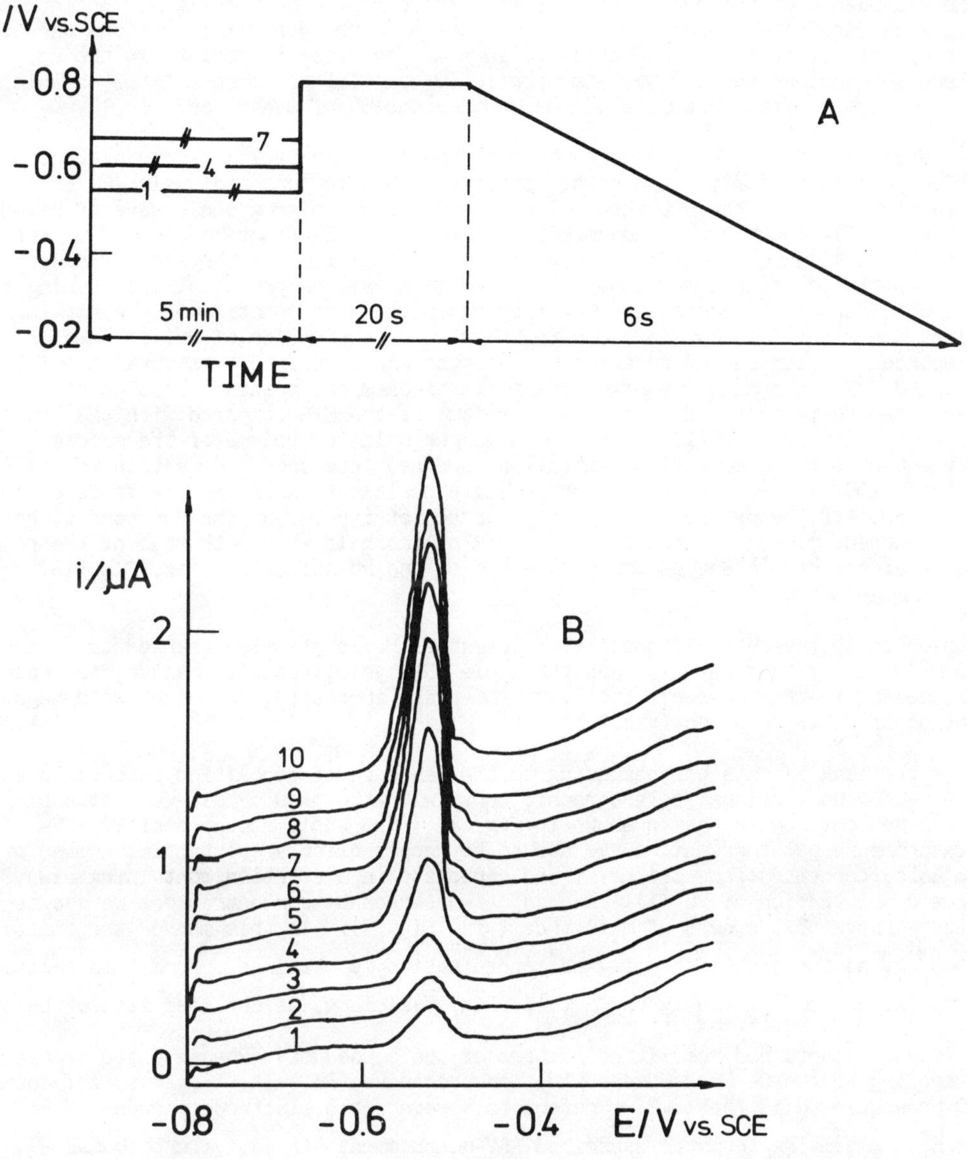

Fig. 1. (A.&B.) Determination of anodic stripping peak current-deposition potential relationship (ASV-PCDPR).
A. Course of the polarizing voltage.
B. Family of ASV-curves for various cathodic deposition potentials. $6x10^{-9}$ M Pb, 0.59 M $NaClO_4$, 0.10 M $NaHCO_3$, 0.01 M NaCl, pH 7.

was produced by cathodic deposition for 20 min of a thin mercury film (thickness 1000 Å) on the glassy carbon substrate from a 10^{-5}M $Hg(NO_3)_2$ solution. The electrode was rotating with 1500 rpm during the film formation.

The investigated samples had a Pb(II)-concentration of only 6.10^{-9}M. To follow the dependence of the half-wave potential shift $\Delta E_{1/2}$ on the ligand concentration, i.e. the carbonate concentration, a sequence of carbonate concentrations was adjusted, as will be pointed out later, in a manner close to reality in the sea, where interaction between the atmospheric CO_2 partial pressure and the oceanic carbonate buffer system prevails and determines the pH of sea water.

For each adjusted carbonate concentration the following voltammetric procedure had to be performed. At first a rather positive cathodic potential value E_d in the potential range of the hypothetical conventional dc-polarographic wave of reversible Pb(II)-reduction was adjusted, e.g. -0.53 V(SCE) (compare Fig. 1C). For a deposition time t_d of 5 min. Pb was deposited as amalgam on the working electrode under potentiostatic control of the cathodic deposition potential E_d. During this cathodic deposition period the solution was stirred by rotation of the working electrode with 1500 rpm, in order to speed up mass transfer of Pb(II) towards the electrode. After t_d the rotation was stopped and a quiescent interval t_r of 20 s elapsed. In this manner a significant electrochemical preconcentration of the deposited Pb is achieved in the small volume of the MFE compared with the low bulk concentration level Pb(II) in the much larger solution volume of the sample. Subsequently the electrode potential was scanned into anodic direction with a scan rate of 100 mV s^{-1}. Due to the progressive positivation of the electrode potential an aliquot of the deposited Pb is reoxidized (stripped) and the recorded corresponding current yields a characteristic peak at the half wave potential of the reversible electrode process of Pb (II)-reduction and Pb-oxidation (viz. Fig. 1B, voltammogram 1).

Subsequently the cathodic deposition potential E_d is stepwise changed to an about 30 mV more negative E_d-value and the whole ASV-cycle, described above, is repeated. Proceeding in this manner a family of ASC-peaks increasing in height with negativation of E_d is obtained (viz. Fig. 1B).

From the peak heights corresponding to the sequence of deposition potentials E_d the needed conventional polarographic curve with the pseudo-half-wave potential $E^*_{1/2}$ could be constructed, as is shown in Fig. 1C. The whole sequence of 10 ASV-measurements polarographic curve has to be repeated for a substantial number of adjusted concentrations of the ligand carbonate in a solution containing always the same concentration of Pb(II). The adjusted carbonate concentrations were extended over a range of 5 orders of magnitude (viz. Fig. 7). In this manner the shift of $\Delta E^*_{1/2}$, as function of the ligand concentration is obtained. From this relationship between $\Delta E^*_{1/2}$ and lg $c_{CO_3^{2-}}$ (viz. Fig. 7) follow, as outlined later, the stability constants and the ligand numbers of the consecutive Pb-carbonato complexes according to the De Ford-Hume evaluation procedure (Ref. 14, 15, 20). A fundamental prerequisite is that $\Delta E^*_{1/2}$ refers to a reversible electrode process of Pb(II). $\Delta E^*_{1/2}$ resulting from the described ASV-measurements is equivalent to $\Delta E_{1/2}$, which would be obtained immediately from the potential shift of the corresponding reversible polarographic dc-wave measurable immediately for sensitivity reasons only at substantially higher Pb(II)concentrations, i.e. $\geq 10^{-5}$M Pb(II). With the outlined ASV-procedure, utilizing the ASV-PCDPR-approach, these speciation studies can be extended down to ultra trace levels of Pb(II) due to the substantially larger sensitivity of ASV (Ref. 21, 22), which is particularly high in connection with the MFE (Ref. 1-4, 19).

As a large number of runs is necessary to establish in this manner the shift of $E^*_{1/2}$ with the free ligand concentration the recording of the Pb-peaks as function

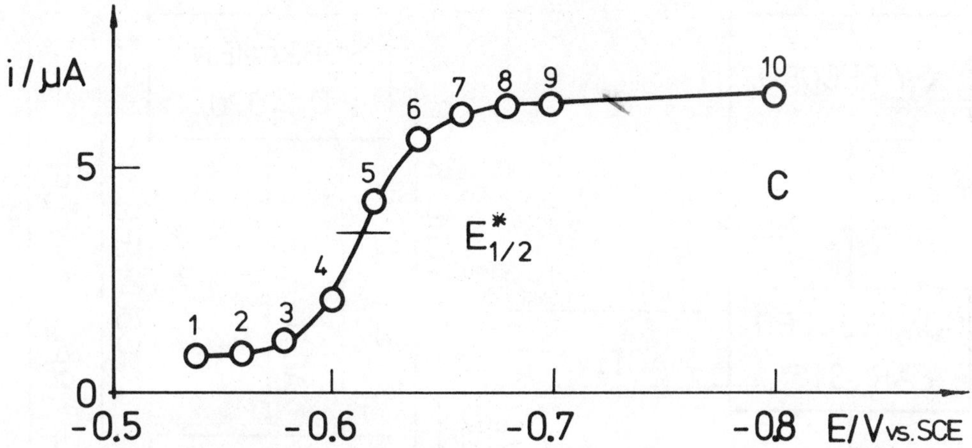

Fig. 1. (C.) Determination of the pseudo half-wave potential $E^*_{1/2}$ from the reconstructed current-potential curve (analogue of conventional dc-polarogram).

of the varied deposition potentials E_d was automated by linking to the polarograph a homemade programming device described in detail elsewhere (Ref. 23). Here only one block diagram of the automatic device is shown in Fig. 2. The Polarographic Analyzer, PAR Mod. 174A, generates the polarizing voltage, ensures the potentiostatic control and performs the recording of the current-potential curves on a X-Y-recorder Hewlett-Packard, Mod. 7004B. The Automated Electroanalysis Controller, PAR Mod. 315, controls the series of steps of the voltammetric analysis from the deaeration of the solution with 99.999% N_2 for 20 min to the cathodic deposition of Pb, the subsequent adjustment of the deposition period $t_d + t_r$ and of the following stripping period during which the automatic recording of the current-potential-curve is performed. A Digitimer, Mod. D 4030, in connection with a homemade interface (Ref. 23) enables the automatic stepwise change of the deposition potential E_d and the recyclization of the system. The auxiliary electrode in the voltammetric cell was a Pt-coil. As reference electrode served a saturated Ag/AgCl electrode connected to the cell via a salt bridge, filled with a saturated solution of NaCl.

All chemical used were Merck, suprapur. Temperature was adjusted to 25°C.

By the beforementioned voltammetric methods the complex equilibria of Pb(II) carbonates existing in sea water could be established for the first time. To characterize the complex species of Pb(II) and to identify the occuring carbonato complexes of Pb(II), the stability constants of Pb(II)-complexes have been determined at a realistic total Pb(II)-concentration in artificial sea water containing all significant components of genuine sea water, and also in model solutions of an ionic strength of 0.7 M, containing Na^+, H^+, ClO_4^-, HCO_3^-, CO_3^{2-} and H_2CO_3 as components.

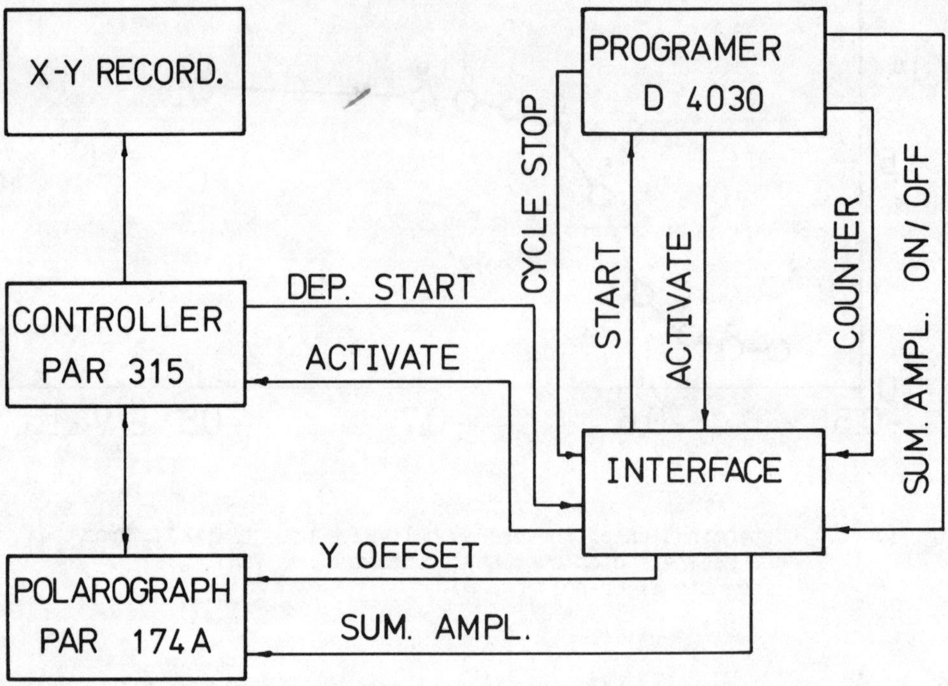

Fig. 2. Block diagram of the device for automatic registration of ASV-PCDPR.

INVESTIGATIONS ON THE NATURE OF THE EFFECTIVE CARBONATO LIGAND

At first exploratory experiments have been carried out by single sweep voltammetry at the HMDE in $NaClO_4$-$NaHCO_3$-model solutions with at a rather high Pb(II)-concentration of 4×10^{-6}M to check, whether carbonate and/or bicarbonate are involved as ligands in the complex equilibria of Pb(II). Two series of measurements have been performed, one at pH 6.5 and the other at pH 7.5, respectively. The total concentration of carbonate was stepwise increased and the reversible cathodic half-peak potential $E_{p/2}$ of Pb(II) reduction was measured. The respective HCO_3^-- and CO_3^{2-}-concentrations c in these experiments have been evaluated from the lg c-pH diagram of the carbonato system shown in Fig. 3. At the lower pH of 6.5 the ratio of $c_{CO_3^{2-}}/c_{HCO_3^-}$ is lo times lower than at pH of 7.5. This fact enables to distinguish between the eventually occurring complex species with the ligands HCO_3^- and CO_3^{2-}. The experimental set up for those measurements is shown in Fig. 4. The required pH was established in the voltammetric cell containing a solution with a certain $NaHCO_3$-concentration by bubbling a mixture of N_2 and CO_2 through two washing bottles prior to the voltammetric cell. The pH was measured in the second washing bottle, filled also with sample solution, before and after the voltammetric determination. For solutions with a low carbonate content it was difficult to maintain a constant pH in the described manner, since a very low partial pressure of CO_2 was necessary. Therefore, in such cases N_2 used for deaeration was passed through a different first washing bottle containing a suspension of magnesium hydroxy carbonate in borate buffer of a given pH. The partial pressure of CO_2 in the N_2 could be easily adjusted by varying the pH of the buffer and the flow rate of N_2. The total carbonato concentration of the solution in the voltammetric vessel was determined by potentiometric titration with 0.1 M HCL solution.

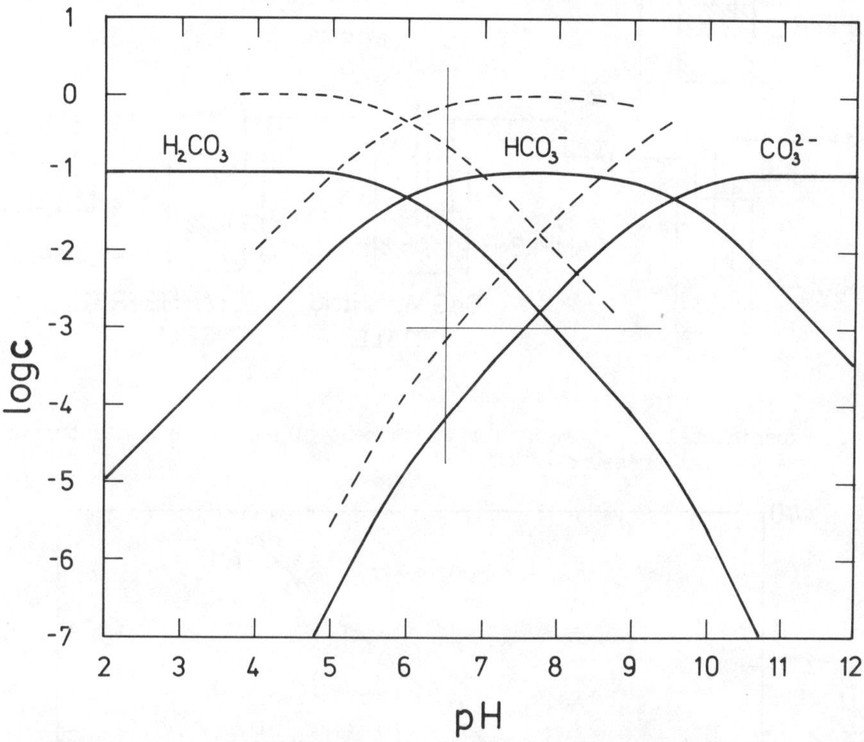

Fig. 3. pH dependence of components of carbonato system.

The plots of the obtained $E_{p/2}$-values (half peak potentials) of the single sweep voltammetric curves versus the logarithm of the ligand concentration are shown in Fig. 5 for the bicarbonate ligand and in Fig. 6 for carbonate ligand, respectively, at the two pH values 7.5 and 6.5. The concentrations $c_{HCO_3^-}$ and $c_{CO_3^{2-}}$ were calculated from the experimentally determined total carbonato concentration and the measured pH with the aid of the first and the second dissociation constant of carbonic acid. The curves corresponding to pH 7.5 and 6.5, respectively, differ by about 26 mV in Fig. 5. On the other hand, the curves in Fig. 6 coincide for both pH values. Therefore, it is to be concluded that under the adjusted experimental conditions complexation of Pb(II) only by carbonate takes place, whereas the Pb-bicarbonato and Pb-hydroxo complex equilibria can be neglected.

DETERMINATION OF THE STABILITY CONSTANTS OF Pb-CARBONATO COMPLEXES

For the determination of the stability constants again the two voltammetric methods, single sweep voltammetry and the outlined ASV-PCDPR-procedures were applied, respectively, at the two Pb(II)-concentration levels, 4×10^{-6}M and 6×10^{-9}M, to measure

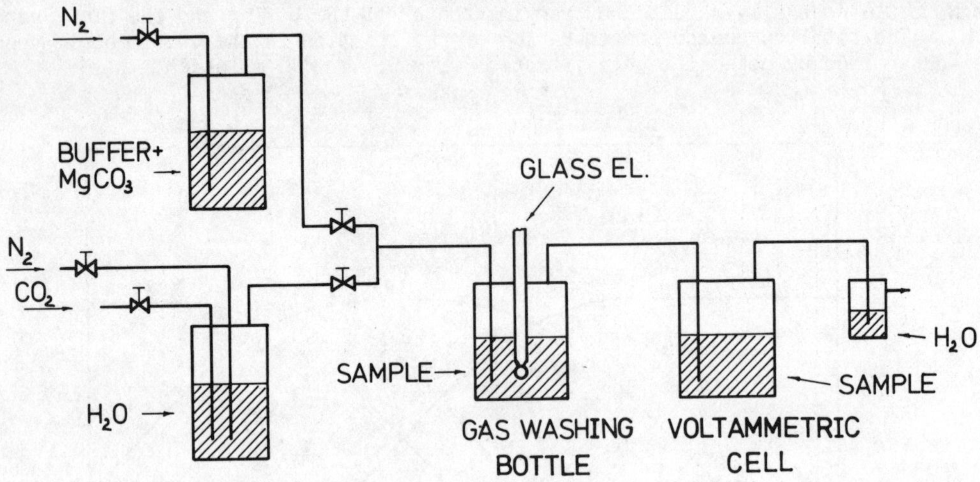

Fig. 4. Experimental arrangement for adjustment of carbonate concentration via partial pressure of CO_2.

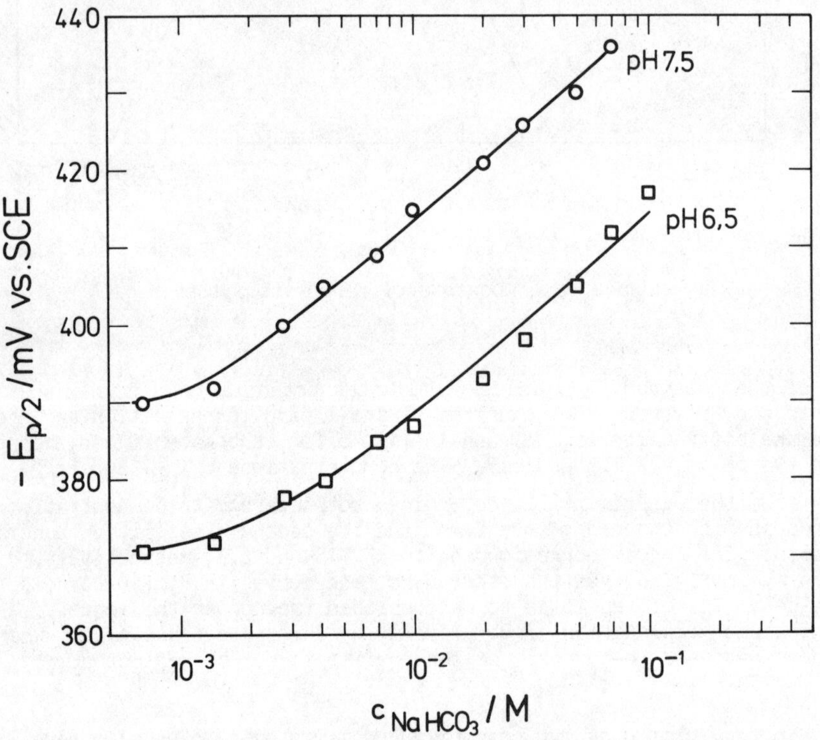

Fig. 5. Dependence of the half peak potential $E_{p/2}$ of Pb(II)-reduction on concentration of bicarbonate. $4x10^{-6}M$ Pb(II), 0.7 M $NaClO_4$ + $NaHCO_3$, single sweep voltammetry at HMDE, curve 1:pH 7.5, curve 2:pH 6.5.

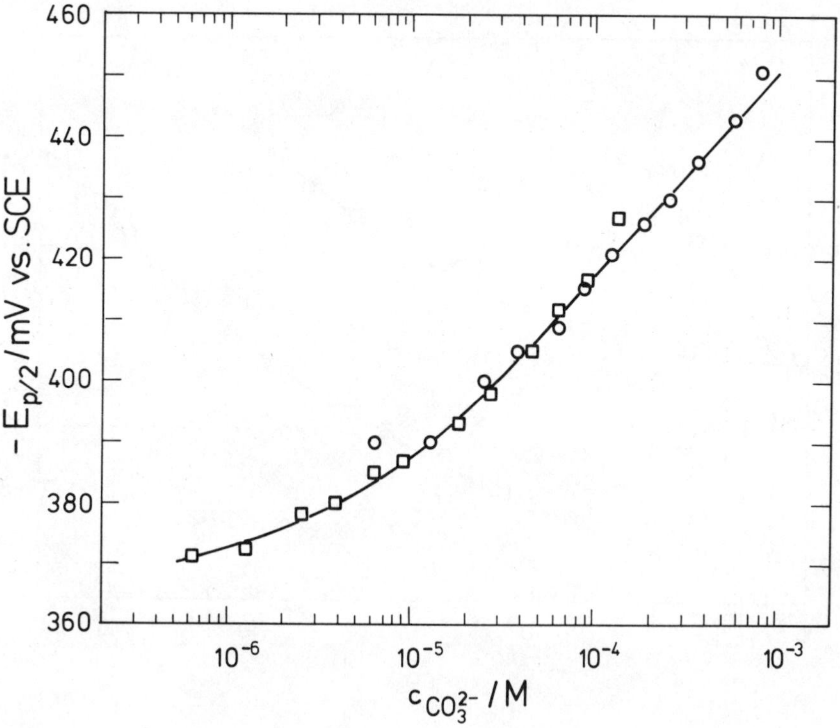

Fig. 6. Dependence of halfpeak potential $E_{p/2}$ of Pb(II)-reduction on concentration of carbonate; other conditions as in Fig. 5.

the function of $\Delta E_{p/2}$ and $\Delta E^{*}_{1/2}$, respectively, on the carbonate concentration in the $NaClO_4$ -$NaHCO_3$-model solution (viz. Fig. 7.).

The evaluation was performed using the DeFord and Hume method (Refs. 14, 15, 20) based on the half-wave potential shift $\Delta E_{1/2}$ as function of the ligand concentration in case of a consecutive complex formation as described in eq. (1).

$$\Delta E_{1/2} = E^{c}_{1/2} - E_{1/2} = - \frac{RT}{nF} \ln \sum_{j=o}^{N} \beta_{Me(L)_j} [L]^j \tag{1}$$

$E_{1/2}$ is the reversible half-wave potential of reduction of the uncomplexed metal ion Me(II), $E^{c}_{1/2}$ is the reversible half-wave potential of reduction of the metal when complexation occurs in presence of ligand concentration $[L]$, j is the ligand number, the term RT/nF has its usual significance and $\beta_{Me\,L_j}$ is the overall stability constant of the complex species $[MeL_j]$ defined by eq. (2).

$$\beta_{Me(L)_j} = \frac{[Me(L)_j]}{[Me][L]^j} \tag{2}$$

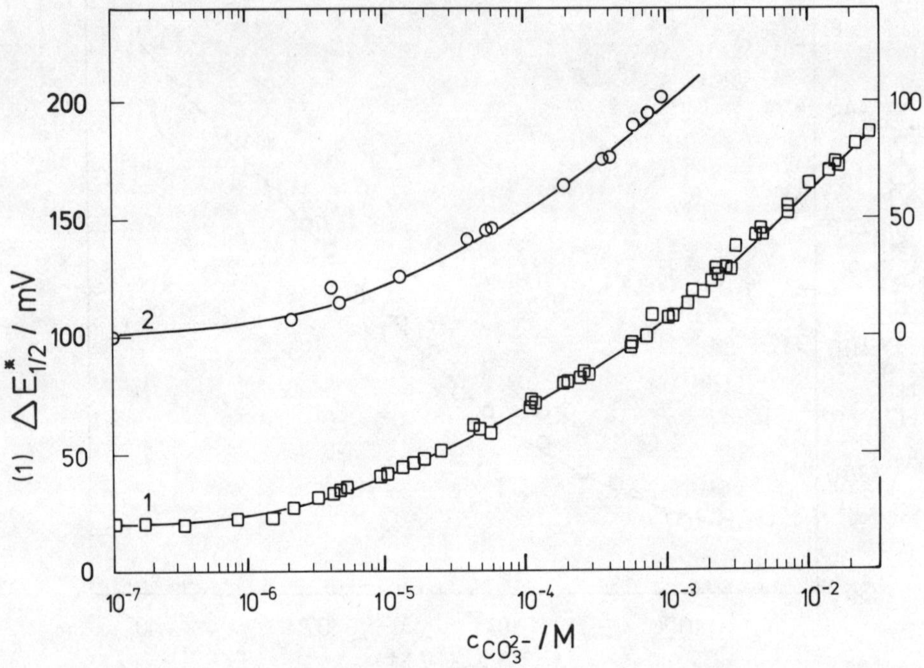

Fig. 7. Dependence of the shifts of pseudo half-wave potential $\Delta E^*_{1/2}$ or $\Delta E_{p/2}$, respectively, of Pb(II)-reduction on concentration of carbonate; $4x10^{-3}$M to $7x10^{-1}$M $NaHCO_3$; I 0.7, if necessary adjusted by addition of $NaClO_4$; curve 1: $6x10^{-9}$M Pb(II), anodic stripping voltammetry (ASV) at rotating mercury film electrode (MFE); curve 2: $4x10^{-6}$M Pb(II), single sweep voltammetry at HMDE.

The function $F_o(L)$ following from eq. (1) is defined in a simplified form by eq. (3).

$$F_o(l) = \exp\left\{(nF/RT)\ (E_{1/2} - E^c_{1/2})\right\} = 1 + \sum_{j=o}^{N} \beta_{Me(L)_j} [L]^j \qquad (3)$$

This equation was solved on a computer with available computer programs described elsewhere (Ref. 24).

Both curves in Fig. 7 have similar shapes and enable the calculation of the stability constants according to the DeFord-Hume-method in its computerised version. The evaluated stability constants are presented in Table 1. Furthermore, these investigations have established that two carbonato complexes of Pb(II), $[PbCO_3]^o$ and $[Pb(CO_3)_2]^{2-}$, exist.

The value of the stability constant depends on the values of the dissociation constants of carbonic acid K_1 and K_2 used in the calculation. To enable a comparison with literature data the stability constants have been determined with respect to the "ionic medium" pH-scale (Ref. 26) and the "infinite dilution" pH-scale (Ref. 27).

In the former case pH is defined so that the activity coefficient of hydrogen ions approaches zero, if the H^+-concentration approaches zero in the pure ionic medium. In the operational "infinite dilution" activity scale pH is defined by a standard dilute buffer solution. The dissociation constants K_1 and K_2 have been taken from Dyrssen and Hanson (Ref. 26) for the "ionic medium" and from Pytkowicz and Hawley (Ref. 27) for the "infinite dilution" scale. Both values had been determined in approximtely 0.7 M NaCl solution.

TABLE 1 Stability Constants of Carbonato Complexes of Pb(II) at I 0.7 for two pH-Scales

Activity Scale	Pb(II)-Concentration M	β_{PbCO_3}	$\beta_{Pb(CO_3)_2^{2-}}$
"ionic medium"	10^{-6}	$(2.34 \pm 0.12)\ 10^5$	$(2.37 \pm 0.08)\ 10^8$
	10^{-9}	$(2.20 \pm 0.31)\ 10^5$	$(5.4 \pm 0.9)\ 10^8$
"infinite dilution"	10^{-6}	$(4.19 \pm 0.22)\ 10^5$	$(6.25 \pm 0.28)\ 10^8$
	10^{-9}	$(3.71 \pm 0.53)\ 10^5$	$(1.8 \pm 0.3)\ 10^9$

It is evident from Table 1 that no significant difference exists between the stability constants obtained at two significantly different concentration levels of Pb(II). This confirms the reliability of the data obtained at the with respect to sea water rather realistic level of 6×10^{-9}M Pb(II) from the measurement of pseudo-half wave potentials by the ASV-PCDPR-approach.

EXPERIMENTAL VERIFICATION OF A CHEMICAL MODEL FOR TRACE METAL SPECIATION IN SEA WATER

It had now to be proved that the stability constants determined for the two Pb-carbonato complexes in the $NaClO_4$-$NaHCO_3$-model solution having an ionic strength of 0.7 equivalent to the sea water value, are also valid in sea water. In this context it has to be recalled that the range of carbonate concentrations adjusted (viz. Fig. 7) corresponds necessarily to a range of pH-values as the concentration ratio of HCO_3^- and CO_3^{2-} is continously changed.

Therefore in artificial sea water, prepared according to (Ref. 28), the dependence of $\Delta E^*_{1/2}$ on pH has been measured by the outlined ASV-PCDPR-procedure (viz. experimental points in Fig. 8).

The pH was again adjusted via the $NaHCO_3$-concentration and the respective partial pressure of CO_2 in the N_2 used for deaeration. In this manner the action of the carbonate buffer system operative in the oceans has been simulated over an extended pH-range.

The experimental data have been compared with the theoretical values of $\Delta E^*_{1/2}$ computed from the sea water model of Pytkowicz and Hawley (Ref. 27). In this computation also the stability constant and the concentration of the other labile Pb(II) complexes relevant for the speciation of Pb(II) complexes relevant for the speciation of Pb(II) in sea water have to be taken into account; i.e. the chloro- and hydroxo complexes of Pb(II). The needed stability constants (viz. Table 2) have been taken from literature values determined in aqueous media at approximately the same ionic strength as in sea water. For the Pb(II)-carbonato complexes the values determined in the $NaClO_4$-$NaHCO_3$-model solution based on the "infinite dilution pH-scale" (viz. Fig. 7 and Table 1) have been inserted.

TABLE 2 Stability Constants used in Computation of the Theoretical Dependence of Half-Wave Potential of Pb(II) on pH at I of 0.7

Metal-Ligand	1	2	Ref.
$Pb-Cl^-$	8.6	22.4	(29), in 0.7 M NaCl
$Pb-OH^-$	7.94×10^6	6.31×10^{10}	(30), in 1 M KNO_3
$Pb-CO_3^{2-}$	4.19×10^5	6.25×10^8	this work, in artificial sea water, (infinite dilution pH-scale)
pK_w	13.77	-	(26)

The theoretical calculation of the $\Delta E^*_{1/2}$-dependence on pH had to be based on the relation for the reversible polarographic half-wave potential shift as function of the ligand concentrations in a multiligand system taking into account all considered kinds of labile Pb(II)-complexes, i.e. chloro-, hydroxo- and carbonato complexes.

The stability constants of the particular complexes are given for a series of complexes $[Me(L_i)_j]$ with the ligands $L_1, L_2 \ldots L_i$ by eq. (4).

$$\beta_{Me(L_i)_j} = \frac{[Me(L_1)_j]}{[Me][L_1]^j}, \quad \beta_{Me(L_2)_j} = \frac{[Me(L_2)_j]}{[Me][L_2]^j} \quad \text{etc.} \tag{4}$$

The concentration of the potential determining free metal ions $[Me^{2+}]$ will then follow from eq. (5). Charges of the species are omitted for simplicity.

$$[Me] = \frac{\sum_{i=1}^{N} \sum_{j=1}^{M} [Me(L_i)_j]}{(1+ \sum_{i=1}^{N} \sum_{j=1}^{M} \beta_{Me(L_i)_j} [L_i]^j)} \tag{5}$$

This leads to a general equation (6) for the shift of the half wave potential $\Delta E_{1/2}$, which is analogous to the equation derived for a single ligand by v. Stackelberg and v. Freyhold (Ref. 31) and by Lingane (Ref. 32).

$$\Delta E_{1/2} = - \frac{RT}{nF} \ln \left\{ 1 + \sum_{i=1}^{N} \sum_{j=1}^{M} \beta_{Me(L_i)_j} [Li]^j \right\} \tag{6}$$

In the calculations of the $\Delta E^*_{1/2}$ we have used the sum of the free CO_3^{2-} and the $NaCO_3^-$, both calculated on the basis of the sea water model (Ref. 27) - and not the free CO_3^{2-} as the ligand. This was necessary because the stability constants of the carbonato complexes of lead were determined in solutions containing Na^+ ions and include the formation of $NaCO_3^-$.

The resulting theoretical curve and the experimental data are compared in Fig. 8. The excellent agreement between the experimental data and the theoretical curve originating from the chemical model of sea water by Pytkowicz and Hawley (Ref. 27) established that also in sea water the carbonato system speciates Pb(II) in form of the two carbonate complexes $[PbCO_3]$ and $[Pb(CO_3)_2]^{2-}$, which had been identified and of which the stability constants had been determined in the $NaClO_4$-$NaHCO_3$-model solutions having a sea water analogous ionic strength of 0.7. Among the labile Pb(II)-complexes present in sea water the rather stable carbonato complexes are obviously prevailing. The most abundant dissolved Pb(II)-species will be $[PbCO_3]$ followed by $[PbOH]^{+}$. A further small amount of Pb(II) will be present as $[Pb(CO_3)_2]^{2-}$. Altogether about 50 % of the dissolved Pb(II) occurs thus in the sea complexed by carbonate. Despite the high Cl^--concentration in sea water the contribution of chlorocomplexes remains rather small due to the low stability of these species (viz. Table 2). The established kind of Pb-speciation by carbonato complexes will be therefore predominant in large parts of the open oceans, which are essentially marine deserts and have thus a too low level of dissolved organic material with strong chelating properties to cause significant Pb(II)-chelation. This is with respect to Pb(II) an important general conclusion from the viewpoint of chemical oceanography of trace metals.

It has to be added that after the completion of the reported investigation on the stability constants of Pb-carbonato species very recent studies have revealed that

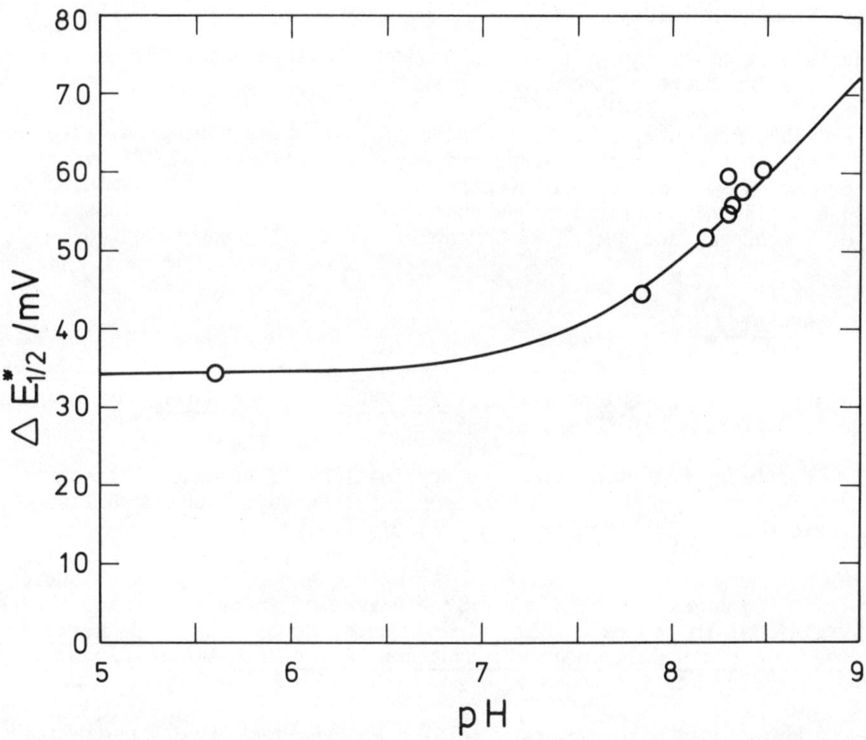

Fig. 8. Comparison between based on Pytkowicz-Hawley sea water model theoretically computed pH-dependence of $\Delta E^{*}_{1/2}$ (curve) and experimental data (points) in artificial sea water.

ion-pairing between the major cationic and anionic components of sea water salinity deserved attention. As will be pointed out elsewhere (Ref. 33), the inclusion of the ion pairing effects into the consideration will lead to some refinement of the values of the stability constants of the respective trace metal complexes. This will be also the case for the Pb-carbonato complexes in sea water. Nevertheless, the important general conclusion to be drawn from the good agreement with the model of Pytkowicz and Hawley (viz. Fig. 8) that the predominant chemical species of dissolved Pb(II) in sea water are the studied carbonato complexes will not be affected by this refined consideration taking into account ion pairing.

In general, this study emphasizes that the voltammetric approach permits in a rather unique manner, due to its particular methodological properties, to obtain even at the realistic trace level data of general validity on trace metal speciation by labile complexes in sea water from measurements based on the ASV-PCDPR relation. As has been demonstrated in a further contribution for Pb(II) (Ref. 34) and elsewhere for Cd(II) (Ref. 35) different voltammetric and polarographic methods provide as well reliable access to the quantitative elucidation and characterisation of the general principles and mechanism of trace metal chelation by organic ligands in the sea. These kinds of chelates of natural and anthropogenic origin are likely to have significance in regions of the sea containing higher amounts of dissolved organic matter (DOM) due to pronounced biological activity, as in upwelling areas, and/or the presence of polluting chelating compounds, which are to be expected particularly in coastal waters and estuaries.

ACKNOWLEDGEMENTS

This study is part of the joint research project "Environmental Research in Aquatic Systems" of the Institute of Chemistry, Institute 4 Applied Physical Chemistry, Nuclear Research Centre (KFA), Juelich, and the Center for Marine Research of the "Ruđer Bošković" Institute, Zagreb. Support of this study during the attachment of Dr. L. Sipos to the Institute 4 Applied Physical Chemistry of KFA by the International Bureau of Nuclear Research Centre, Juelich (KFA), on the basis of the bilateral scientific and technical cooperation between the Socialist Federative Republic of Yugoslavia and the F. R. Germany is gratefully acknowledged.

REFERENCES

(1) H. W. Nürnberg and P. Valenta, Polarography and Voltammetry in Marine Chemistry, in E. D. Goldberg (Ed.), The Nature of Sea Water, Dahlem-Konferenzen, Berlin, 1975, pp. 87-136.

(2) H. W. Nürnberg, Potentialities and Applications of Advanced Polarographic and Voltammetric Methods in Environmental Research and Surveillance of Toxic Metals, Electrochim. Acta 22, 935 (1977).

(3) H. W. Nürnberg and M. Stoeppler, Voltammetric and Spectroscopic Studies on Toxic Trace Metal Levels and Characterization in Sea Water and Marine Organisms, in Lecons et Sèminaires, Quatrième Session de E4 Ecole Europèenne d'Etè d'Environnement, Systèmes Marins, Louvain-la-Neuve, 1977 A. Sanfeld, Ed. (in press).

(4) H. W. Nümberg (1978), Potentialities and Applications of Advanced Polarographic and Voltammetric Methods in Aquatic and Marine Trace Metal Chemistry, in Proc.Int.Symp. "Structure and Dynamics in Chemistry", Uppsala 1977, Acta Universitatis Upsaliensis, Sympos. Univ. Upsaliensis Annum Quingentesium Celebrantis 12, 270-307, Almquist and Wiksell International, Stockholm.

(5) E. D. Goldberg, Minor Elements in Sea Water, in J. P. Riley and G. Skirrow (Eds.), Chem. Oceanography, Academic Press, London, 1965 , 163-196.

(6) W. Stumm and H. Bilinski, Trace Metals in Natural Waters; Difficulties and Interpretation Arising from our Ignorance of their Speciation, in S. H. Jenkins (Ed.), Proc. Water Poll. Res. Conf.,Israel, Pergamon Press, Oxford, 1972, pp. 39-49.

(7) A. Zirino and S. Yamamoto, A pH-Dependence Model for the Chemical Speciation of Copper, Zinc, Cadmium and Lead in Sea Water, Limnol. Oceanogr. 14, 661 (1972).

(8) J. D. Hem and W. H. Drum, Solubility and Occurence of Lead in Surface Water, J. Am. Wat. Wks. Ass. 65, 562 (1974).

(9) H. Bilinski, P. Schindler, W. Stumm, and J. Zobrist, Kupfer und Blei in natürlichen Gewässern, Vom Wasser 43, 107 (1974).

(10) D. Dryssen and M. Wedborg, Equilibrium Calculations of the Speciation of Elements in Sea Water, in E. D. Goldberg (Ed.), Marine Chemistry, Wiley Interscience, New York, 1974, Vol. 5, 181-195.

(11) E. Stumm and A. Brauner, Chemical Speciation, in J. P. Riley and G. Skirrow, Eds., Chemical Oceanography, 2nd Ed., Academic Press, London, 1, 173-239 (1975).

(12) S. Ahrland, Metal Complexes Present in Sea Water, in E. D. Goldberg (Ed.), The Nature of Sea Water, Dahlem-Konferenzen, Berlin, 1975, pp. 219-244.

(13) M. Whitfield and D. Turner, Theoretical Studies of the Chemical Speciation of Lead in Sea Water, Proceedings of the International Experts Discussion on the "Lead - Occurrence, Fate and Pollution in the Marine Environment", Rovinj, October 1977, M. Branica and Z. Konrad (eds.), Pergamon Press, 1979, 109-148.

(14) D. R. Crow, Polarography of Metal Complexes, Academic Press, London, 1964.

(15) J. Heyrovsky and J. Kuta, Grundlagen der Polarographie, Akademie Verlag, Berlin, 1965.

(16) L. Sipos, P. Valenta, H. W. Nürnberg, and M.Branica, Chemical Species of Lead and Cadmium on Sea Water, Marine Chem. (to be submitted).

(17) S. Bubić and M. Branica, Voltammetric Characterization of Ionic State of Cadmium Present in Sea Water, Thalassia Jugosl. 9, 47 (1973).

(18) M. Branica, D. M. Novak, and S. Bubić, Application of Anodic Stripping Voltammetry to Determination of the State of Complexation of Trace Metal Ions at Low Concentration Levels, Croat. Chem. Acta 49, 539 (1977).

(19) H. W. Nürnberg, P. Valenta, L. Mart, B. Raspor, and L. Sipos, Applications of Polarography and Voltammetry to Marine and Aquatic Chemistry, Fresenius Z. Anal. Chem. 282, 357 (1976).

(20) D. D. DeFord and D. N. Hume, The Determination of Consecutive Formation Constants of Complex Ions from Polarographic Data, J. Am. Chem. Soc. 73, 5321 (1951).

(21) F. Vydra, K. Stulik, and E. Julàkovà, Electrochemical Stripping Analysis, Ellis Horwood, Chichester, 1976.

(22) M. Branica, L. Sipos, S. Bubić, and S. Kozar, Electroanalytical Determination and Characterization of Some Heavy Metals in Sea Water, Proc. 7th IMR Symp. Gaithersburg 1974, Nat. Bur. Stand. Spec. Publ. 422, 1976, 917-928.

(23) L. Sipos, P. Krumpen, P. Valenta, and H. W. Nürnberg, Chem. Instrumentation (to be published).

(24) I. Piljac, B. Grabarić, and I. Filipović, Improved Technique for Determination of Stability Constants by the Polarographic Method, J. Electroanal. Chem. 42, 433 (1973).

(25) L. Mart, H. W. Nürnberg, and P. Valenta, Comparative Base Line Studies on Pb-Levels in European Coastal Waters, Proceedings of the International Experts Discussion on the "Lead - Occurrence, Fate and Pollution in the Marine Environment",Rovinj, October 1977, M.Branica and Z. Konrad (eds.), Pergamon Press, 1979, 155-179.

(26) D. Dryssen and I. Hansson, Ionic Medium Effects in Sea Water - A Comparison of Acidity Constants of Carbonic Acid and Boric Acid in Sodium Chloride and Synthetic Sea Water, Marine Chem. 1, 137 (1972).

(27) R. M. Pytkowicz and I. E. Hawley, Bicarbonate and Carbonate Ion-Pairs and a Model of Sea Water at 25°C, Limnol. Oceanogr. 19, 223 (1974).

(28) D. N. Kester, J. W. Duedall, D. N. Connors, and R. M. Pytkowicz, Preparation of Artificial Sea Water, Limnol. Oceanogr. 12, 176 (1967).

(29) L. Sipos (1978), unpublished work

(30) L. G. Sillèn and A. E. Martell, Stability Constants of Metal Ion Complexes, Chem. Soc. Spec. Publ. 17, 1971.

(31) M. V. Stackelberg and H. V. Freyhold, Polarographische Untersuchungen an Komplexen in wässriger Lösung, Z. Elektrochem. angew. physik Chem. 46 120 (1940).

(32) J. J. Lingane, Interpretation of the Polarographic Waves of Complex Metal Ions, Chem. Rev. 29, 1 (1941).

(33) B. Raspor, L. Sipos, H. W. Nürnberg, and R. M. Pytkowicz, Interaction of Metal Complexes with Coulombic Ion-Pairs in High Salinity Aqueous Media, Marine Chem . (in press).

(34) B. Raspor, H. W. Nürnberg, P. Valenta, and M. Branica, The Chelation of Lead by Organic Ligands in Sea Water, Proceedings of the International Experts Discussion on the "Lead - Occurrence, Fate and Pollution in the Marine Environment", Rovinj,October 1977, M. Branica and Z. Konrad (eds.), Pergamon Press, 1979, 181-195.

(35) B. Raspor, P. Valenta, H. W. Nürnberg, and M. Branica, The Chelation of Cadmium with NTA in Sea Water as a Model for the Typical Behaviour of Trace Heavy Metal Chelates in Natural Waters, The Science of the Total Environment 9, 87 (1978).

(36) L. Meites, H. W. Nürnberg, P. Zuman, et al., Classification and Nomenclature of Electroanalytical Techniques, Pure Appl. Chem. 45, 81 (1976).

A Sensitive Technique for the Analysis of Tetraalkyllead Compounds

Y. K. Chau, P. T. S. Wong and P. D. Goulden
Canada Centre for Inland Waters, Burlington, Ontario, Canada

ABSTRACT

The gas chromatograph-atomic absorption system provides a sensitive, element- and speciation-specific technique for the analysis of tetraalkyllead compounds in gaseous and liquid samples. The system inherits the separation capability of gas chromatography and the specificity of element detection of atomic absorption. The atomic absorption instrument can be operated in either the flame or the furnace mode with detection sensitivity at the subnanogram level expressed as the element. It has been applied to the analysis of tetramethyllead in the headspace of biological systems to study biological methylation of lead.

INTRODUCTION

The combination of a gas chromatograph and an atomic absorption spectrophotometer provides a sensitive element- and speciation-specific system for the analysis of tetraalkyllead compounds in gaseous and liquid samples. The atomic absorption unit is employed as an element-specific detector after the chromatographic separation. When a furnace is used for atomization, subnanogram detection can be achieved. This technique was specially developed for the study of biological conversion of lead compounds to tetramethyllead in sediment systems. It can be readily applied to the analysis of these compounds in the atmosphere.

EXPERIMENTAL

The GC-AAS System

Any gas chromatograph (GC) instrument with temperature programming capability and atomic absorption spectrophotometer (AAS) with background correction accessory can be used to set up the system. The interfacing of the two instruments was made by means of a stainless steel tubing (2 mm o.d.) connected to the column outlet of the GC by Swagelock joints and leading to the silica furnace of the AAS (Fig. 1). The transfer line was wrapped with heating tapes and was heated to ca. 80 - 90°C. A 4-way valve was installed between the carrier gas inlet and the column injection port so as to facilitate the introduction of gaseous samples into the GC column through a sample trap. The recorder (10 mV) was equipped with an electronic integrator to measure the peak areas, and was simultaneously actuated with the sample introduction so that the retention time of each component was recorded for

for peak identification.

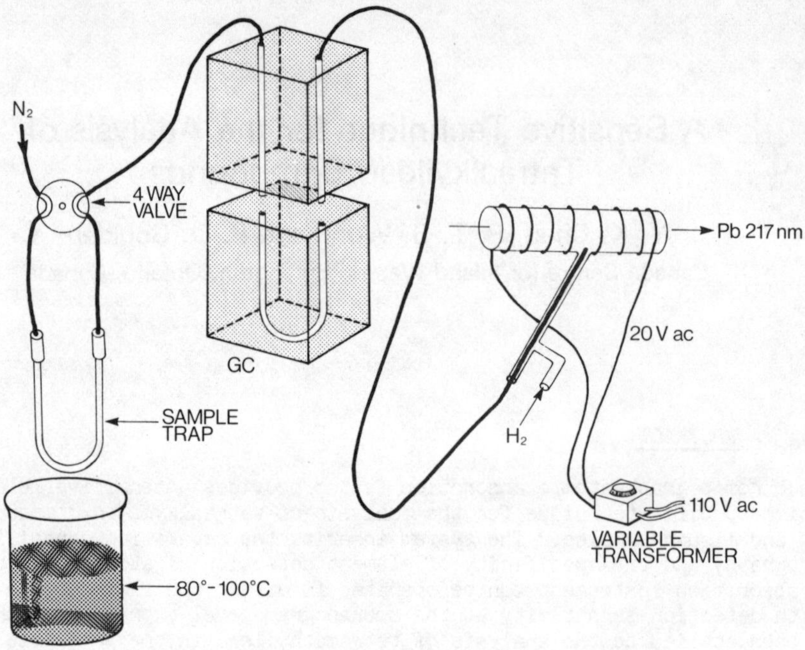

Fig. 1. Interfacing of the GC-AAS system (furnace).

The Silica Furnace

The furnace was constructed from silica tubing, 7 mm i.d. 4 cm long with open ends (Fig. 2). The effluent from the GC was introduced through a side arm to the centre

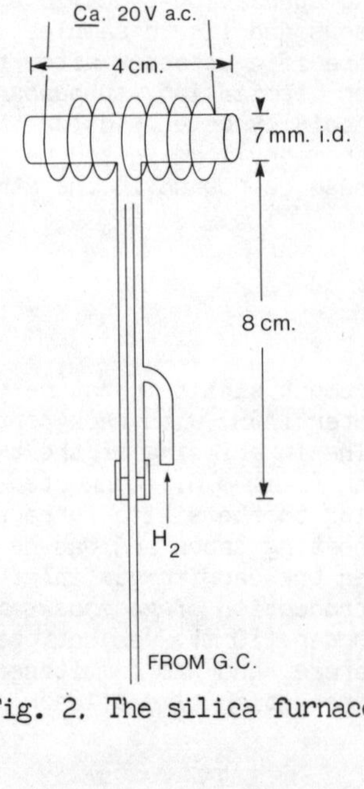

Fig. 2. The silica furnace.

of the furnace. Hydrogen was added at the same point at a flow rate of 135 ml min^{-1}; the burning of hydrogen improved the sensitivity. The main furnace was wrapped with asbestos and resistance wire (No. 26 Chromel wire) to give a resistance of ca. 5 ohms. The voltage applied to the furnace was ca. 20V a.c. regulated by a variable transformer so that the furnace temperature with the hydrogen burning was about 980 - 1000°C.

Sampling

The sampling device for gaseous samples was a glass U-tube, 6 mm o.d. 26 cm long, packed with 3% OV-1 on Chromosorb W, 80/100 mesh, immersed in a dry ice-methanol bath (ca. -70°C). The gaseous sample or the atmosphere to be analyzed was sucked through the trap and metered by a peristaltic pump (130 - 150 ml min^{-1}). The cold trap was found very effective in retaining all the alkyllead compounds under study. After sampling, the trap was connected to the 4-way valve, warmed to ca. 80 - 90°C in a beaker of hot water, and the sample was introduced to the GC column by the carrier gas.

For liquid samples, such as lead alkyls absorbed in solvents, or standard solutions in solvent, they can be directly injected to the GC column through the normal injection port without the use of the sample trap.

Instrument Parameters

The following instrument parameters are based on a system using a MicroTek 220 Gas Chromatograph and a Perkin-Elmer 403 Atomic Absorption Spectrophotometer.

Glass column, 1.8 m long, 6 mm o.d. packed with 3% OV-1 on Chromosorb W, 80/100 mesh; carrier gas, N_2 at 70 ml min^{-1}; injection port temperature, 150°C; temperature programs, initial 50°C for 2 min, programmed at 15°C min^{-1} until 150°C.

Electrodeless Discharge Lamp (Perkin-Elmer), Pb line 217 nm; special band width, 0.7 nm; scale expansion, 4X(0.25A Full Scale); furnace gas, 135 ml H_2 min^{-1}. The deuterium background corrector was used.

Standards

Tetramethyl- and tetraethyl-lead (Alfa Chemicals, Beverly, Mass.) and mixed alkyllead compounds: Me_3EtPb, Me_2Et_2Pb, $MeEt_3Pb$ (Ethyl Corporation, Ferndale, Mich.) were diluted with benzene to prepare working standards (2 ng/μl). The purity of these compounds were assessed by gas chromatography; standardization was done by atomic absorption determination of the lead content as described in a previous study (Ref. 1).

RESULTS AND DISCUSSION

The constant temperature furnace (ca. 1000°C) has many advantages over the programmed heating units. It maintains a uniform temperature throughout for good reproducibility in atomization; the furnace is continuously self-cleaning while in use; it is always ready to receive the analytes in the effluent stream of the GC column. The hydrogen in the furnace burns in a small jet which enhances the sensitivity. Furthermore, as the conventional flame is not used, the electrometer of the AAS can be operated at much lower gains, which give very steady baselines at high scale expansion. The lower limit of detection of the method depends on the scale expansion of the instrument. Under the present conditions, 0.1 ng Pb give a signal 3x of the baseline noise. For analyses where such sensitivity is not a critical consideration, the system can be operated at the microgram level by using the conventional flame burner (air-acetylene) instead of the furnace. The interfacing is identical except that the transfer line is connected to the nebulizer of the burner system, and no hydrogen is used. The interior of the nebulizer, generally coated with plastic materials, must be shielded by a glass liner to prevent the absorption of organolead on the walls (Fig. 3). The system using a burner has a sensitivity of 0.1 μg Pb (Ref. 1).

The identification of compounds was based on retention times established from authentic compounds, with element-specific confirmation to eliminate possible artifacts.

The cold trap technique for collection of volatile organometallic compounds has been previously discussed (Ref. 1, 2). The precision of the method was evaluated by replicate analyses (10 runs) of a synthetic sample containing ca. 5 ng (as Pb) of each of the five tetraalkyllead compounds. The relative standard deviation for each compound was in the range 10 - 15%.

The method was calibrated by injecting a known amount of a standard (usually 10-20 ng Pb) and obtaining the peak area. As atomic absorption was used as the detector, the response was related directly to the element. All the five tetraalkyllead compounds gave similar calibration curves. The response was linear up to at least 200 ng Pb above which overlapping of peaks occurred. If only a single compound was present (e.g., tetramethyllead), the calibration was linear up to at least 2000 ng. For analysis at the microgram level, the flame AAS mode is more suitable. Figure 4 illustrates typical recorder tracings of a mixture of the five tetraalkyllead compounds, using the furnace in the AA unit.

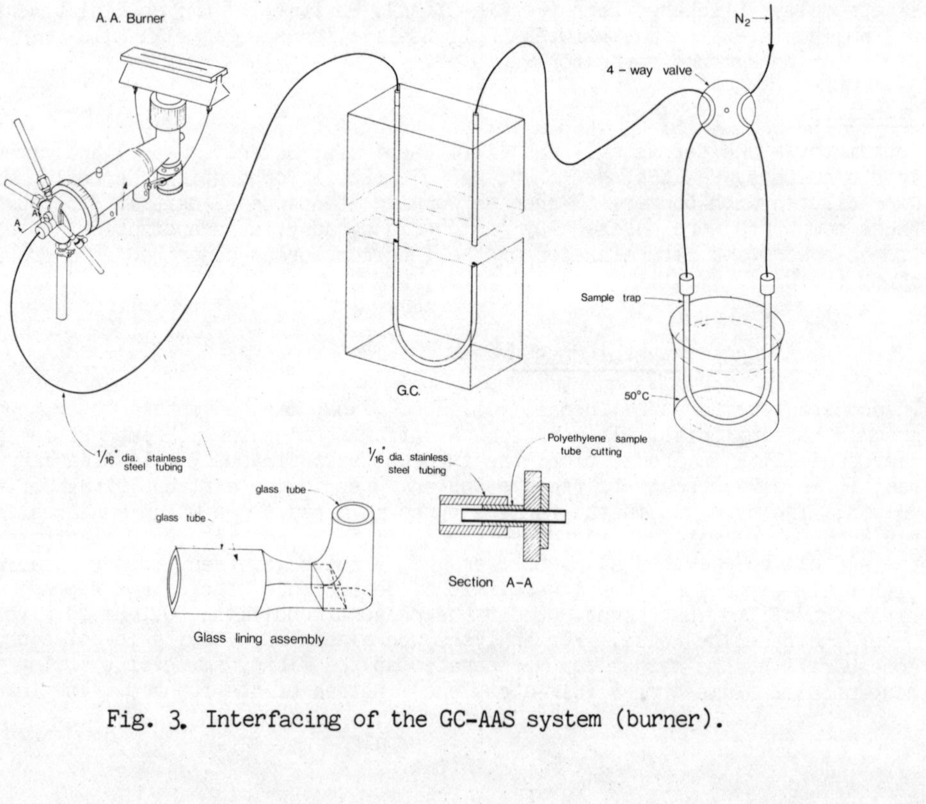

Fig. 3. Interfacing of the GC-AAS system (burner).

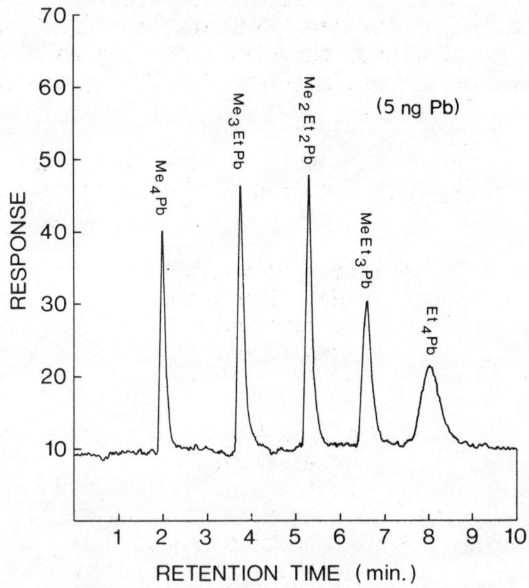

Fig. 4. Recorder tracings of a mixture of tetraalkyllead.

Solvents such as chloroform, carbon tetrachloride, hexane and benzene interfere because of their non-specific absorption at the lead resonance line. Although these solvent peaks generally emerge well before the lead peaks, the use of the background corrector is recommended.

REFERENCES

(1) Y. K. Chau, P. T. S. Wong and H. Saitoh, Determination of tetraalkyllead compounds in the atmosphere, J. Chrom. Sci. 14, 162-164 (1976).

(2) Y. K. Chau, P. T. S. Wong and P. D. Goulden, Gas chromatography-atomic absorption spectrometry for the determination of tetraalkyllead compounds, Anal. Chim. Acta 85, 421-424 (1976).

The Determination of Tetraalkyllead Compounds and their Degradation Products in Natural Water

F. G. Noden

Associated Octel Company Limited, Ellesmere Port, U.K.

ABSTRACT

An analytical scheme used in the study of the behaviour of tetraalkyllead compounds when exposed to sea water is presented. The methods cover the determination of tetraalkyllead compounds and their degradation products in the presence of each other at concentrations in the $\mu g\ l^{-1}$ range.

A choice of analytical techniques comprising mass spectrometry, atomic absorption and colorimetric chemical methods is described.

INTRODUCTION

The Organo lead compounds Tetraethyl lead (TEL) and Tetramethyl lead (TML) are used world wide as "antiknock" additives during the manufacture of gasoline to be used as fuel for spark ignition internal combustion engines. TEL was introduced into gasoline technology in 1923 and since that time a great deal of knowledge and information has been accumulated to ensure the safe manufacture, transport and use of these toxic materials. Until recently however little was known about the behaviour of lead alkyls when exposed to sea water. The accumulation of reliable data on the solubility of tetraalkyl lead compounds in water has proved to be a difficult task for a variety of reasons:

(a) The low solubility of the compounds, necessitating the development of sensitive analytical methods for measurement. The steady state $(C_2H_5)_4Pb$ concentration in sea water in contact with TEL Compound is approximately 100 $\mu g l^{-1}$ and that for $(CH_3)_4Pb$ in contact with TML Compound is of the order of 9,000 $\mu g\ l^{-1}$.

(b) The effect of such factors as the salinity of the aqueous phase the presence of air and organic liquids (dichloroethanes and dibromoethane) blended with the tetraalkyl lead compounds in "antiknock compounds".

(c) The formation of an emulsion in the aqueous phase, and the presence of very finely divided solid particles, especially in sea water.

In aqueous solution degradation of tetraalkyl lead compound takes place by the following stages:

$$R_4Pb \longrightarrow R_3Pb^+ \longrightarrow R_2Pb^{2+} \longrightarrow Pb^{2+}$$

Thus in order to obtain accurate values for the organic lead content of water that has been in contact with tetraalkyl lead compounds it was necessary to develop a scheme of analysis for the determination of the tetraalkyl lead compounds and their degradation products in the presence of each other at concentrations down to the parts per billion ($\mu g\ l^{-1}$) range.

Several of the methods have been used in our laboratories for several years. Although we have introduced atomic absorption techniques and mass spectrometry in order to achieve increased sensitivity and speed of analysis we have always kept in mind the need for methods that can be used in laboratories that do not have the resources for sophisticated equipment.

Within the described scheme it is possible to carry out a resolution of alkyl lead compounds using only spectrophotometric methods.

BASIS OF THE ANALYTICAL SCHEME

The scheme is based on the following analytical principles:

1. The separation of TEL and/or TML from their ionised degradation products by exaction with n-hexane. Concentration of the TEL and TML is also achieved at this stage.

2. The quantitative conversion of R_4Pb and R_3Pb^+ compounds to R_2Pb^{2+} on reaction with iodine monochloride as described by Moss and Browett (Ref. 1).

3. The spectrophotometric determination of R_2Pb^{2+} as the dialkyllead dithizonate following reaction with diphenylthiocarbazone (dithizone) according to the method of Cremer (Ref. 2) or by an alternative method using 4-(2-pyrydylazo resorcinol)(PAR) according to the method of Pilloni and Plazzogna (Ref. 3).

4. The extraction of trialkyl lead as undissociated trialkyl lead halide into toluene from a solution containing sodium chloride and potassium iodide.

   The analytical scheme is set out in diagram 1. An alternative procedure for the separation of tetraalkyl lead is indicated by the dotted line. The analytical scheme is usually applied to samples of volume 50 ml except in the case of the total lead determination where the sample volume may be varied depending on the choice of method for completing the determination.

DETERMINATION OF LEAD IN SEA-WATER

Collection and Storage of the Samples

The parts of the sampling equipment in contact with the sample must be lead-free. It is recommended that prior to the sampling operation the sampling vessel should be filled with a control sample of water taken at a point away from the sampling location. The control sample is analysed following collection and again after 24 hours. No increase in the lead content should be observed.

Immediately after collection the sample must be transferred to a borosilicate glass bottle with a ground glass stopper. The bottle must be completely filled otherwise loss of tetraalkyl lead compounds may take place via the air space. The usual precautions for pre-cleaning the bottles with nitric or chromic acid or alternatively with a recommended surface active agent must be taken.

Experience has shown that samples to be used for the determination of organic lead are best stored under neutral conditions in the dark. Storage tests carried out on

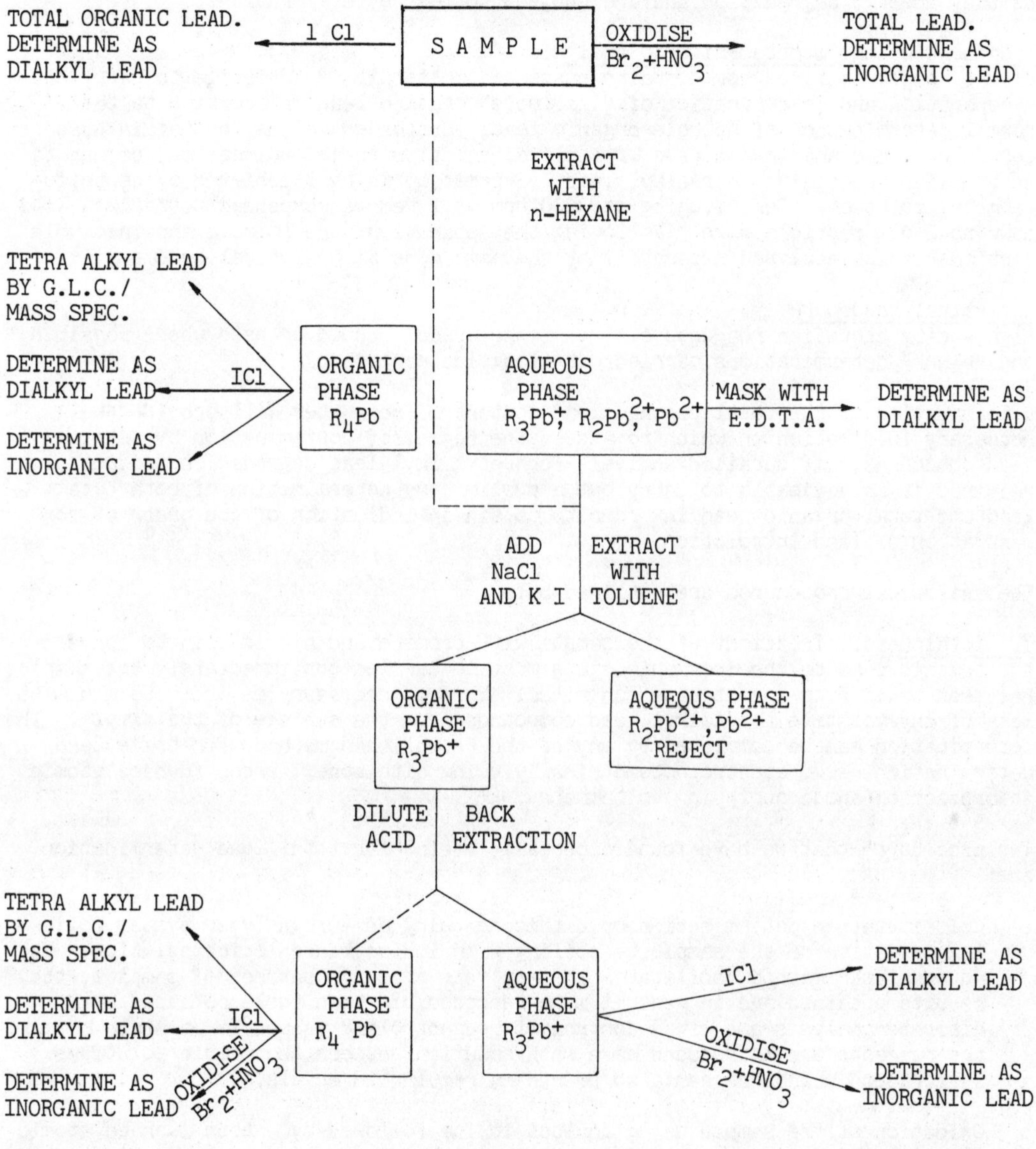

DIAGRAM 1. The Determination of Tetra Alkyl Lead Compounds and their Degradation Products in Natural Waters

using sea-water containing triethyl lead chloride and trimethyl lead chloride =10 μg Pb l^{-1} showed 90% recovery after storage for periods of up to 6 weeks.

Samples containing tetra alkyl lead compounds must not be stored for long periods as they undergo degradation and are subject to loss by evaporation.

Treatment of the Sample Prior to Analysis

If the sample is cloudy or contains suspended matter it is advisable to carry out a separation and determination of "insoluble" organic lead in addition to the normal determination of soluble organic lead. Suspended solids may retain adsorbed tetraalkyl lead and in the case of a cloudy solution the cloudiness may be due to an emulsion containing tetraalkyl lead. Separation is best achieved by centrifuging the solution. Centrifuging at 3000 rpm will remove suspended tetraalkyl lead down to 1 μ m particle size. Following the separation, the soluble and insoluble portions can be analysed separately by the same general analytical scheme.

Analytical Procedure

High purity certified reagents e.g. (Aristar Grade) should be used where possible and "blank" determinations carried out alongside each sample.

Determination of the total organic lead content of sea-water will provide the necessary information on which to assess the degree of contamination by tetraalkyl lead compounds. If detailed analysis for tetraalkyl lead degradation products is required it is advisable to carry out a preliminary determination of both total lead and total organic lead in order to obtain an indication of the order of concentration of lead in solution.

The analytical procedures are outlined below.*

1. Total Lead. Treatment of the sample with bromine and nitric acid to convert the organic lead to the inorganic state must be carried out immediately the sample has been taken from the storage container. This is necessary in order to minimise loss of any volatile tetraalkyl lead compounds from the surface of the sample. The determination can be completed by any of the established methods for trace lead determination, e.g. spectrophotometrically using dithizone, carbon furnace atomic absorption or anode stripping voltammetry.

Two procedures that we have found especially useful for total lead determination are:

a) Colorimetetric determination by dithizone using an autoanalyser system following oxidation of the sample by boiling with iodine monochloride solution. The autoanalyser method facilitates the analysis of large batches of samples, the results obtained are in general more reproducible than those obtained by discrete analysis and close control of reagent blank values is possible because the reagents are introduced as a wash solution between the sample solutions. Calibration standards can also be run at regular intervals.

b) Oxidation of the sample using aqueous iodine followed by carbon furnace atomic absorption. Prior to injection into the carbon furnace the solution is treated with diluent containing ascorbic acid (to reduce free iodine) plus phosphoric acid and ammonium molybdate (to overcome matrix effects of salts present in the sample).

2. Total Organic Lead. The sample is treated with sufficient iodine monochloride,

* Full details of individual metods can be made available on request.

1.0 M ICl in 5 M HCl, to give a final solution 0.1 M wrt ICl and 0.5 M wrt HCl. In this solution the R_4Pb and R_3Pb^+ compounds are converted to the dialkyl lead state. The dialkyl lead ion is stable in ICl solution for at least 7 days at room temperature.

There is a choice of methods for completing the determination and these are described in Section 4.

3. Separation and Determination of Tetraalkyl Lead. Good separation of tetraethyl lead from their degradation products in sea-water can be achieved by extraction with n- hexane followed by water washing of the hexane extract. This is important in cases where a mass spectrometer is not available for the determination of the tetraalkyl lead and it is necessary to complete the determination of lead in the dialkyl or inorganic state. Determination as inorganic lead will give erroneous results if organically combined lead other than tetraalkyl lead is present.

a) Determination of tetra alkyl lead by mass spectrometer following separation from hexane by gas liquid chromatography is the recommended method because it allows the differentation of tetraethyl lead and tetramethyl lead and has good sensitivity. The limit of detection for lead alkyls when present together in a sample is as follows:

$(CH_3)_4Pb$	-	4 μg l^{-1}
$(CH_3)_3Pb\ (C_2H_5)$	-	4 "
$(CH_3)_2Pb\ (C_2H_5)_2$	-	6 "
$(CH_3)\ Pb\ (C_2H_5)_3$	-	8 "
$(C_2H_5)_4Pb$	-	8 "

These limits were achieved with a VG Micromass 12B instrument using single ion monitoring. The fragment ion is m/e 208 being common to all the tetraalkyl leads. The intensity of this ion is about 50% of that of the individual ions but varies for individual tetraalkyl lead compounds.

4. Determination of Dialkyl Lead. As mentioned previously we have used two colorimetric reagents for the determination of dialkyl lead and have used them in both manual and continuous flow systems.

The various ways of completing the determination are shown in Table 1.

Dithizone forms complexes with trialkyl lead, dialkyl lead and inorganic lead with absorption maxima at 435 nm, 485 nm and 520 nm respectively but inorganic lead, and other possible interfering metals present in sea-water can be effectively masked by EDTA.

In methods 1 and 2 the dialkyl lead dithizonate is converted to inorganic lead by treatment with dilute nitric acid/hydrogen peroxide solution whereas in the automatic method the dialkyl lead dithizonate is measured. In these procedures a high result for dialkyllead will be obtained if trialkyllead is also present therefore ideally they should only be applied where it is desired to determine the sum of di- and tri- alkyl lead after conversion of trialkyl lead to dialkyl lead state.

It is possible to correct for this error which is consistent and amounts to 10% of the trialkyl lead content of the solution. However, in some cases the ratio of R_3Pb^+ to R_2Pb^{2+} in solution is too high to allow the calculation of accurate values for dialkyl lead content by applying the correction.

TABLE 1 Determination of Dialkyl Lead

	Reagent	Method of Determination	Normal Range
1	Dithizone	Back extraction of dialkyl lead dithizonate into dilute nitric acid followed by spectrophotometric determination as inorganic lead using dithizone.	5 to 200 $\mu g\ l^{-1}$
2	Dithizone	Back extraction of dialkyl lead dithizonate into dilute acid followed by carbon furnace atomic absorption.	2 to 40 $\mu g\ l^{-1}$
3	Dithizone	Automatic Spectrophotometric, continuous flow system.	2 to 100 $\mu g\ l^{-1}$
4	PAR	Manual Spectrophotometric	20 to 20,000 $\mu g\ l^{-1}$
5	PAR	Automatic Spectrophotometric	100 to 20,000 $\mu g\ l^{-1}$

PAR reagent does not react with trialkyl lead therefore the reagent can be made specific for dialkyl lead after masking inorganic lead. We have found that in this determination CDTA is a more effective masking agent than EDTA.

Unlike dithizone PAR forms water soluble lead complexes. The wavelength for maximum absorption of the dimethyl and diethlyllead complexes with PAR are practically identical and the same applies in the case of complexes with dithizone.

For calibration purposes the most convenient organic lead standard is trimethyl lead chloride. This is treated with iodine monochloride to give a solution containing dialkyl lead and which is stable for at least 7 days at room temperature.

5. Separation and Determination of Trialkyl Lead. The determination of trialkyl lead is normally carried out following the removal of any tetraalkyl lead from the sample but the analytical scheme provides for an alternative method of separation if necessary.

In the normal procedure the sample after removal of tetraalkyl lead is extracted with toluene, the trialkyl lead is quantitatively transferred to the toluene together with any tetra alkyl lead not previously removed. Dialkyl lead and inorganic lead remain in the aqueous phase. Quantitative extraction of triethyl lead takes place if the aqueous phase is saturated with sodium chloride (Ref. 4) but only 20% extraction of trimethyl lead takes place under the same conditions. However, it was found that by the addition of sufficient potassium iodide to give a 40% solution complete extraction of trimethyl lead could be achieved. However, under these conditions there was some extraction of diethyl lead (about 20%). This determination was carried out with solutions containing 100 $\mu g\ l^{-1}$ of diethyl lead but in our experience there is little evidence of the presence of a significant amount of dialkyl lead during degradation of tetra alkyl lead compounds in sea-water. When tetraalkyl lead accompanies trialkyl lead in the toluene extract separation can be achieved by back extraction with 1% v/v nitric acid solution. Tetra alkyl lead remains unchanged in the toluene phase whereas the trialkyl lead is extracted by the acid and can be determined either as dialkyl lead following reaction with iodine monochloride or as inorganic lead following oxidation with bromine and nitric acid.

6. Direct Determination of Inorganic Lead. The effective masking of inorganic lead by EDTA (as applied in the determination of dialkyl lead) can be used as a basis for the determination of inorganic lead in the presence of tetra, tri and di alkyl lead compounds.

The sample is buffered to pH 11.5 and reacted with dithizone solution in chloroform. On shaking with EDTA the lead dithizonate is decomposed and the inorganic lead content is obtained by measurement of the light absorbance of the inorganic phase at a wavelength of 530 nm before and after the reaction with EDTA.

RECENT DEVELOPMENTS

1. Application of Polarographic Techniques. We have recently used differential pulse polarography (after anode stripping from a glassy carbon electrode) for the determination of ionic species of lead in sea-water. We have in the main followed the procedure used by Magjer and Branica (Ref. 5) for the determination of labile lead, i.e., without acidification of the sample.

Using this procedure we have measured:

$R_3Pb^+ + R_2Pb^{2+} + Pb^{2+}$ (at - 1.0V plating potential)

$R_2Pb^{2+} + Pb^{2+}$ (at - 0.7V plating potential)

R_3Pb^+ can be determined by difference.

It is necessary to add a reducing agent, ascorbic acid, to stabilise the organic lead species in this system in order to obtain reproducible results. The possibility of masking inorganic lead by the addition of EDTA is being investigated. If this is successful it should facilitate the separate determination of R_2Pb^{2+} and also provide a means of monitoring trace amounts of R_3Pb^+ and R_2Pb^{2+} in the presence of relatively large amounts of inorganic lead.

To date we have worked at concentrations down to 0.2 $\mu g l^{-1}$ for inorganic lead and 1 $\mu g l^{-1}$ for organic lead.

2. Determination of Tetraalkyl Lead by "Head Space" Analysis
By the application of "Head Space" analysis in conjuction with mass spectrometry we have achieved lower detection limits than is possible using solvent extraction. The solution is allowed to reach equilibrium with its vapour in a closed vessel at 30°C.

Taking 0.5 cm^3 of vapour for analysis the following detection limits were attained for the determination of all 5 species. The values represent the concentration in water.

$(CH_3)_4$ Pb = 0.2 $\mu g l^{-1}$

$(CH_3)_3$ Pb (C_2H_5) = 0.3 "

$(CH_3)_2$ Pb $(C_2H_5)_2$ = 0.4 "

$(C_2H_5)_4$ Pb = 1.0 "

Individual detection limits may be improved by adjusting the operating conditions for a particular lead alkyl. There is scope for lowering the detection limit further by working at a higher temperature.

Some Typical Analyses of Sea-Water Sample Following Contact with Tetraalkyl Lead Compounds. The data shown in Table 2 was obtained from the analysis of sea-water following contact with tetraalkyl lead compounds under a variety of experimental conditions.

TABLE 2 Typical Analysis of Sea-Water Following Contact with Tetra Alkyl Lead Compounds

Details of Sample(1)	Lead Content $\mu g l^{-1}$ as Pb					
	Total Lead	Total Organic Lead	R_4Pb	R_3Pb^+	R_2Pb^{2+}	$Pb^{2+(2)}$
After stirring with pure TEL for 2 days	718	661	626	39	< 5	57
After stirring with TEL-CB for 18 hours	332	301	300	40	< 5	31
After stirring with TEL-CB for 1 hour	133	113	92	31	< 5	20
Sample 3 after storage, then 37 $\mu g l^{-1}$ R_2Pb^{2+} added	99	75	27	25	34	24
After stirring with TEL-CB for 2.5 hours	44	39	26	14	< 2	3^+
After stirring with TEL-CB for 2 hours	17	12	6	8	< 2	3^+
After stirring with TML-CB for 17 hours	936	925	762	137	5	11
After stirring with TML-CB for 1 hour	393	355	302	35	< 5	38
Aged sample after contact with TML-CB	294	199	2	200	< 5	95
After stirring with TML-CB for 1 hour	45	41	39	4	< 2	4^+
Method Reference in Text	1a	2 & 4.3	3a	5 & 4.3	4.3	See (2).

(1) Varying ratios of lead alkyl Compound and sea-water were used during these experiments.

(2) The results of Pb^{2+} marked $^+$ were obtained by anodic stripping polarography. All other values were calculated as the difference between the total lead and total organic lead contents.

ACKNOWLEDGEMENTS

The analytical work described in this report was carried out by members of the Analytical Section of the Research and Development Department.

REFERENCES

(1) R. Moss and E. V. Browett, Analyst 91, 428 (1966).

(2) J. E. Cremer, Brit. J. Ind. Med. 16, 191 (1959).

(3) P. Pilloni and G. Plazzogna, Anal. Chim. Acta 35, 325 (1966).

(4) W. Bolanowska, J. Ind. Med. 25, 203 (1968).

(5) T. Magjer and M. Branica, Croat. Chem. Acta 49, (1) L1 - L5 (1977).

Studies of Lead in Water and Biological Materials using X-ray Emission Spectroscopy*

V. Valković

Rice University, Houston, Texas 77001, U.S.A. and
"Rudjer Boskovic" Institute, Zagreb, Yugoslavia

ABSTRACT

Advantages and limitations of Pb-determination in water and biological materials using X-ray emission spectroscopy are summarized. Sample excitations by X-ray tube, radioactive sources and charge-particle beams are discussed. X-ray emission spectroscopy is very useful in the determination of the origin of lead in the marine environment. This can be accomplished by the simultaneous measurement of several trace element concentration levels, i.e., elemental fingerprints. Sample preparation techniques for water and biological material are described. Continuous monitoring of Pb-level is also possible.

INTRODUCTION

Lead is present in all organs and tissues of mammals, although it is not essential to their nutrition. The toxicology of lead has been studied extensively. Organic lead poisoning from contact with tetra-alkyl lead compounds represents a special problem. The major portion of organic lead accumulates in the brain due to a special affinity between organic lead and the lipid of nerve tissues. According to Schroeder (Ref. 1) tetraethyl lead is a hundred or more times more toxic than inorganic lead. Lead can interfere with the synthesis of heme, thereby altering the urinary or blood concentration of enzymes and intermediates in heme synthesis or their derivates.

The major stationary sources of lead induced into the atmosphere are the primary and secondary smelting of lead, smelting of brass, storage battery manufacturing, paint manufacturing and insecticide manufacturing. The major source of atmospheric lead is the variety of lead-based antiknock additives in the gasoline used in internal combustion engine. From the atmosphere lead is transferred to waters, soils and plants (Ref. 2). There is a clear evidence of lead pollution caused by motor vehicles close to busy roads. It seems that there is an increase in lead concentrations in snow from the Greenland ice cap which could be correlated with the emission of lead aerosols from automobile exhausts. Chester and Stoner (Ref. 3) carried out a sampling programme in which particulates were collected from the lower atmo-

* Supported in part by R. A. Welch Foundation and National Science Foundation, Washinton, D. C.

sphere over large tracts of the world's oceans. The study of Pb concentrations along the length of Atlantic Ocean indicated that human activities are very important in the distribution of this element in the marine atmospheric particulates.

All this has generated great interest in the Pb concentration levels in the environment and in the living beings.

A variety of methods has been used for the monitoring of lead in water, air, and biological material. In this paper we shall describe the use of X-ray emission spectroscopy for lead determination.

EXPERIMENTAL METHODS

Fundamentals

X-rays are produced as a result of the removal of at least one inner-shell electron. Several processes might result in the removal of an electron from the inner shell in the atom. The most important are scattering processes in which the incoming charged particle or photon hits an electron of an K- or L-shell and transfers part of its kinetic energy to the electron. As a result, a vacancy is created which can be filled with an electron from one of the outer shells; and the energy is emitted in the form of an X-ray photon.

In recent years the detection of characteristic X-ray for trace element analysis has increased significantly, primarily because high resolution Si(Li) detectors have made energy dispersive analysis possible. X-rays from X-ray tubes and radioactive sources have been used to excite characteristic X-rays in different samples. Because of large cross sections for X-ray production, charged particles can also be used to produce characteristic X-rays from elements only present in trace amounts in target materials. Figure 1 indicates the principle of X-ray emission spectroscopy (Ref. 4). The principle of X-ray spectroscopy is the excitation of the target by a beam of photons or charge particles, and subsequent detection of the X-rays produced in the target.

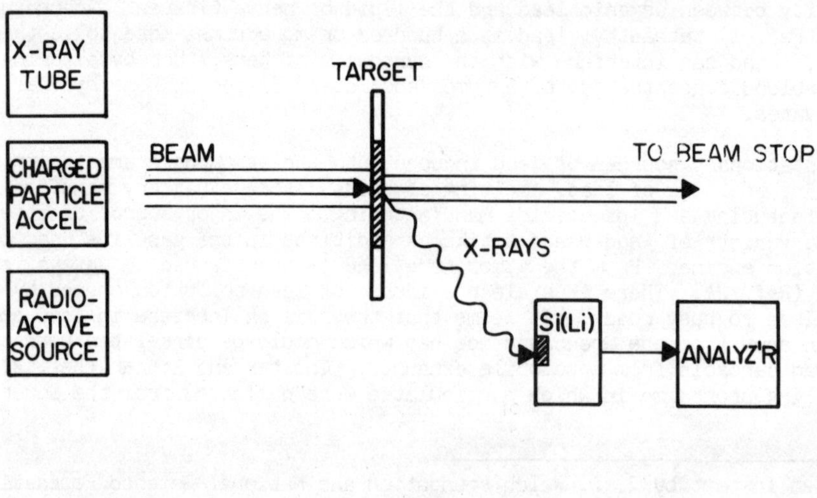

Fig. 1. The principle of X-ray emission spectroscopy.

Semiconductor detectors used for X-ray detection have the first stage of preamplification cooled. In order to prevent contamination of the detector surface by condensation of impurities, the detector is protected by a part sealed by a beryllium. The absorption of X-rays in this window determines the detector efficiency for lower X-ray energies; while the detector size and geometry define the efficiency for high X-ray energies. Figure 2 shows the calculated efficiency for a Si(Li) detector for different values of Be window thickness. The energy interval 5-20 keV embraces the K-lines of medium heavy elements, the L-lines of heavy elements, and the M-lines of super heavy elements. Figure 3 shows the measured X-ray spectrum from Ni-Pb solution deposited on Al-formvar backing and bombarded by 3 MeV protons. Nickel K_{α} (E=7.5 keV) and K_{β} (E=8.3 keV) and lead L-lines (Ll = 9.2 keV, L_{α} = 1o.5 keV, L_{β}, L_{γ}) are seen in the spectrum. The energy of lead K-lines is too high (72 - 85 keV) to be seen with Si(Li) detector.

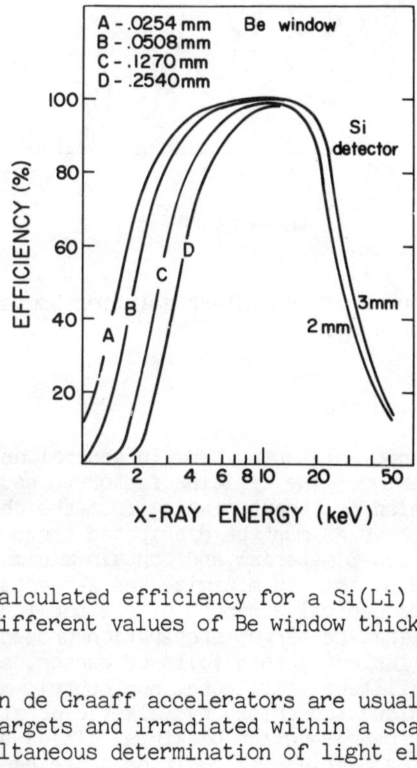

Fig. 2. Calculated efficiency for a Si(Li) detector for different values of Be window thickness.

As a proton source, Van de Graaff accelerators are usually used. Samples are prepared in the form of targets and irradiated within a scattering chamber. Such a set-up allows the simultaneous determination of light elements using backscattering, medium and heavy elements using characteristic X-rays and hydrogen determination using the measurements of p-p coincidences. The block diagram for such a set-up is shown in Fig. 4.

Sample Preparation

Trace elements to be determined are always mixed with the bulk of the material in the sample, the so-called matrix. Matrix effects (i.e., interference due to internal absorption and enhancement of the fluorescent radiation, and changes in emitted intensity due to specimen heterogeneity and finite particle size) are central problems in excitation by X-ray tube or radioactive sources (Ref. 6 and 7). Charged-particle induced emission has an advantage with respect to these techniques in that, because of large cross sections, relatively thin targets can be used.

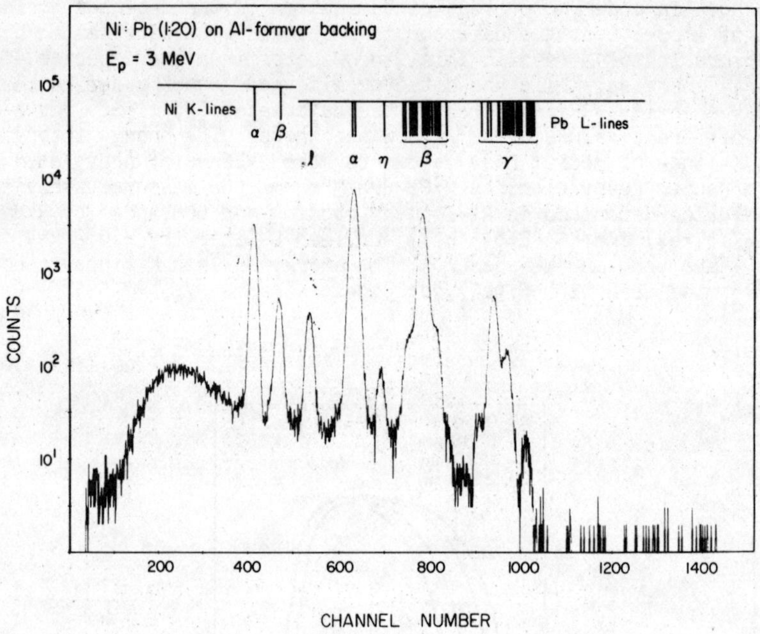

Fig. 3. X-ray spectrum from Ni-Pb solution bombarded with 3 MeV protons.

Thus, matrix effects can be ignored.

Thin biological targets are not self-supporting in general and must be prepared as deposits on some kind of backing. The limiting factor to count rate in the charged-particle induced X-ray emission is target heating from the charged-particle beam. In vacuum, heat input from the beam must be dissipated through the target backing. The backing must consist of low-Z material and contain minimal amounts of higher-Z impurities. Backings should be thin to minimize the X-ray background and thus to increase sensitivity. The mechanical strength of the backing should be such to withstand irradiation, handling and target preparation procedures. Many materials have been used for backings including thin foils of carbon, collodion, formvar, kapton, mylar and different filters (millipore, nuclepore), and others (Ref. 8). The target preparation techniques described by Valković et al. (Ref. 9) involve the fabrication of aluminized formvar sandwich backings. These backings possess both good mechanical and thermal properties, in addition to being thin and made of low-Z material. The X-ray spectrum resulting from the bombardment of such a backing with a proton beam does not show any characteristic X-rays and is dominated by a hump resulting from secondary electron induced bremsstrahlung. The bremsstrahlung hump decreases rapidly above an X-ray energy of $E_x \approx (4m/M)E$ which is the maximum energy transfer from a projectile of mass M and energy E to a free electron of mass m. This is the reason that the bombarding energy for beams of protons is usually chosen between 2 to 3 MeV.

Once the backing is chosen and the proton energy determined one should optimize the other experimental parameters in order to determine the maximum beam current giving results of acceptable reproducibility and the volatilization of trace elements due to beam induced target damage. As an example, Fig. 5 shows the results of a

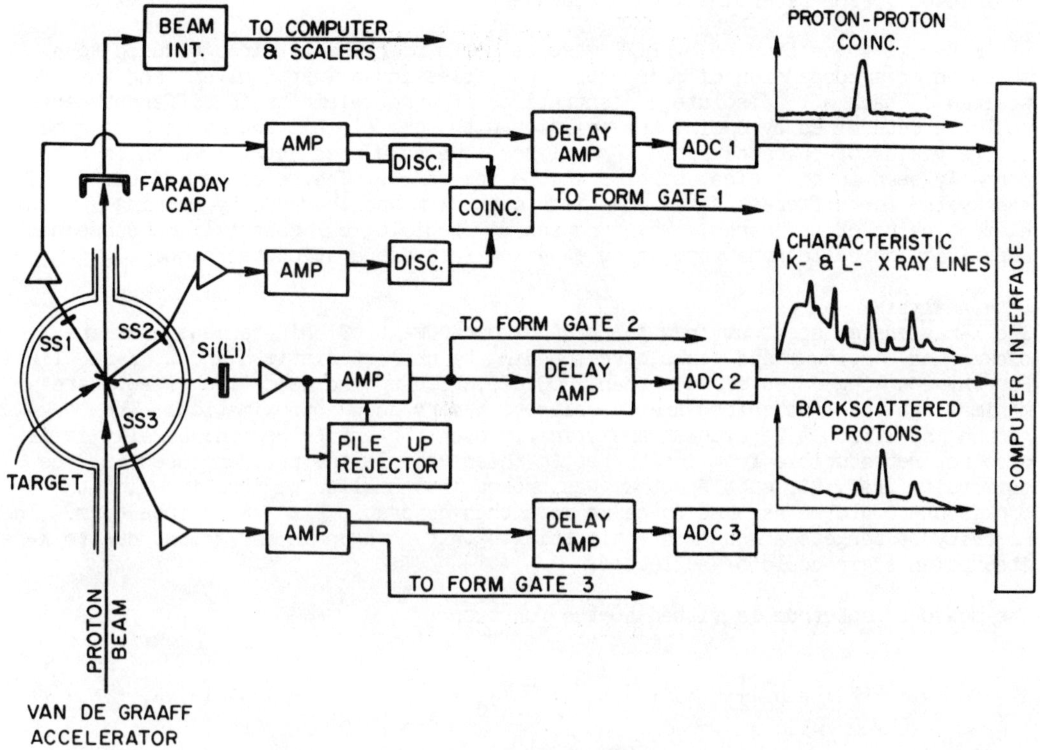

Fig. 4. Experimental set-up for the simultaneous measurement of proton induced X-rays, proton backscattering and proton-proton coincidences.

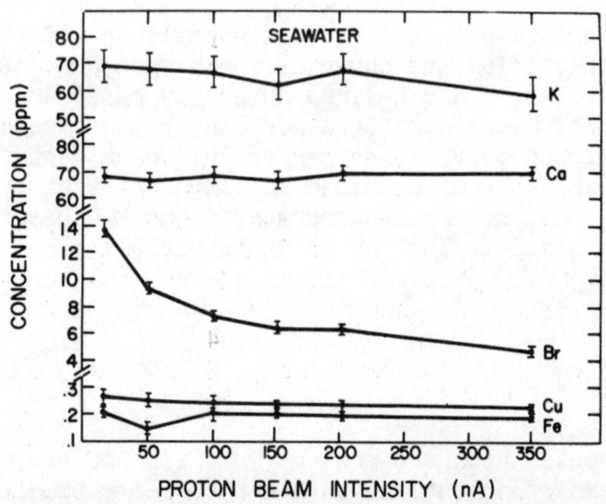

Fig. 5. Concentrations of different trace elements determined from the exposure of sea water targets to a 3-MeV proton beam as functions of beam intensity.

volatilization study for a sea water target (Ref. 10). Concentrations of different trace elements determined from the exposure of sea water targets to a 3 MeV proton beam as functions of beam intensity. It is obvious that the beam current should be restricted for Br determination in seawater.

Any material to be studied by PIXE must be selfsupporting or first reduced to a solution or a suspension of microscopic particles in an inert solvent and then deposited on backing. Absolute concentrations of trace elements in different targets could be determined by doping the solution with some element and by measuring relative yields of characteristic X-ray lines. We have used yttrium which is not normally seen in biological samples in the ppm range. The relative efficiency of the system for different X-ray energies to yttrium was measured by running yttrium-element pairs of standards. We found ashing of biological material and subsequent deposition on cellophane tape to be easy and fast preparation technique.

Data Analysis

The X-ray energy spectrum from the proton bombardment of thin targets contains background continuum and peaks corresponding to characteristic K and L X-ray lines. For the counting rates of less than 1000 c/s, pile-up and dead-time effects are minimal and the line shapes are Gaussian to a very good approximation. The background is difficult to express analytically over the entire region; usually it is also not reproducible from one target to the next. In the program used at Rice University (Ref. 9), each spectrum was broken into smaller regions in which the background could be assumed to be no more than quadratic plus an inverse term. The majority of targets are sufficiently thin so that changes in background due to X-ray absorption edges could be neglected.

The measured spectrum is fitted to the function

$$F(x_j) = N_o + N_1 x_j + N_2 x_j^2 + N_3/x_j + \sum_{\text{all peaks}} A_i \left[\exp\left[-(x_i - x_j)^2/\Gamma^2\right]\right]$$

where area is defined by

$$P_i = \pi^{1/2} A_i \Gamma_i .$$

Analysis of data was performed in two steps. A small set of standard targets containing combinations of yttrium and other elements were run. Evaluation of resulting spectra for peak widths, K and L-lines intensity ratio and relative efficiency for each element to yttrium was done by a non-linear least squares routine in which all parameters were allowed to vary independently. Each X-ray spectrum measured during the proton bombardment of a sample was then analyzed by on-line program (a linear least-square routine) which generated K and L Gaussian functions for each element present in the sample. The resulting output was given directly in ppm concentrations.

RESULTS

X-ray Emission Spectroscopy of Water

Concentrations of elements which are present in fresh and sea waters in the level of ppm or less have been of scientific interest for some time. Water pollution has received much attention since, over the years, water quality has deteriorated in many areas of the world. Increasing concern has been connected with the quantities of different metals, expecially heavy ones such as mercury and lead.

Water targets are easily prepared by simply allowing a few drops to dry on an aluminum formvar backing. Such a simple target preparation allows the detection of elements present in water with concentrations $\geq$ 0.1 g/ml. As an example, Fig. 6 shows the X-ray spectrum from a clean sea water target; peaks associated with K, Ca, Br, and Sr are present (Z is dopant). In some cases pollutants can be determined using this simple method. Figure 7 shows the X-ray spectrum from polluted harbour sea water; Fe and Cu are present in high enough concentration to be easily determined. For analysis of dissolved trace metals present in water in smaller concentrations, one often uses a method which involves the formation of insoluble metal chelates in coordination with dithiocarbamate (ammonium pyrolidine dithiocarbamate or diethyl dithiocarbamate), filtration through a membrane filter, and the analysis of the precipitate. Simultaneous analysis of most of the transition elements is possible, but alkali and alkaline-earth metals are excluded. The following elements can be separated by dithiocarbamate: V, Cr, Mn, Fe, Co, Ni, Zn, Ga, As, Mo, Rh, Pd, Ag, Cd, In, Sn, Sb, Pt, Au, Hg, Tl, Pb, and U (Ref. 11). The dithiocarbamate precipitation technique for target preparation offers several advantages over procedures based upon evaporation, among them the uniformity of distribution of elements on the target, the simplicity which reduces both handling time and contamination hazard, detection limit of 10^{-3} ppm and lower, and high efficiency within a wide pH range. Figure 8 shows the X-ray spectrum from a sea water target prepared on millipore filter from 0.1 liter of sea water. The target was bombarded with only 1-2 nA beam of 3 MeV protons.

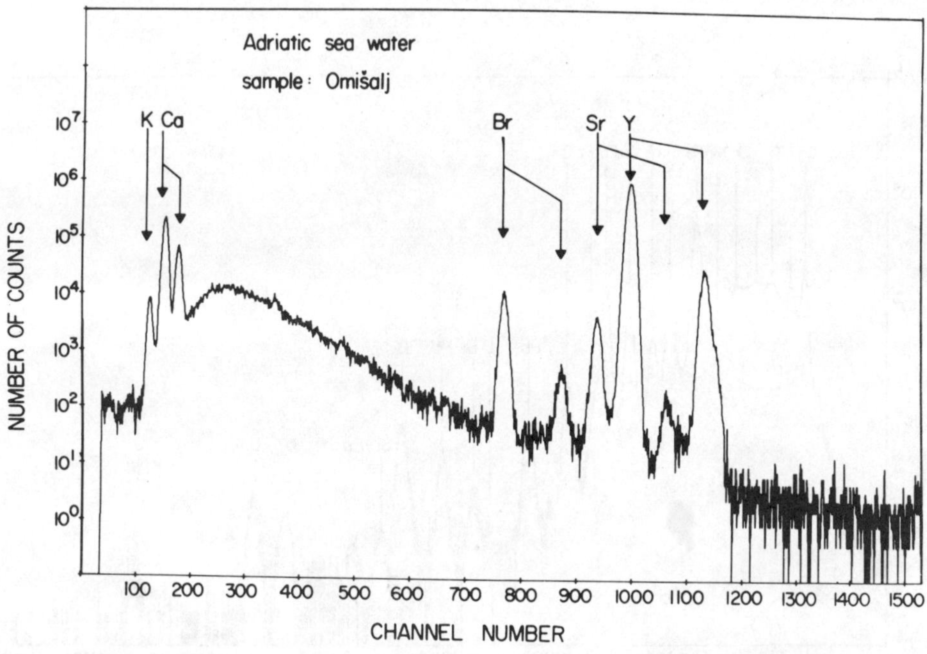

Fig. 6. X-ray spectrum resulting from proton bombardment of a clean sea water target.

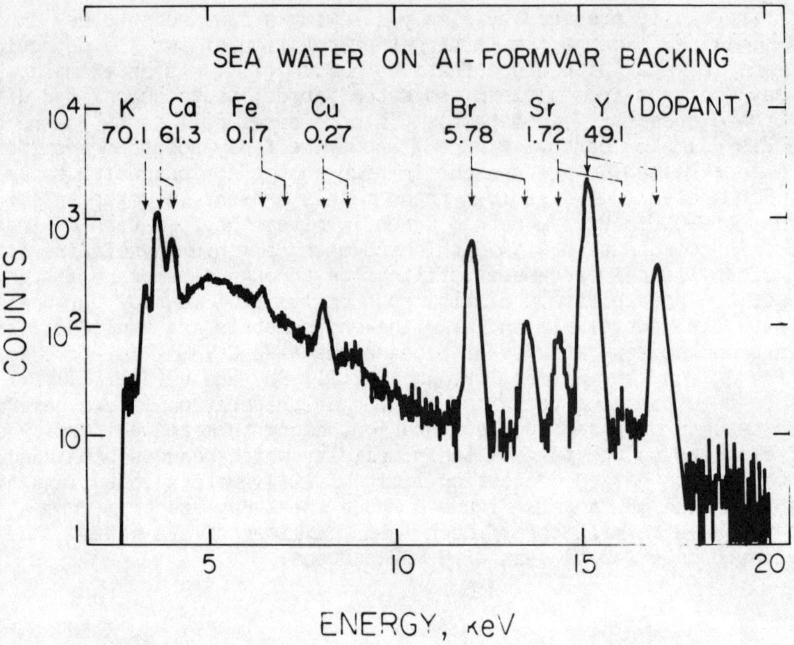

Fig. 7. X-ray spectrum resulting from proton bombardment of polluted harbour sea water target.

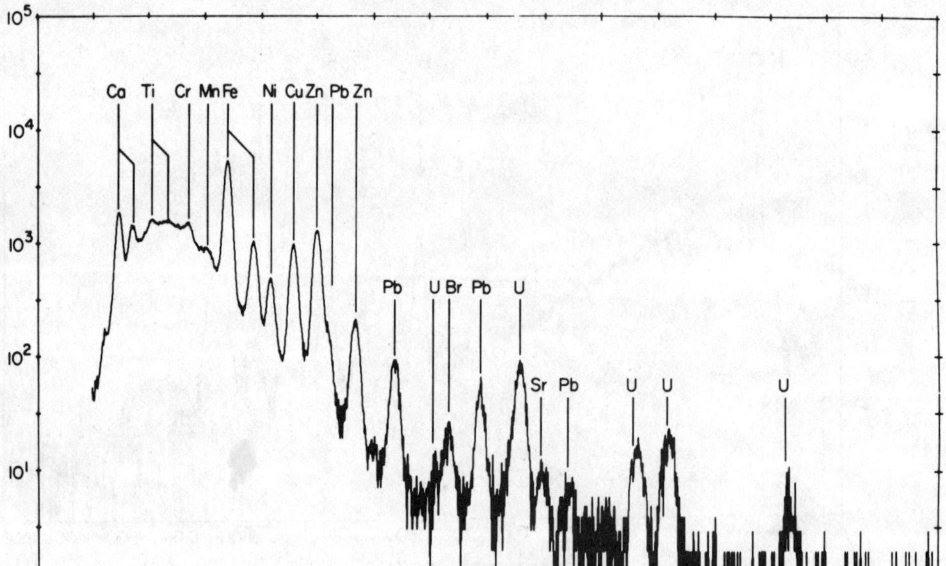

Fig. 8. X-ray spectrum from a sea water target prepared on millipore filter by dithiocarbamate precipitation technique.

Lead in Biological Materials

Lead can be found in a variety of biological materials using the detection method of characteristic X.rays.

The sources of lead in the environment are varied. The use of fossil fuels and their products is a widespread source. Lead can be detected in both crude oil and coal using proton induced X-ray emission spectroscopy. As an sample, Fig. 9 shows the X-ray spectrum from the bombardment of a target made from the ash of South Louisiana crude oil. Crude oils of different origin have different amounts of Pb.

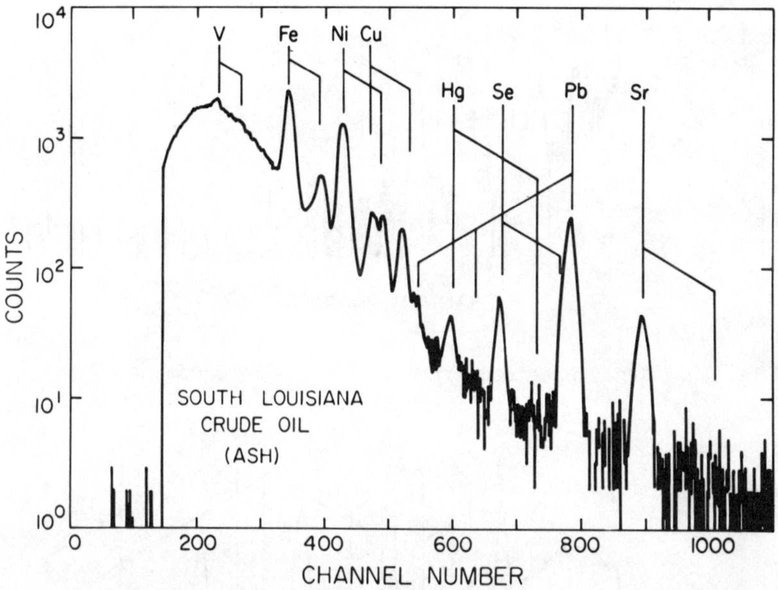

Fig. 9. X-ray spectrum from the ash target of South Louisiana crude oil.

The same is valid for coals (see Fig. 10). It should be mentioned that X-ray emission spectroscopy allows the determination of Pb within the "elemental fingerprints" of other elements. This is very useful in following the destiny of Pb in fossil fuels and products. The use of fossil fuels (and their products with Pb-compound additives) is responsible for the major part of the atmospheric load. X-ray emission spectroscopy is an accepted method for the determination of the Pb content of air particulates. Figure 11 shows the X-ray spectra of filter paper before and after air had been pumped through it. Pb:Br ratio indicates the source of Pb to be automobile traffic.

From the atmosphere, Pb is transferred to soil, waters and plants. It is even incorporated into the structure of tree rings as a permanent record of atmosphere and soil loadings. We have investigated variations in Pb concentration across tree rings in an effort to establish variations in the atmospheric load. This is still in progress; Fig. 12 shows the X-ray spectrum obtained by the proton bombardment of the target made from ash deposited on cellophane tape. Single ring variations can be established using such an approach.

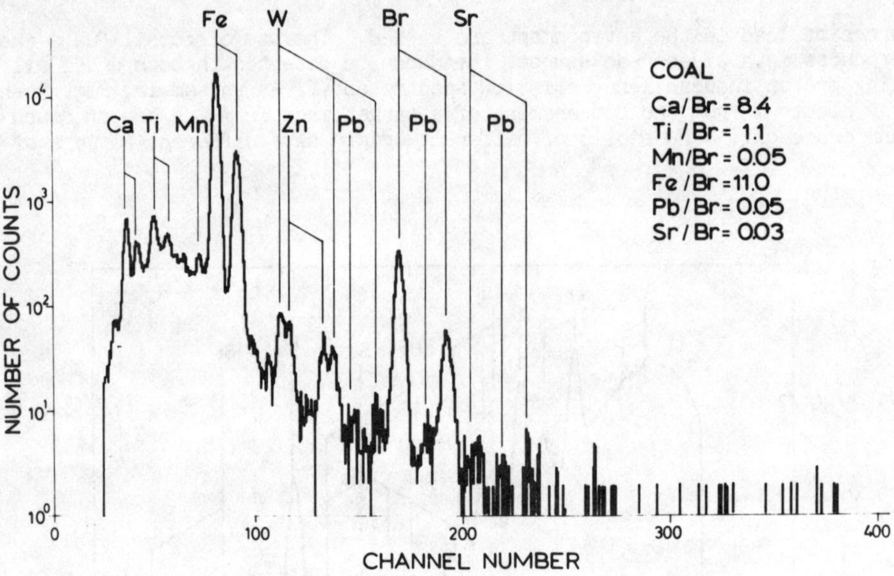

Fig. 10. X-ray spectrum from the proton bombardment of coal.

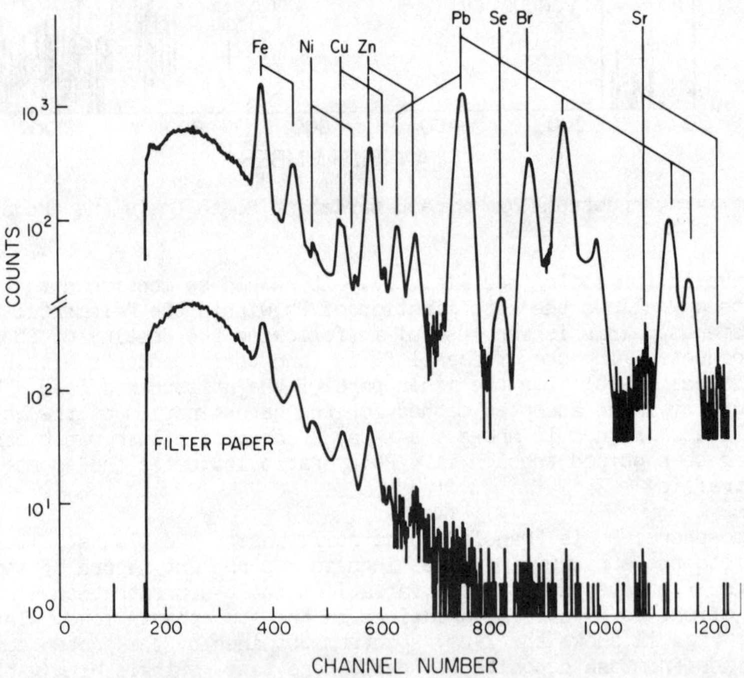

Fig. 11. X-ray spectra from filter paper before and after air was pumped through it.

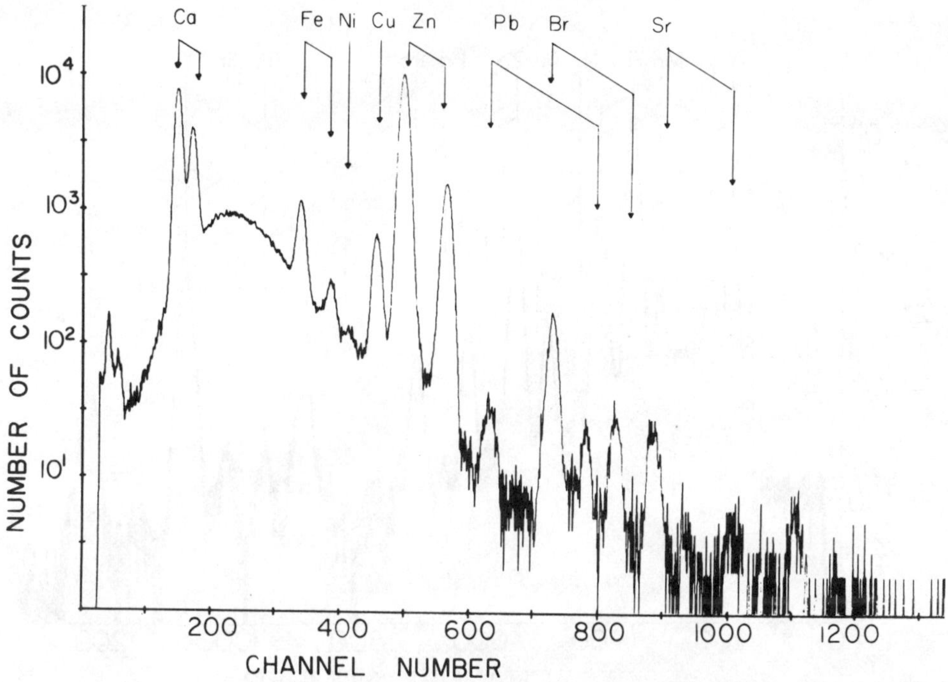

Fig. 12. X-ray spectrum from target made of tree ring ash deposited on cellophane tape.

Animals living in a Pb contaminated environment reflect this exposure to some extent by elevated Pb levels in their tissues. It has been suggested (Ref. 12) that elemental analysis of honey can be used as an indicator of Pb pollution. We have found significant amounts of Pb in honey originating from locations near the congested areas. Figure 13 shows the X-ray spectrum from one of the measured honey samples, with Pb concentration of 0.8 % in ash.

The biological experiments on the effects of pollutants are often performed by introducing controlled amounts of pollutant in the environment in which given species live. This is customary procedure with metals and sea life. X-ray emission spectroscopy allows the determination of many metals in different parts of the body simultaneously. Figure 14 shows the measured X-ray spectrum from fathead living in a Pb enriched environment. The sample was homogenized before the target preparation and yttrium was introduced as the internal standard.

Hair as an Indicator of Pb Exposure

It has been suggested by several authors that hair can be used as an indicator of exposure to environmental pollutants (see Ref. 13 for the list of publications relevant to this subject). The concentrations of most trace elements should increase with increasing distance from the scalp if the exposure of hair to the elements of the environment is constant. X-ray emission spectroscopy allows simultaneous measurement of several elements (including Pb) on single hairs. Figure 15 shows the X-ray spectra from the bombardment of human head hair. Lead concentration can be determined from one of many Pb-lines. It should be mentioned that one of the Pb line coincides with the As K_α line and that Pb can interfere with As determination in hair. X-ray emission spectroscopy can be used for the measurement of trace element ratios along human hair. It has been shown (Ref. 14) that the elements

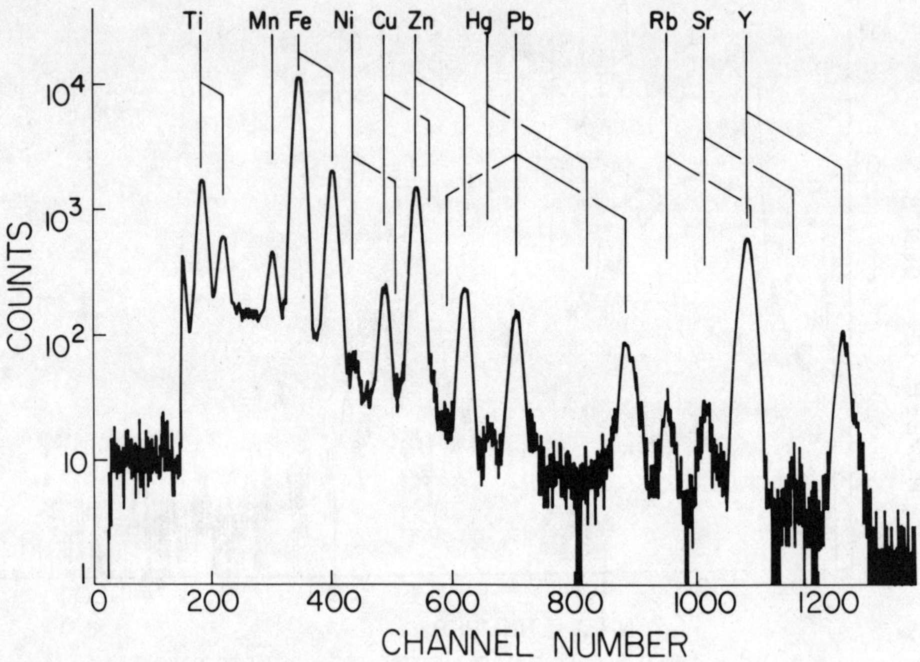

Fig. 13. X-ray spectrum from the proton bombardment of honey (ash target).

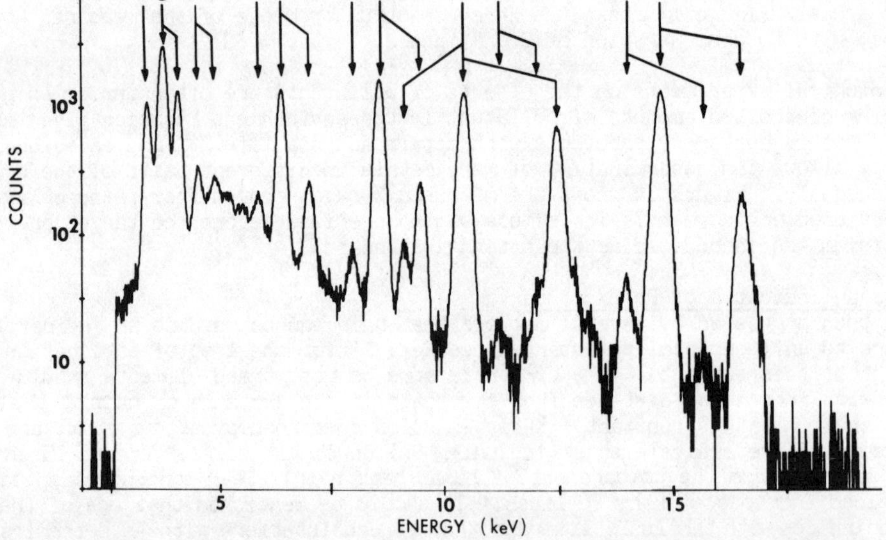

Fig. 14. X-ray spectrum of target made by homogenized fathead (solution dried on Al-formvar backing).

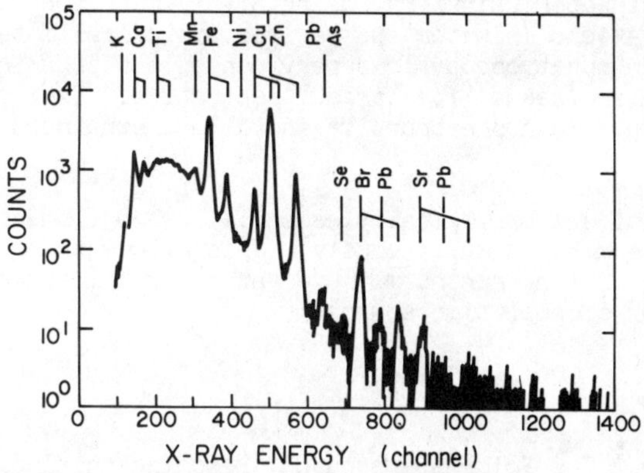

Fig. 15. X-ray spectrum from head hair.

whose concentrations increase along the hair can be identified as pollutants in the area. For widespread pollutants (such as lead) even the medium value of a group of subjects can be used as a measure of the exposure. Figure 16 (from Ref. 15) shows a median concentration ratio values for Cu/Zn, Pb/Zn, Sr/Zn, and As/Zn as a function of the distance from the scalp. It is seen that the value of the Cu/Zn concentration ratio is remarkably constant. The medians for Pb/Zn, Sr/Zn and As/Zn increase along the hair indicating the presence of Pb, Sr and As as pollutants in the environment.

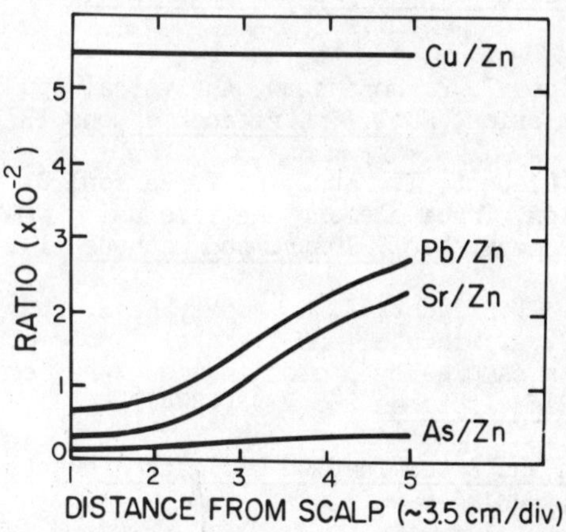

Fig. 16. Median concentration ratio values for Cu/Zn Pb/Zn, Sr/Zn and As/Zn as a function of the distance from the scalp.

CONCLUSION

Detection of the characteristic X-ray using Si(Li) detectors allows the determination of Pb concentrations in water and various biological materials. Proton induced X-ray emission spectroscopy is a very convenient technique since only small amounts of material are needed for target preparation. Excitation by radioactive sources and X-ray tube is nondestructive and allows continuous monitoring in the field.

The concentrations of all biologically essential trace elements and many toxic elements can be determined simultaneously. This allows the study of complex interdependences as well as fingerprint measurements. Such measurements are very useful in the determination of pollution source.

REFERENCES

(1) H. A. Schroeder, The Poisons Around Us, Bloomington, Indiana University Press, 1974.

(2) V. Valković, Trace Element Analysis, Taylor and Francis, London, 1975.

(3) R. Chester and J. H. Stoner, Nature, London 246, 138, 1973.

(4) V. Valković, X-ray emission spectroscopy, Contemp. Phys. 14, 415 (1973).

(5) V. Valković, Nuclear Microanalysis, Garland Publishing, New York, 1977.

(6) J. R. Rhodes, Energy Dispersion X-ray Analysis, (J. C. Russ, ed.), ASTM Publication 485, 243 (1971).

(7) J. C. Russ, Elemental X-ray Analysis of Materials, EXAM Methods, EDAX Int. Inc., 1972.

(8) S. A. E. Johansson and T. B. Johansson, Analytical application of particle induced X-ray emission, Nucl. Instr. and Methods 137, 473 (1976).

(9) V. Valković, R. B. Liebert, T. Zabel, H. T. Larson, Đ. Miljanić, R. M. Wheeler, and G. C. Phillips, Trace element analysis using proton induced X-ray emission spectroscopy, Nucl. Instr. and Methods 114, 573 (1974).

(10) M. E. Alexander, E. K. Biegert, J. K. Jones, R. S. Thurston, V. Valković, R. M. Wheeler, G. A. Wingate, and T. Zabel, Trace element analysis of seawater and fish samples by proton induced X-ray emission spectroscopy, Inter. J. App. Rad. Isotopes 25, 229 (1974).

(11) M. Pinta, Detection and Determination of Trace Elements, (English translation), original: Dunod, Paris, 1962.

(12) S. C. Tong, R. A. Morse, C. A. Bache, and D. J. Lisk, Elemental Analysis of Honey as an Indicator of Pollution, Arch. Environ. Health 30, 329 (1975).

(13) V. Valković, Trace Elements in Human Hair, Garland Publishing Inc., New York, 1977.

(14) V. Valković, D. Rendić, and G. C. Phillips, Elemental ratios along human hair as indicators of exposure to environmental pollutants, Environ. Science and Technology 9, 1150 (1975).

(15) D. Rendić, S. Holjević, V. Valković, T. H. Zabel, and G. C. Phillips, Trace-element concentrations in human hair measured by proton-induced X-ray emission, Journal of Investigative Dermatology 66, 371 (1976).

The Theoretical Studies of the Chemical Speciation of Lead in Seawater

M. Whitfield and D. R. Turner
Marine Biological Association of the United Kingdom,
The Laboratory, Citadel Hill, Plymouth PL1 2PB, U. K.

ABSTRACT

Stability constants relevant to the inorganic speciation of lead in sea water are critically reviewed and a pH-dependent model for lead speciation is derived from selected constants. The model confirms the predominance of the $PbCO_3^0$ species at pH 8 and introduces a mixed ligand complex, $Pb(Cl, CO_3)^-$, which is also important at this pH. A simple general relationship is derived between the thermodynamic activity of lead in sea water and the characteristics of added organic complexing agents (e.g. EDTA, NTA, citric acid and humic acid). The importance of considering competition between trace metal ions in such calculations is briefly discussed. The speciation model is used, in conjunction with a newly derived theory for plating at the rotating disc electrode, to define the electrochemical availability of lead in sea water and to estimate the electrochemical lability of individual complexes.

1. INTRODUCTION

The low levels of dissolved (<0.45 micron) lead in natural sea water and the ease with which samples are contaminated have given rise to considerable uncertainties in the figures for total lead concentration. Although there are signs that these difficulties are now being overcome (Ref. 1-3), any attempts to split the total lead concentrations in natural open ocean samples into subfractions representing different chemical forms are still met with some scepticism. In discussing the speciation of lead we will therefore focus on theoretical approaches that may help clarify some of the physico-chemical problems that must be solved once good analytical data become more universally available. We will first make a critical re-assessment of models describing the inorganic speciation of lead in sea water. Once a reasonable model has been established we will use it to look at aspects of lead speciation that are likely to be significant in interpreting the biological effects of lead in sea water. In particular we will look at (i) the thermodynamic activity of lead in sea water and its modification by organic complexing agents and (ii) the influence of speciation on the electrochemical availability of lead as sensed by anodic stripping voltammetry. It should be emphasised that alkyl lead compounds, probably the most toxic forms of lead encountered in the marine environment, will not feature in the equilibrium calculations although they may be sensed by electrochemical measurements.(Ref 4). Their persistence in the marine environment can only be modelled when the rates of their release and of their uptake or hydrolysis have been measured. Their toxicity, however, is likely to be more closely correlated with their thermodynamic activity (as measured by their partial pressures) rather than with their total concentration.

2. PREVIOUS WORK ON MODELS OF INORGANIC SPECIATION

Lead is present at such low concentrations in sea water (10^{-10} M or less) that the inorganic complexes it forms will have no noticeable effect on the free concentrations of the ligands involved. It is therefore possible to define these free concentrations using a suitable chemical model for sea water and to use such values as inputs for the lead speciation calculations. To establish a speciation model for lead against this background we must identify those equilibria that are likely to make a significant contribution to the solution chemistry of the metal. An equilibrium constant is then assigned to each reaction and this constant is suitably corrected so that it is relevant to the temperature, pressure and composition of the sea water ionic medium (Ref. 5). The discussion of lead speciation (Table 1) has been conventionally confined to conditions typical of tropical surface waters (25 °C, 1 atm. pressure, 35 % salinity) since most of the available thermodynamic data refer to these conditions. Given the solution composition and the relevant stability constants, lead speciation can be calculated using either a general computer programme (Ref. 6 and 7) or simple combinations of equilibrium data (Ref. 8 and 9, see Section 3(ii)).

TABLE 1 Previous Theoretical Studies of the Inorganic Speciation of Lead in Sea Water (35 ‰ Salinity) at 25 °C and 1 Atmosphere Pressure

Species	% of total lead				
	a	b	c	d	e
Pb^{2+}	2	5	4	1	1̶
$PbCl^{+}$	11	19	13	9	2
$PbCl_2^{o}$	3	42	8	12	1
$PbCl_3^{-}$	2	9	3	5	1̶
$PbCl_4^{2-}$	1̶	4	1̶	2	1̶
$PbCl_6^{4-}$	-	-	-	-	29
$Pb(Cl,OH)^{o}$	-	9	-	-	-
$PbOH^{+}$	1	10	30	1	1̶
$Pb(OH)_2^{o}$	1	-	2	1̶	1̶
$PbCO_3^{o}$	80	1̶	40	68	65
$Pb(CO_3)_2^{2-}$	-	-	-	-	2
$PbHCO_3^{+}$	1̶	1	-	1̶	1̶
$PbSO_4^{o}$	1̶	1̶	-	2	1̶
% uncharged	84	51	50	82	66

a (Ref. 9) b (Ref. 8) c (Ref. 10) d (Ref. 6)
e (Ref. 7) 1̶, considered but < 1 %; -, not considered

A number of general points of agreement arise from previous studies of lead speciation in sea water (Table 1). (a) Most models consider $PbCO_3^o$ as the dominant complex at pH 8 and all conclude that uncharged species account for more than 50 % of the dissolved lead at this pH. This finding may prove significant if it can be confirmed that such uncharged complexes tend to form colloidal aggregates in solution. (b) All the models agree that $PbHCO_3^+$ and $PbSO_4^o$ are minor species at pH 8. (c) Ionic lead contributes only a few percent to the total lead in all cases - a point we will return to when discussing the thermodynamic activity of lead.

Apart from these general points, the various models show a considerable divergence in the details of lead speciation which reflects the spread in the stability constants employed (Table 2).

TABLE 2 Stability Constants Used in Previous Calculations of Lead Speciation

Species	$\log_{10} \beta_j^*$ (X) (μ = 0.72)			
	a	b	c	d
$PbCl^+$	1.10	0.88	0.8	1.60
$PbCl_2^o$	1.04	1.49	?	1.78
$PbCl_3^-$	0.94	1.09	?	?
$PbCl_4^{2-}$	0.64	0.94	?	?
Pb(Cl,OH)	-	6.2	-	-
$PbOH^+$	5.70	6.0	6.8	?
$Pb(OH)_2^o$	10.16	-	-	?
$PbCO_3^o$	6.51	3.0	6.2	7.4
$Pb(CO_3)^{2-}$	-	-	-	9.89
$PbHCO_3^+$	2.40	2.2	-	?
$PbSO_4^o$	1.71	1.0	-	?
$PbCl_6^{4-}$	-	-	-	3.02

a Zirino and Yamamoto (Ref. 9). The constants used by Long and Angino (Ref. 11) are identical except for log $\beta^*_1(CO_3^{2-})$ = 5.31 and log $\beta^*_2(CO_3^{2-})$ = 9.65.

b Dyrssen and Wedborg (Ref. 8)

c Stumm and Brauner (Ref. 10)

d Lu and Chen (Ref. 7)

We will now re-assess the inorganic speciation models for lead and establish a useful model for further calculations.

3. A REASSESSMENT OF INORGANIC SPECIATION MODELS

(i) Activity Coefficient Corrections

To establish a suitable model, stability constants determined in simple supporting electrolytes must be adjusted to make them relevant to the sea water ionic medium. For the reaction

$$Pb^{2+} + jL^{p-} \rightleftharpoons Pb\,L_j^{(jp-2)-} \tag{1}$$

the thermodynamic stability constant ($\beta_j^o(L)$), relative to a reference state at infinite dilution in pure water is given by

$$\begin{aligned}\beta_j^o(L) &= a(PbLj)/a(Pb).\ a(L)^j \\ &= \beta_j^*(L).\ \Gamma j(L)\end{aligned} \tag{2}$$

Symbols are defined in Appendix 1. Charges are omitted for convenience. $\beta_j^o(L)$ is dependent on temperature and pressure only. The stoichiometric stability constant ($\beta_j^*(L)$) is defined by

$$\beta_j^*(L) = [PbLj] \,/\, [Pb]\,[L]^j \tag{3}$$

and the activity coefficient quotient ($\Gamma j(L)$) by,

$$\Gamma j(L) = g(PbLj)/g(Pb).g(L)^j \tag{4}$$

$\beta_j(L)$ can be corrected from the value determined in the simple supporting electrolyte (subscript 1) to the value relevant to the sea water ionic medium (subscript 2) using an equation of the form,

$${}_2\beta_j^* = {}_1\beta_j^*\,.\,{}_1\Gamma_j/{}_2\Gamma_j \tag{5}$$

It is assumed that the free single-ion activity coefficients (g(Pb), etc.) are dependent only on the ionic strength (μ) of the medium and a suitable conventional equation is used to define the relationship between g and μ (Ref. 12 and 13). The most widely used relationship is the Davies equation (Ref. 14),

$$\log g(X) = 0.5108\ Z_X^2\ [0.3\mu - \mu^{\frac{1}{2}}/(1+\mu^{\frac{1}{2}})] \tag{6}$$

This equation provides only a crude approximation to g(X) in sea water since it was originally developed for calculations up to 0.1 M ionic strength (Ref. 14). However, it gives values for g(Pb) that are in reasonable agreement with those calculated from the experimental mean-ion activity coefficient of lead perchlorate ($\gamma\pm(Pb(ClO_4)_2)$ using the MacInnes convention (Fig. 1, Ref. 15 and 16). According to this convention

$$\gamma_{\pm}^{2}(KCl) = g(K).\ g(Cl) \tag{7}$$

since KCl is not associated. Similarly,

$$\gamma_{\pm}^{3}(Pb(ClO_4)_2) = g(Pb).\ g(ClO_4)^2 \tag{8}$$

The conventional single ion activity coefficient of lead can therefore be calculated from,

$$g(Pb) = \gamma_{\pm}^{3}(Pb(ClO_4)_2)\ .\ \gamma_{\pm}^{4}(NaCl)/\gamma_{\pm}^{4}(NaClO_4)\ .\ \gamma_{\pm}^{2}(KCl) \tag{9}$$

Values of $\gamma_{\pm}$ can be obtained from Robinson and Stokes (Ref. 17). An alternative procedure frequently used (e.g. Ref. 11 and 18) is the supporting electrolyte convention of Helgeson (Ref. 19). For sea water the equation may be written as

$$\log g(X) = 0.038\mu - 0.5108\ Z_X^2\mu^{\frac{1}{2}}/(1+0.329\ \mathring{a}_X\mu^{\frac{1}{2}}) \tag{10}$$

For lead $\mathring{a}_X = 4.5$. Values of g(Pb) calculated from eq. 10 are considerably lower than those calculated from eq. 6 or 9 (Fig. 1).

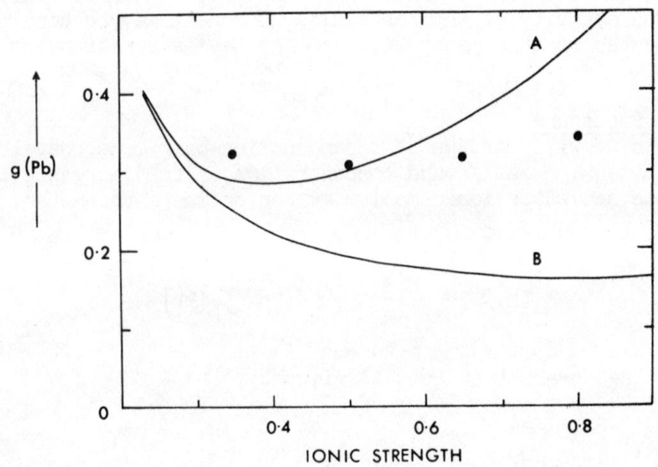

Fig. 1. Conventional free-ion activity coefficients for lead.

A - Davies convention (eq. 6)
B - Supporting electrolyte convention (eq. 10)
● - Mean-salt convention (eq. 9)

Even so the use of different conventions for correcting ${}_1\beta_j$ (eq. 5) will result in relatively small shifts in ${}_2\beta_j$. Take the extreme example of the formation of a neutral complex such as

$$Pb^{2+} \quad + \quad 2Cl^- \rightleftharpoons PbCl_2^{\circ} \tag{11}$$

If ${}_1\beta_2^*$ is determined in a medium of 0.1 M ionic strength and is corrected to a seawater ionic strength of 0.72 M eq. 6 will give

$$\log {}_2\beta_2^* \quad = \quad \log {}_1\beta_2^* \quad + \quad 0.644 - 0.753$$

$$= \quad \log {}_1\beta_2^* \quad - \quad 0.109$$

whereas eq. 10 (with $\mathring{a}_X = 3$ for Cl^-) will give

$$\log {}_2\beta_2^* \quad = \quad \log {}_1\beta_2^* \quad - \quad 0.478$$

A difference of 0.37 logarithmic units. If the simplified medium had been at an ionic strength of 1 M the difference between the two conventions would have been 0.21 units. Such shifts are small when compared with the variation in the ${}_1\beta_j^*$ values themselves (Table 2). They will certainly have only a minor effect on the relative proportions of the various complexes, although their influence on the actual value of the activity of any particular component may be considerable. Calculations presented in this paper will use the Davies equation.

(ii) Calculation Procedure

In our calculations we will use the pH-dependent ion-pair model developed previously (Ref. 20) to establish the background concentration of the inorganic ligands (Table 3). In this sea water ionic medium we can define a parameter, $\sigma_{L,j}$, so that (from eq. 3),

$$\sigma_{L,j} \quad = \quad \beta_j^*(L) \quad . \quad (L)^j \quad = \quad (PbL_j)/(Pb) \tag{12}$$

The percentage of lead present as PbL_j is given by,

$$\%(PbLj) \quad = \quad 100 \times (PbLj)/(Pb)_T \tag{13}$$

where the total lead concentration, $(Pb)_T$, is defined by,

$$(Pb)_T \quad = \quad (Pb) \;\; (1 + \Sigma_{Pb}) \quad = \quad Pb \; . \; \overline{\alpha}_{Pb} \tag{14}$$

TABLE 3 Free Ligand Concentrations for the Sea Water Model[a] (S = 35 ‰, μ= 0.72 M)

A. Major ions at pH 8

ION	Cl^-	OH^-	SO_4^{2-}	Na^+	Mg^{2+}	Ca^{2+}
[X]	0.5662	1.62×10^{-6}	0.0147	0.478	4.89×10^{-2}	9.66×10^{-3}

B. Components of CO_2-system as f(pH)

pH	2	2.5	3	3.5	4	4.5	5	5.5
$p(CO_3)$[b]	14.52	13.52	12.52	11.52	10.52	9.53	8.56	7.64
$p(HCO_3)$	6.74	6.24	5.74	5.24	4.74	4.26	3.79	3.37

pH	6	6.5	7	7.5	8	8.5	9	9.5	10
$p(CO_3)$	6.82	6.15	5.57	5.06	4.58	4.16	3.87	3.70	3.63
$p(HCO_3)$	3.05	2.87	2.80	2.78	2.80	2.89	3.09	3.42	3.85

[a] Using the stability constants and activity coefficients given by Whitfield (Ref. 20) and the sea water stoichiometry given by Millero (Ref. 21).

[b] $p(X) = -\log_{10}[X]$

Here Σ_{Pb} indicates the sum of <u>all</u> of the σ-values for the inorganic complexes of lead. $\bar{\alpha}_{Pb}$ is equivalent to the overall side reaction coefficient of Ringbom (Ref. 22, see also Ref. 23). From eq. 12,

$$[PbL_j] = [Pb]\sigma_{L,j} \tag{15}$$

so that, substituting (15) and (16) into (13) gives

$$\%\ [PbLj] = 100 \times \sigma_{L,j}/\bar{\alpha}_{Pb} \tag{16}$$

Similarly the percentage of free metal ion at equilibrium can be calculated from,

$$\%\ [Pb] = 100/\bar{\alpha}_{Pb} \tag{17}$$

In all cases we will only calculate the species distribution to the nearest percent since uncertainties in the stability constants, and indeed uncertainties in the value of detailed speciation charts, render finer resolution meaningless.

(iii) Selection of Stability Constants

(a) Chloride complexes. There is some disagreement between the published models as to whether $PbCl^+$ or $PbCl_2^o$ is the dominant chloride complex of lead in sea water (Table 1). In particular the published diagrams of lead speciation as a function of pH (Ref. 9 and 10) suggest that $PbCl^+$ is the dominant form. A survey of the original data in the compilations of Sillen and Martell (Ref. 22 and 23; see Table 4) indicates that, apart from data at μ = 2 M, which show a very wide spread, all of the measurements confirm that $PbCl_2^o$ is more stable than $PbCl^+$. If the data are corrected to μ = 0.72 M using eq. 6 this conclusion is confirmed for $\mu \leqslant 1$ M (Table 5). At higher ionic strengths the Davies equation is meaningless. The mean of the values corrected form media where $\mu \leqslant$ 1 M are reasonably close to the values adopted by Dyrssen and Wedborg (Ref. 8; see Table 2) and will be used in further calculations.

TABLE 4 Stepwise Stability Constants[a] for Lead Chloride Complexes Simplified Media at Different Ionic Strengths

μ	n	log K_1^*(Cl)	log K_2^*(Cl)	log K_3^*(Cl)	log K_4^*(Cl)
0	2	1.35 ± 0.35	0.67 ± 0.69	-0.25 ± 0.21	-0.23 ± 0.11
1	4	1.04 ± 0.26	0.45 ± 0.39	0.00 ± 0.39	-0.04 ± 0.15
2	3	1.27 ± 0.17	-0.06 ± 0.18	0.07 ± 0.34	-
3	6	1.16 ± 0.06	0.56 ± 0.07	0.23 ± 0.10	-0.87 ± 0.40
4	6	1.26 ± 0.03	0.53 ± 0.04	0.47 ± 0.09	-0.43 ± 0.29
Mean values		1.22 ± 0.12	0.43 ± 0.29	-0.05 ± 0.39	-0.39 ± 0.36

[a] $K_j^*(Cl) = [PbCl_j]/[PbCl_{j-1}][Cl]$. Values from Sillen and Martell (Ref. 24 and 25).

(b) Sulphate complexes. The $PbSO_4^o$ complex is rather weak and makes only a minor contribution to the speciation of lead in sea water (Table 1). The stability constants estimated by Zirino and Yamamoto (Ref. 9) and by Dyrssen and Wedborg (Ref. 8) are in reasonable agreement with one another (Table 2) and we will adopt the mean of these values (log $\beta_1^*(SO_4)$ = 1.4).

(c) Hydroxide complexes. Estimates for β_1^*(OH) vary by more than an order of magnitude (Table 6). Baes and Mesmer (Ref. 26), as part of a critical review encompassing the hydrolysis of all cations, concluded that the hydrolysis of lead was

TABLE 5 Mean Values of Overall Stability Constants for Chloride Complexes (cf Table 4) Corrected to $\mu = 0.72$ M Using the Davies Equation

Initial μ	$\log \beta_1^*(Cl)$	$\log \beta_2^*(Cl)$	$\log \beta_3^*(Cl)$	$\log \beta_4^*(Cl)$
0[1]	0.85	1.28	1.03	0.80
1[1]	0.95	1.36	1.36	1.32
2	0.74	0.42	0.49	-
3	0.12	0.16	0.39	-0.48
4	-0.33	-0.59	-0.12	-0.55
Mean value[1]	0.90	1.32	1.20	1.06

TABLE 6 Stability Constants for Lead-Hydroxy Complexes[a]

Reference	$\log \beta_1^*(OH)$	$\log \beta_2^*(OH)$	$\log \beta_3^*(OH)$
Zirino and Yamamoto (Ref. 9)	5.70	10.16	13.16
Dyrssen and Wedborg (Ref. 8)	6.0	-	-
Stumm and Brauner (Ref. 10)	6.0	-	-
Schindler (Ref. 27)	6.03	10.38	-
Baes and Mesmer (Ref. 26)	5.61	9.99	13.31
Sillen and Martell (Ref. 25)			
μ = 1 M	6.60	10.34	12.76
μ = 0.3 M	6.03	10.43	20.42[b]
Bilinski et al. (Ref. 28)	6.90	10.80	-
Mean values	6.21 ± 0.49	10.35 ± 0.27	13.08 ± 0.28

[a] Adjusted to $\mu = 0.72$ M where necessary using eq. 6.

[b] Omitted from average.

best described by equations of the form

$$\log \beta_j^*(OH) = \log \beta_j^{o}(OH) + a\,\mu^{1/2}/(1 + \mu^{1/2}) + b\,[Pb]_T \qquad (18)$$

where a = -1.022 for n = 1 or 2 and a = 0 for n = 3. The final term is negligible if $(Pb)_T = 10^{-10}$ M. Values given by Baes and Mesmer (Ref. 26) for $\beta^o_j(OH)$ refer to the reaction

$$Pb^{2+} + jH_2O \rightleftharpoons Pb(OH)_j^{(2-j)+} + jH^+ \tag{19}$$

and must be converted by the addition of the reaction

$$jH^+ + jOH^- \rightleftharpoons jH_2O \tag{20}$$

to give the formation of the hydroxy complex by direct addition of the ligand to the metals,

$$Pb^{2+} + jOH^- \rightleftharpoons Pb(OH)_j^{(2-j)+} \tag{21}$$

Values of log $\beta^*(OH)_j$ adjusted to μ = 0.72 M and corrected in this way, assuming log β^* for reaction 20 is 13.79 (Ref. 29), are consistently lower than experimental values quoted by other workers (Table 6). Consequently, the overall mean values of $\beta^*(OH)_1$ and $\beta^*(OH)_2$ (Table 6) will be used in subsequent calculations. The influence of variations in $\beta^*(OH)_j$ on lead speciation may be judged from Table 7. There is clearly a need for a more rigorous investigation of this parameter since its effects on lead speciation become dominant at high pH.

TABLE 7 Influence of $\beta^*_j(OH)$ on Lead Speciation in Sea Water

Species	% of total lead					
	pH 8			pH 10		
	a	b	c	a	b	c
Pb^{2+}	2	2	2	-	-	-
$PbCl^+$	7	8	8	-	-	-
$PbCl_2^o$	10	12	13	-	1	1
$PbCl_3^-$	4	5	5	-	-	-
$PbCl_4^{2-}$	2	2	2	-	-	-
$PbSO_4^o$	1	1	1	-	-	-
$PbCO_3^o$	53	63	65	9	23	36
$Pb(CO_3)_2^{2-}$	2	3	3	4	9	13
$PbOH^+$	20	5	1	38	20	8
$Pb(OH)_2^o$	-	-	-	48	44	30
$Pb(OH)_3^-$	-	-	-	2	4	10

a $\beta^*_j(OH)$ values of Bilinski et al.(Ref.28); b Mean values of $\beta_j(OH)$ from Table 6; c $\beta^*_j(OH)$ values calculated from eq. 18.

All other constants are as listed in Table 12. Mixed ligand complexes are omitted.

(d) Carbonate complexes. Recent re-assessments of the speciation of lead in carbonate media (Ref. 28, 30 and 31) indicate that there is no convincing evidence for the formation of $PbHCO_3^+$ and $Pb(HCO_3)_2^0$ species. Even if the $PbHCO_3^+$ species is included its σ-value (eq. 12) is so small that it makes a negligible contribution to lead speciation at all pH values. It will therefore be omitted here.

The value of $\beta_1^*(CO_3^{2-})$ used in previous sea water models has varied by more than four orders of magnitude (Table 8). If on the basis of more recent experimental data we discount the low value used by Dyrssen and Wedborg (Ref. 8) we still find a range of nearly two orders of magnitude which can introduce serious discrepancies into the speciation picture (Table 9) since $PbCO_3^0$ is a dominant complex at normal sea water pH values (7.5 to 8.5). The most careful and convincing determinations of $\beta_1^*(CO_3^{2-})$ and $\beta_2^*(CO_3^{2-})$ have been made by voltammetric techniques using differential pulse polarography (DPP) and anodic stripping voltammetry (ASV) (Ref. 2, 28, 30 and 31). The measurements made by workers using both techniques are in encouragingly close agreement (Table 10). The similarity of the DPP and ASV results confirms that the reduction of lead is not subject to any kinetic control (Ref. 31) and suggests that it is reasonable to use the Lingane equation (eq. 49) to analyse the experimental data. We will adopt the mean of the voltammetric estimates of $\beta_1^*(CO_3)$ and $\beta_2^*(CO_3)$ using both DPP and DPASV in this paper (see Table 10). These values are close to the mean values of the stability constants used in previous speciation calculations (Table 8).

TABLE 8 Stability Constants for Lead-Carbonate Complexes[a]

Reference	$\log \beta_1^*(CO_3)$	$\log \beta_2^*(CO_3)$
Schindler (Ref. 27)	5.72	9.40
Lu and Chen (Ref. 7)	7.4	9.89
Dyrssen and Wedborg (Ref. 8)	3.0[b]	-
Zirino and Yamamoto (Ref. 9)	6.51	-
Stumm and Brauner (Ref. 10)	6.2	-
Sipos et al. (Ref. 30)	5.62	8.80
mean values	6.29 ± 0.72	9.36 ± 0.55

[a] Corrected to $\mu = 0.72$ where necessary using eq. (6).

[b] Omitted in calculating mean.

[c] Nürnberg et al. (Ref. 2) give values of 5.32 and 8.73 respectively in perchlorate media using the same technique.

(e) Mixed ligand complexes. Dyrssen and Wedborg (Ref. 8) pointed out that in a multi-ligand system where stepwise complexation of the metal occurs, mixed ligand complexes should be statistically favoured. Lead forms di-complexes with a number of monodentate ligands (e.g. Cl^-, OH^-, CO_3^{2-}) and for the formation of a mixed ligand complex e. g.

TABLE 9 Influence of $\beta_j^*(CO_3^{2-})$ on Lead Speciation in Sea Water

Species	% of total lead pH 8 a	b	c	pH 10 a	b	c
Pb^{2+}	3	2	1	-	-	-
$PbCl^+$	12	8	2	-	-	-
$PbCl_2^o$	17	12	3	1	1	-
$PbCl_3^-$	7	5	1	-	-	-
$PbCl_4^{2-}$	3	2	1	-	-	-
$PbSO_4^o$	1	1	-	-	-	-
$PbCO_3^o$	47	63	88	14	23	55
$Pb(CO_3)_2^{2-}$	4	3	2	9	9	13
$PbOH^+$	7	5	1	22	20	9
$Pb(OH)_2^o$	-	-	-	49	44	21
$Pb(OH)_3^-$	-	-	-	4	4	2

a $\beta_1^*(CO_3) = 6.92 \times 10^5$; $\beta_2^*(CO_3) = 2.00 \times 10^9$. Bilinski and Schindler as reported by Bilinski et al.(Ref. 28).

b $\beta_1^*(CO_3) = 1.32 \times 10^6$; $\beta_2^*(CO_3) = 2.09 \times 10^9$. Mean of values from Table 10.

c $\beta_1^*(CO_3) = 6.61 \times 10^6$; $\beta_2^*(CO_3) = 6.61 \times 10^9$. Baranova (Ref. 32).

All other constants are listed in Table 12. Mixed ligand complexes are omitted.

TABLE 10 Stability Constants for Lead-Carbonate Complexes Determined by both DPP and DPASV[a]

Reference	Method	$\log \beta_1^*(CO_3)$	$\log \beta_2^*(CO_3)$
Ernst et al. (Ref. 31)	DPP	6.07	-
	DPASV[b]	6.17	-
Bilinski et al. (Ref. 28)	DPP	5.97	8.97
	DPASV[b]	6.27	9.67
mean values [c]		6.12 ± 0.13	9.32 ± 0.49

a Corrected from original data in 0.1 M KNO_3 to an ionic strength of 0.72 M using eq. 6.

b Using shifts in the potential of the stripping peak and the Lingane equation (Ref. 33).

c Inclusion of the values reported by Sipos (Ref.30) would give 6.02 ± 0.25 and 9.15 ± 0.46, respectively.

$$Pb^{2+} + Cl^- + OH^- \rightleftharpoons Pb(Cl,OH)^o \qquad (22)$$

the stability constant, $\beta_2^*(Cl,OH)$ can be calculated from

$$\beta_2^*(Cl,OH) = 2\,(\beta_2^*(Cl).\ \beta_2^*(OH))^{1/2} \qquad (23)$$

By analogy with eq. 12 we can write

$$\sigma_{Cl,OH} = \beta_2^*(Cl,OH).\ [Cl][OH] \qquad (24)$$

and thus calculate the contribution of this complex to the speciation of lead via eq. 16. Dyrssen and Wedborg (Ref. 8) considered only the $Pb(Cl, OH)^o$ complex but the recognition that carbonate forms strong complexes with lead (Tables 8 and 10) suggests that the species $Pb(OH,CO_3)^-$ and $Pb(Cl,CO_3)^-$ should also be included. The $Pb(Cl,CO_3)^-$ species in particular makes a significant contribution to lead speciation at pH 8 (Table 11) and all mixed ligand complexes become more important at higher pH values. All three species will be included in subsequent calculations.

Table 11 Influence of Mixed Ligand Complexes on Lead Speciation in Sea Water

Species	% of total lead			
	pH 8		pH 10	
	a	b	a	b
Pb^{2+}	2	2	-	-
$PbCl^+$	8	7	-	-
$PbCl_2^o$	12	11	1	-
$PbCl_3^-$	5	5	-	-
$PbCl_4^{2-}$	2	2	-	-
$PbSO_4^o$	1	1	-	-
$PbCO_3^o$	63	55	23	15
$Pb(CO_3)_2^{2-}$	3	2	9	6
$PbOH^+$	5	4	20	13
$Pb(OH)_2^o$	-	-	44	29
$Pb(OH)_3^-$	-	-	4	3
$Pb(OH,Cl)^o$	-	2	-	6
$Pb(CO_3,Cl)^-$	-	12	-	25
$Pb(CO_3,OH)^-$	-	1	-	3

a Without mixed ligand complexes; b Including mixed ligand complexes; Other stability constants as listed in Table 12.

(iv) Inorganic speciation model for lead. The selected stability constants are collected in Table 12. The distribution of the various lead species as a function of pH is summarised in Table 13 and Figure 2. The picture that emerges is quite complex, particularly in the natural sea water pH range where four complexes each contribute more than 5 % to the total lead concentration. The recalculated model confirms the general points made in Section 2 but, in our present state of ignorance over the biological significance of the individual complexes, it is not profitable to present a more detailed interpretation of the model. It is possible, however, to use the model to look at two aspects of lead speciation in sea water that might have a direct bearing on interpretations of the biochemical and geochemical behaviour of lead via (i) the thermodynamic activity of lead, and (ii) the electrochemical availability of lead.

TABLE 12 Stability Constants Selected for the Sea Water Speciation Model (μ = 0.72, t = 25 °C)

Ligand	log $\beta_1^*(X)$	log $\beta_2^*(X)$	log $\beta_3^*(X)$	log $\beta_4^*(X)$
Cl^-	0.90	1.32	1.20	1.06
OH^-	6.21	10.35	13.08	-
CO_3^{2-}	6.12	9.32	-	-
SO_4^{2-}	1.4	-	-	-
(Cl^-, OH^-)	-	6.14	-	-
(Cl^-, CO_3^{2-})	-	5.62	-	-
(OH^-, CO_3^{2-})	-	10.14	-	-

4. THERMODYNAMIC ACTIVITY OF LEAD IN SEA WATER

(i) Inorganic Model

It is well established that activities rather than concentrations are required to predict the status of a chemical system at equilibrium (Ref. 13 and 34), see also discussion of eq. 2. It is also true, although less widely appreciated, that where gradients or discontinuities in solution composition occur (e.g. at the sediment-water or membrane-water interface) activities rather than concentrations are required for the quantitative interpretation of reaction kinetics (Ref. 35) and of diffusion phenomena (Ref. 36). Consequently, for both inorganic (Ref. 37) and organic components (Ref. 38), activities frequently show more coherent correlations than concentrations when used to interpret toxicity data. As a result there is considerable interest in attempts to define activities in multicomponent systems and in procedures for measuring these activities directly.

Ion-selective electrodes respond to ion activities rather than to ion concentrations (Ref. 39) and the potential of a cell composed of a lead and a chloride selective electrode in a sample, X, is given by

$$E_X - E_X^o = (2f)^{-1} \log a_{Pb,X} \cdot a_{Cl,X}^2 \qquad (25)$$

TABLE 13 Speciation of Lead in Sea Water at 25 °C and 1 Atmosphere Pressure [a]

Species	% of total lead			
	pH 4	pH 6	pH 8	pH 10
Pb^{2+}	6	6	2	-
$PbCl^{+}$	27	27	7	-
$PbCl_2^{o}$	40	40	11	-
$PbCl_3^{-}$	17	17	5	-
$PbCl_4^{2-}$	7	7	2	-
$PbSO_4^{o}$	2	2	1	-
$PbCO_3^{o}$	-	1	55	15
$Pb(CO_3)_2^{2-}$	-	-	2	6
$PbOH^{+}$	-	-	4	13
$Pb(OH)_2^{o}$	-	-	-	29
$Pb(OH)_3^{-}$	-	-	-	3
$Pb(Cl,OH)^{o}$	-	-	2	6
$Pb(CO_3,OH)^{-}$	-	-	1	25
$Pb(CO_3,Cl)^{-}$	-	-	10	3

[a] 35 ‰ using stability constants given in Table 12.

In a standard medium the cell potential is given by

$$E_S - E_S^{o} = (2f)^{-1} \log a_{Pb,S} \cdot a^2_{Cl,S} \tag{26}$$

If the standard and the sample have the same temperature and major ion composition and lead is present as a minor component (less than 10^{-4} M) then

$$E_X - E_S = (2f)^{-1} \log [Pb]_X / [Pb]_S \tag{27}$$

For lead the cell can be standardised in an organic-free sea water with the same salinity as the sample but with the pH adjusted to 5 to 6 where lead speciation depends only on the concentration of chloride and sulphate (Fig. 2). $[Pb]_S$ can be

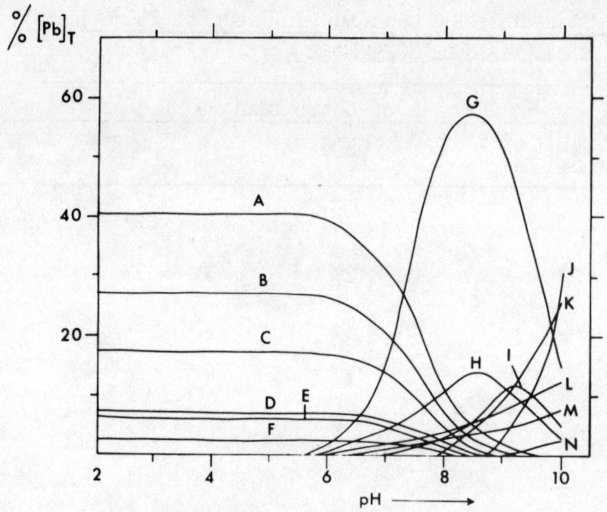

Fig. 2. Distribution of the inorganic species of lead in 35 ‰ salinity sea water at 25 °C and 1 atmosphere pressure for the model defined in Table 12.

Key to species identified:

A, $PbCl_2^{o}$: B, $PbCl^{+}$: C, $PbCl_3^{-}$: D, $PbCl_4^{2-}$: E, Pb^{2+}: F, $PbSO_4^{o}$: G, $PbCO_3^{o}$

H, $Pb(Cl,CO_3)^{-}$: I, $Pb(CO_3)_2^{2-}$: J, $Pb(OH)_2^{o}$: K, $Pb(OH,CO_3)^{-}$: L, $PbOH^{+}$

M, $Pb(OH,Cl)^{o}$: N, $Pb(OH)_3^{-}$.

estimated from the inorganic speciation model and variations in $[Pb]_X$ can be monitored. If the samples and standards all have the same salinity than g(Pb) will be constant, since by definition it is sensitive only to changes in ionic strength, and variations in $[Pb]_X$ will be proportional to variations in the lead activity. Measurement of this kind have shown useful correlations in toxicity tests with copper but are restricted to metal concentrations greater than 10^{-6} to 10^{-7} by the response characteristics of the electrodes.

At low metal concentrations we must resort to the use of speciation models to calculate [Pb] and relate this to the lead activity via conventional g(Pb) values. The values of -log [Pb] calculated from the various models are in good agreement with one another (Table 14) considering the wide range of stability constants employed (Table 2). In particular note the agreement between the estimates from the models of Dyrssen and Wedborg (Ref. 8) and Stumm and Brauner (Ref. 10) despite their widely different speciation pictures (Table 1). Even large changes in the stability constant of the dominant lead carbonate species cause only relatively small changes in the free lead concentration. It should be emphasised again that although the conventions used to define g(X) values will have only a small effect β^{*}-values (and hence on the species distribution) they can have a considerable influence on individual a(X) values. If lead activity is the decisive factor in defining the biological availability of lead it seems likely that the fine details of the speciation models will be of little significance. If, however, certain complexes, because of their size or charge, are selectively transported across cell membranes then the detailed species distribution will be important. In this context

TABLE 14 The Free Concentration of Lead in Sea Water

A. Calculated from different models at pH 8.0

Source	a	b	c	d	e	f	g
- log (Pb)	11.70	11.30	11.40	12.00	11.82	12.19	11.59

B. Calculated from model defined in Table 12

pH	2-5	6	7	8	9	10
- log (Pb)	11.22	11.23	11.33	11.82	12.55	13.32

a Zirino and Yamamoto (Ref. 9)
b Dyrssen and Wedborg (Ref. 8)
c Stumm and Brauner (Ref. 10)
d Florence and Batley (Ref. 6)
e Model defined in Table 12
f As e but with β^{*}_{j} (CO_3^{2-}) values from Baranova (Ref. 32), see Table 9
g As e but with β^{*}_{j} (CO_3^{2-}) from Sipos et al.(Ref. 30)

it is interesting to note that in the present model uncharged species are dominant in the lead system at all pH values (Fig. 2).

(ii) Influence of Organic Complexing Agents

Organic complexing agents can be used either to enhance or to supress the availability of metals to phytoplankton cultures (Ref. 40 and 41) and their behaviour is most aptly summarised by the parameter $\sigma_{Pb,L}$ since this specifies the ratio of complexed to free metal (eq. 12). To demonstrate this we have chosen four organic acids (citric acid, humic acid, EDTA and NTA) that form complexes with lead whose stability constants range over twelve orders of magnitude (Table 15). The stability constants for all except the humic acid complexes were taken from the compilation of Sillen and Martell (Ref. 24 and 25) and were adjusted to an ionic strength of 0.72 using eq. 6. The humic acid constants were taken in the main from the experimental work of Mantouva et al. (Ref. 42) and were adjusted to an ionic strength of 0.72 M using eq. 6 on the assumption that humic acid is dibasic. The value for the lead-humic acid stability constant was taken from the measurements of Stevenson (Ref. 43) and similarly adjusted to $\mu = 0.72$ M. The ratios of the stability constants estimated here for humic acid are close to those observed for the other organic ligands (Table 16) suggesting that they are of the right order.

Provided that the total ligand concentration remains less than the total concentration of the metals to be complexed the speciation of each ligand in sea water can be calculated from relationships analogous to eq. 17. From such a calculation (Table 17) it is immediately obvious that in sea water the divalent cations (Mg^{2+} and Ca^{2+}) buffer all four ligand systems very effectively. As a consequence

TABLE 15 Stability Constants (Expressed as $\log \beta^*_j (X)$) for Organic Complexing Agents Adjusted to $\mu = 0.72$ Using the Davies Equation

Complex	EDTA	NTA	Citric Acid	Humic Acid
HL	10.13	9.63	5.58	a
H_2L	16.19	11.93	9.77	a
H_3L	18.79	14.01	12.82	b
H_4L	20.75	b	b	b
KL	1.05	a	a	a
NaL	1.88	1.46	0.50	a
Na_2L	2.63	b	a	a
NaHL	10.91	a	a	a
MgL	8.42	6.80	3.20	3.23
MgL_2	b	10.10	b	b
MgHL d	13.21	a	7.97	b
MgH_2L	a	a	10.67	b
CaL	10.30	6.68	3.25	3.35
CaL_2	b	10.86	b	b
CaHL d	13.44	a	7.55	b
CaH_2L	a	a	10.88	b
PbL	18.03	11.60	6.30	7.49
PbHL d	20.54	a	11.87	b
CuL	18.52	12.80	c	8.79
CuL_2	b	a	c	b
CuHL d	24.04	a	c	b

a not available; b not applicable; c not considered; d refers to reaction $M^{m+} + H^+ + L^{n-} \rightleftharpoons MHL^{(m+1-n)+}$

TABLE 16 Stability Constant Ratios for Organic Complexing Agents (μ = 0.72 M)

Compound	$\log \beta_1$*(PbL)	Ratio of stability constant [a] of ligand with metal M			
		Na	Mg	Ca	Cu
EDTA	18.0	0.10	0.47	0.57	1.03
NTA	11.6	0.13	0.59	0.58	1.10
Humic acid	7.5	-	0.43	0.45	1.17
Citric acid	6.3	0.08	0.51	0.52	0.91

a expressed as $\log \beta_1$* (ML)/$\log \beta_1$*(PbL)

TABLE 17 Characteristics of Organic Complexing Agents in Sea Water (pH 8)

Parameter	EDTA	NTA	Citric acid	Humic acid
σ H,1 [a]	134.9	42.7	3.8×10^{-3}	-
σ H,2	1.55	8.3×10^{-5}	5.9×10^{-7}	-
σ H,3	6.2×10^{-6}	1.02×10^{-6}	6.6×10^{-12}	-
σ Na,1	36.3	13.8	1.51	19.0 [e]
σ Na,2	97.5	-	-	-
σ Mg,1	1.29×10^{7}	3.09×10^{5}	77.5	83.0
σ Ca,1	1.93×10^{8}	4.62×10^{4}	17.2	21.6
Σ L [b]	2.06×10^{8}	3.55×10^{5}	96.2	123.6
$(L)_T$* [c]	1.92×10^{-10}	8.92×10^{-7}	4.67×10^{-5}	3.96×10^{-6}
b [d]	3.43×10^{16}	6.53×10^{10}	-	6.39×10^{6}
$(Cu)_T$*	6.0×10^{-10}	5.5×10^{-7}	-	2.0×10^{-6}

a Defined by eq. 12 b Defined by eq. 14 c Defined by eq. 32
d Defined by eq. 34 e Using $\log \beta_1$*(Na) = 1.16 (Table 16)

the distribution of the ligands between the major cations is virtually independent of pH over the range 6 to 10. This effect has been confirmed experimentally for metal complexes in sea water (Ref. 2 and 44). The ratio ($\bar{\alpha}_L$) of the total to the free ligand concentration in this sea water ionic medium is given by (cf eq. 14)

$$\bar{\alpha}_L = (L)_T/(L) = 1 + \Sigma_L \tag{28}$$

Values of Σ_L (analogous to Σ_{Pb}, eq. 14) are given in Table 17.

The influence of an organic ligand, L, on the chemistry of lead will be controlled in turn by the influence of the major inorganic components in the system on the availability of free lead and free ligand. These effects have been summarised by the terms Σ_{Pb} (eq. 14) and Σ_L (eq. 28) which do <u>not</u> include the σ-values for the complex. When the sea water mixture contains <u>both</u> lead and the organic ligand then we must rewrite eq. 14 and 28 to give respectively

$$[Pb]_T = [Pb]\ \bar{\alpha}_{Pb} + [PbL]$$

or $$[Pb]_T/[Pb] = \bar{\alpha}_{Pb} + \sigma_{Pb,L} \quad (29)$$

and $$[L]_T/[L] = \bar{\alpha}_L + \sigma_{L,Pb} \quad (30)$$

Rearranging eq. 29 and taking logarithms we find that

$$-\log [Pb] = \log(\bar{\alpha}_{Pb} + \sigma_{Pb,L}) - \log [Pb]_T \quad (31)$$

This equation provides a direct relationship between [Pb], which is proportional to the thermodynamic activity of lead, and $\sigma_{Pb,L}$, which summarises the characteristics of the organic ligand in the sea water. The free ligand concentration can be calculated from the total ligand concentration using eq. 28 provided that the major cations continue to make a dominant contribution to the ligand speciation. If we plot -log [Pb] versus log $\sigma_{Pb,L}$ we obtain a curve, common to all ligands, which can be divided into two straight line components (Fig. 3). The horizontal component (A, Fig. 3) represents the behaviour of the system at low ligand concentration when $\bar{\alpha}_{Pb} \gg \sigma_{Pb,L}$. Its position on the ordinate is fixed by the inorganic speciation of lead as summarised by $\bar{\alpha}_{Pb}$ (Table 18). In the case of lead, where $\bar{\alpha}_{Pb} = 66.05$, the influence of an added organic ligand on [Pb] conveniently becomes noticeable when $\sigma_{Pb,L} = 1$, that is, from eq. 12 and 28, when

$$[L]_T^* = \bar{\alpha}_L / \beta_1^*(L) \quad (32)$$

The corresponding values of $[L]_T^*$ for the four ligands discussed here are given in Table 17. Typical concentration values for humic acid are around 10^{-5}M (Ref.45) and Gardiner (Ref. 46) has reported EDTA concentrations as high as 10^{-7} M in estuarine and river waters. Further additions of ligand will depress the lead activity until at high ligand concentrations, i.e. about one thousand times the value defined in eq. 32, we have,

$$-\log [Pb] = \log \sigma_{Pb,L} - \log [Pb]_T \quad (33)$$

which corresponds to line B, Fig. 3.

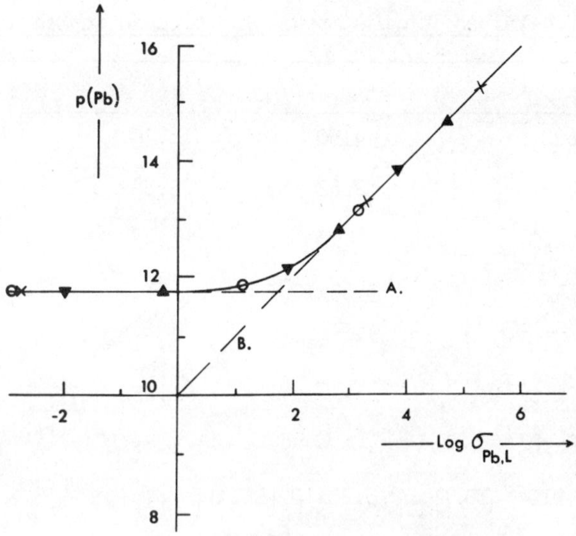

Fig. 3. Relationship between p(Pb) (= $-\log_{10}[Pb]$) and $\log \sigma_{Pb,L}$ (eq. 31).

Data points were calculated using the computer programme HALTAFALL for the following complexes.

Symbol	ligand (L)	p (L)
X	Humic acid	8 to 0
O	Citric acid	7 to 1
▲	EDTA	10 to 5
▼	NTA	8 to 2

Equation 31 therefore provides an immediate and quantitative statement of the observations summarised by Siegel (Ref. 40). If a weak complexing agent, or a low concentration of a strong complexing agent is added to the medium the activity of the metal is unaffected (A, Fig. 3). In fact the buffer capacity of the system for the metal is enhanced somewhat and, if the metal is usually present in a form that is difficult to assimilate (e.g. iron as colloidal ferric hydroxide or lead as colloidal $PbCO_3^{\,o}$), this additional reservoir of available metal may be of vital importance. If the complexation is too strong or the ligand concentration is too high the activity of the metal will be depressed (B, Fig. 3). Equations 31 to 33 enable quantitative limits to be put on these statements (Table 17).

(iii) Competition of Trace Metals for Organic Ligands

As examples we will consider the competition between copper and lead for EDTA and humic acid. Copper itself generally forms more stable complexes with organic ligands than does lead (see Table 15) and it is present in a thousandfold excess over lead in natural sea water. Rather than make a critical survey of the inorganic speciation of copper in sea water we will take the model of Dyrssen and Wedborg (Ref. 8), since this is the only model incorporating mixed ligand complexes, and modify it by using the β^*_j(OH) values of Baes and Mesmer (Ref. 26). Speciation calculations using eq. 15 and 17 indicate a distribution dominated by the hydroxychloride and carbonate complexes (Table 19) with $\overline{\alpha}_{Cu}$ = 96.56 at pH 8.

TABLE 18 σ-Values for the Speciation of Lead in Sea Water at pH 8 [a]

Ligand	L	j= 1	2	3	4
Cl^-	0.5662	4.50	6.70	2.87	1.18
OH^-	1.62×10^{-6}	2.62	5.88×10^{-2}	5.10×10^{-2}	-
CO_3^{2-}	2.63×10^{-5}	34.7	1.44	-	-
SO_4^{2-}	0.0147	3.69×10^{-1}	-	-	-
$(Cl,CO_3)^{3-}$	1.49×10^{-5}	-	6.21	-	-
$(Cl,OH)^{2-}$	9.17×10^{-7}	-	1.25	-	-
$(OH,CO_3)^{3-}$	4.26×10^{-11}	-	5.8×10^{-1}	-	-
EDTA	$4.85 \times 10^{-9}(EDTA)_T$	$5.20 \times 10^{9}(EDTA)_T$	-	-	-
NTA	$2.82 \times 10^{-6}(NTA)_T$	$1.12 \times 10^{6}(NTA)_T$	-	-	-
Citric acid (CA)	$1.04 \times 10^{-2}(CA)_T$	$2.07 \times 10^{4}(CA)_T$	-	-	-
Humic acid (HA)	$8.03 \times 10^{-3}(HA)_T$	$2.48 \times 10^{5}(HA)_T$	-	-	-

a Calculated from eq. 12 using stability constants from Table 12. $\bar{\alpha}_{Pb} = 66.05$

TABLE 19 Inorganic Speciation of Copper in Sea Water at pH 8 [a]

Species	$\log \beta_j^*(L)$	$\sigma_{L,j}$	% of total copper
Cu^{2+}	-	-	1
$CuCl^+$	1.2	8.97	9
$CuCl_2^o$	0.9	2.55	3
$CuOH^+$	5.24	0.28	-
$Cu(OH)_2^o$	9.73	0.01	-
$Cu(OH)_3^-$	13.02	-	-
$Cu(OH,Cl)^o$	7.9	72.84	75
$CuCO_3^o$	5.6	10.91	11

a $\Sigma_{Cu} = 95.56$

To assess the effect of copper on the complexation of lead by the organic ligand we must estimate the correction to $\bar{\alpha}_L$ (eq. 28) which results from the formation of copper-ligand complexes. This will in turn affect the critical ligand concentration (eq. 32) beyond which a rapid decrease in free lead concentration is observed (Fig. 3). For copper complexation

$$\sigma_{L,Cu} = \beta_1^{*}(1).\,[Cu]$$

$$= b\,[Cu]_T \tag{34}$$

where $b = \beta_1^{*}(L)/\,\bar{\alpha}Cu$ (Table 17). Corrected values of $\bar{\alpha}_L$ can then be calculated from

$$\bar{\alpha}'_L = \bar{\alpha}_L + b\,[Cu]_T \tag{35}$$

The corrected value $\bar{\alpha}'_L$ must be used in eq. 28 to calculate [L] from $[L]_T$ and this in turn will affect the value of $\sigma_{Pb,L}$. The presence of copper will shift the critical ligand concentration ($[L]_T^{*}$, eq. 32) by 10 % when

$$[Cu]_T^{*} = \bar{\alpha}_L/10b \tag{36}$$

Values of $[Cu]_T^{*}$ for the various ligands are listed in Table 17. Since copper is normally present in sea water at the 10^{-7} M level, it is likely to compete very effectively with lead for strong complexones such as EDTA but will leave weaker complexes of lead, with ligands such as humic acid, largely unaffected. The procedure outlined here can be used to look at the competition between any mixture of trace metals for organic ligands provided that the trace metal-ligand complexes remain only a small fraction of the total ligand concentration. It is dangerous to ignore competition of this kind when looking at the influence of organic ligands on trace metal speciation.

5. ELECTROCHEMICAL AVAILABILITY OF LEAD IN SEA WATER

(i) Outline of Problem

Natural systems contain such complex mixtures of trace metals and potential organic ligands that the calculation procedures described so far can only provide an approximate guide to the speciation of a particular metal. Usually the organic ligands themselves are unidentified and, even if their identity is known, the stability constants describing their interactions with the appropriate trace metals in the natural medium have rarely been determined. Such speciation models are therefore best suited for interpreting lead activity and toxicity measurements in artificial media where the complexation is dominated by well characterised ligands (Ref. 2 and 44). In natural media an experimental procedure is required that will sense, as far as possible, the concentration of 'free' or 'biologically available' metal in an undisturbed sample. The role of ion-selective electrodes in such measurements has already been briefly discussed. Anodic stripping voltammetry (ASV) also seems to be a promising candidate since it senses only those species that can react at the electrode surface <u>during the time scale of the measurement</u> and is capable of determining lead at natural levels. It is therefore possible to divide the total soluble metal (< 0.45 micron) into 'electrochemically available' and 'electrochemically unavailable' fractions (Ref. 47) and to investigate the relationship between electrochemically and biologically available metal. To date no theoretical treatment has been provided that describes quantitatively the meaning of the term 'electrochemically available' in a complex natural system. The theory of metal reduction at a mercury electrode is largely confined to single-ligand systems (Ref. 49-51) despite a very general mathematical formulation of the transport problem involved in the polarographic reduction of a complex mixture of metal species (Ref. 52).

To clarify the definition of the electrochemically available fraction we will briefly discuss the determination of lead by ASV in the ligand mixture discussed in previous sections (Table 18).

The fraction of metal in the solution sensed by the method depends on the flux of metal to the electrode surface during the plating step. This flux, in turn, defines the magnitude of the peak current observed in the subsequent stripping step. The current measured on stripping, and hence the fraction of the total metal sensed by the method, will be depressed if the solution contains complexes that dissociate slowly in comparison with the time scale from transport across the diffusion layer. A plot of ASV stripping current versus plating potential results in a pseudo-polarographic wave (Fig. 4) which is characterised by a limiting peak current and a half wave potential (Ref. 53 and 54). A theoretical analysis of this curve, which summarises the electrochemistry of the plating step, will provide a quantitative definition of the electrochemically available fraction. In developing such a theory we have chosen the rotating disc working electrode because its well defined hydrodynamic properties simplify the solution of the diffusion equations and provide a constant and easily calculated value of the diffusion layer thickness, δ (Ref. 55). The value of δ, together with the diffusion coefficient D, defines the time scale of the electrochemical measurement (Ref. 56) and provides an important control on the definition of electrochemically available metal. The plating current is assumed to change only slowly so that the equations can be solved at a steady state and then averaged over the plating time (Ref. 54). The theoretical treatment has been considered in detail elsewhere (Ref. 48). The calculations reported here will use the same assumptions and notation as the general theory (Table 20, 21 and Appendix 1) but, for simplicity, we will confine our present discussion to one and two ligand systems.

<u>(ii) Rate Constants for the Dissociation of Lead Complexes</u>

If we consider the reaction

$$Pb + X \underset{k_+}{\overset{k_-}{\rightleftharpoons}} PbX \tag{37}$$

the rate constants for the formation (k_-) and dissociation (k_+) of the complex PbX are related to the stability constant, $\beta_1^*(X)$, by

$$\beta_1^*(X) = k_-/k_+ \tag{38}$$

provided that the ionic medium is homogeneous enough for concentrations rather than activities to be used in describing the reaction. In discussing electrochemical availability we are interested in the relationship between the rate of dissociation of the complex (fixed by k_+ and [PbX]) and the rate of discharge of the metal at the electrode surface, which depends in addition on ε, D and δ (Appendix 1).

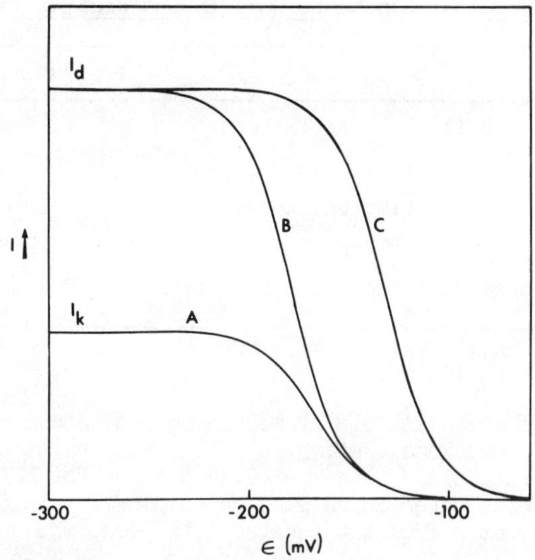

Fig. 4. A. Calculated pseudopolarographic wave for lead forming 1:1 complex with carbonate (eqs. 45-48).

B. As above, but assuming very rapid kinetics (eqs. 41-42).

C. Calculated pseudopolarographic wave for uncomplexed lead (eqs. 39-40).

$\delta = 1.02 \times 10^{-3}$ cm in each case

TABLE 20 Assumptions Made in Deriving the Plating Theory for Anodic Stripping Voltammetry

a. Nature of the electrode

1. The electrode consists of a thin film of mercury on a rotating disc.
2. There are no concentration gradients in the mercury film.

b. The electrode reaction

1. Only uncomplexed Pb^{2+} undergoes reduction at the electrode.
2. The electrode reaction is reversible.
3. There are no effects due to adsorption at the electrode.
4. The diffusion coefficients of Pb^{2+} and its complexes are all the same.
5. The plating current changes only slowly with time.

c. The bulk solution

1. Activity coefficients are set equal to unity.
2. The various species are in thermodynamic equilibrium in the bulk solution.
3. The ligands are either effectively buffered or are present in great excess so that their concentrations are constant throughout.

TABLE 21 Values of Constants Used in the Plating Theory Calculations

Symbol	Definition	Numerical value
D	Diffusion coefficient	$10^{-5} cm^2 s^{-1}$
t	Plating time	300s
l	Mercury film thickness	10^{-4} cm
$(nf)^{-1}$	= RT/nF	12.8 mV
k_-	Rate constant for complex formation	$2 \times 10^8 s^{-1}$

Although several reviews are available of mechanism and rate constant data for metal complex formation reactions (Ref. 57-60), the only rate constants listed for lead are (Ref. 58) for $Pb(HEDTA)^-$ ($\log k_- = 9.3$) and for $Pb(HNTA)^o$ ($\log k_- = 8.2$). For most complexes the rate determining step on complex formation is assumed to be the loss of coordinated water from the aquated metal ion. No values for this water exchange constant (k_{ex}) are available for lead but Wilkins (Ref. 57) noted that, for $M(HEDTA)^-$ complexes, $k_-/k_{ex} \simeq 10$. On this basis we have assumed that $\log k_{ex}(Pb^{2+}) = 8.3$ and have used this value as an order of magnitude estimate of k_- for all lead complexes. It is clear from eq. 38 that this assumption implies that the greater the thermodynamic stability of a complex, the slower it will dissociate and the less likely it is to be electrochemically available.

(iii) Electrochemical Availability in Single-Ligand Systems

Equations derived for the peak current versus plating potential curve (Table 22) are analogous to the corresponding polarographic equations where $\sigma_L >> 1$ or $\sum_n \beta_{L,n} >> 1$ (see Fig. 4). Under these conditions, and in the absence of kinetic control, eq. 42 (Table 22) gives rise to the familiar equation (Ref. 33),

$$\Delta\varepsilon_{\frac{1}{2}} = (nf)^{-1} (\ln \beta_j^{*}(X) + j \ln[X])$$

$$= (nf)^{-1} \ln \sigma_{X,j} \qquad (39)$$

where the complex MX_j predominates. Equation 44 gives the more general equation (Ref. 62),

$$\Delta\varepsilon_{\frac{1}{2}} = (nf)^{-1} \ln \sum_j \sigma_{X,j} \qquad (40)$$

where a series of complexes if formed. Shifts in half wave potential ($\Delta\sigma_{\frac{1}{2}}$) with ligand concentration can be used to determine stability constants in the absence of kinetic control (Ref. 30, 53 and 63). However, when the plating step is dependent on the kinetics of complex dissociation (eq. 45 and 46, Table 22) the Lingane and DeFord-Hume equations will not necessarily apply since the term $\ln (I_k/I_d)$ is itself a function of $[X]$ (eq. 47 and 48, Table 22).

TABLE 22 Description of Pseudopolarographic Waves

Reactions described	Equations		Source of derivation
$M^{n+} \xrightarrow{ne} M(Hg)$	$\varepsilon = \varepsilon_{\frac{1}{2}} + (nf)^{-1} \ln(I_d - I)/I)$	(39)	Zirino &
	$\varepsilon_{\frac{1}{2}} = -(nf)^{-1} \ln Q$	(40)	Kounaves (Ref. 54)
$MX^{(n-m)+} \rightleftharpoons M^{n+} + X^{m-}$	$\varepsilon = \varepsilon_{\frac{1}{2}} + (nf)^{-1} \ln((I_d - I)/I)$	(41)	
(very rapid kinetics)			This work
$M^{n+} \xrightarrow{ne} M(Hg)$	$\varepsilon_{\frac{1}{2}} = -(nf)^{-1} \ln Q - (nf)^{-1} \ln(1 + \sigma X)$	(42)	
$MX_j^{(n-jm)+} \rightleftharpoons M^{n+} + jX^{m-}$	$\varepsilon = \varepsilon_{\frac{1}{2}} + (nf)^{-1} \ln((I_d - I)/I)$	(43)	
(very rapid kinetics)	$\varepsilon_{\frac{1}{2}} = -(nf)^{-1} \ln Q - (nf)^{-1} \ln(1 + \Sigma_X)$	(44)	This work
$M^{n+} \xrightarrow{ne} M(Hg)$	or $\varepsilon_{\frac{1}{2}} = -(nf)^{-1} \ln Q - (nf)^{-1} \ln \bar{\alpha}_X$		
	$\varepsilon = \varepsilon_{\frac{1}{2}} + (nf)^{-1} \ln((I_k - I)/I)$	(45)	
$MX^{(n-m)+} \underset{k-}{\overset{k+}{\rightleftharpoons}} M^{n+} + X^{m-}$	$\varepsilon_{\frac{1}{2}} = -(nf)^{-1} \ln Q - (nf)^{-1} \ln(1+\sigma_X)$		
	$-(nf)^{-1} \ln(I_k/I_d)$	(46)	This work
$M^{n+} \xrightarrow{ne} M(Hg)$	where $I_k/I_d = (1 + \sigma_X V^o)^{-1}$ [a]	(47)	
	and $V^o = (\delta(k_-(1 + \delta_X)/D\beta^*)^{\frac{1}{2}})^{-1}$	(48)	Levich (Ref. 55)
	in the limit of rapid kinetics		
	$V^o = 1$ in the limit of slow kinetics		Hale (Ref. 61)

[a] For intermediate kinetics V^o is tabulated by Hale (Ref. 61) who also provides an approximate function.

To consider kinetic control we must estimate I_k/I_d, the ratio between the kinetically limited maximum peak current (I_k) and the limiting maximum peak current (I_d) which could be observed in the absence of kinetic control (see Fig. 4). Although it is difficult to measure directly, this ratio describes for the system defined in Table 20 the electrochemical availability of the metal in the solution.

Using eq. 46 and 47 (Table 22), values of $\varepsilon_{\frac{1}{2}}$ and I_k/I_d were calculated for the various lead complexes in sea water at pH 8, considering in each case a single 1:1 lead-ligand interaction (Table 23). While single ligand calculations cannot predict quantitatively the behaviour of complex mixtures they can give a qualitative indication of the likely contribution of a particular complex to the electrochemically available metal. Three types of behaviour can be distinguished (Table 23):

1. Complexes labile to ASV where $I_k/I_d \simeq 1$,
2. Complexes non-labile to ASV where I_k/I_d is determined solely by the fraction (θ) of free metal present at equilibrium,
3. Complexes partially labile to ASV where $\theta < I_k I_d < 1$ and I_k/I_d varies with δ so that the plating process is under kinetic control.

There is consequently no clear-cut division between labile and non-labile complexes. For any particular ligand the fraction of electrochemically available metal (I_k/I_d) depends on δ and X , which may be varied experimentally, and on $\beta_1^*(X)$ and k_- which may be subject to some uncertainty. δ defines the time scale of the plating step (Ref. 56) and the more efficiently the system is stirred the smaller will be δ and the smaller the fraction of metal deposited at the electrode if the plating is kinetically controlled (Fig. 5). Changes in the ligand concentration will alter σ_X and around a thousandfold change in concentration is capable of causing a change over from 90 % electroactive metal to 10 % (Fig. 6).

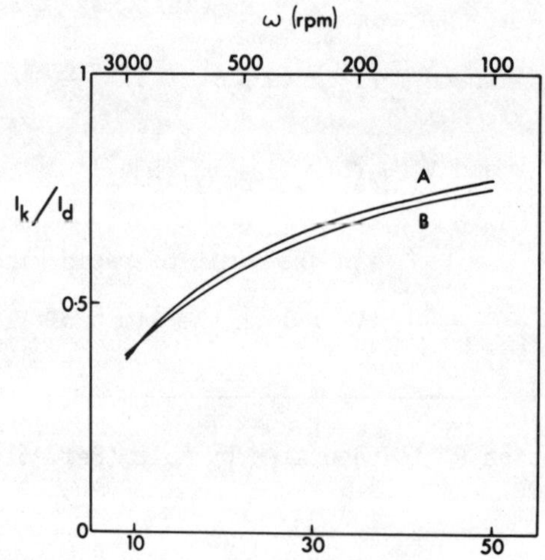

Fig. 5. Calculated dependence of I_k/I_d on σ, illustrated for 1:1 complexation of lead with carbonate (A) and humic acid (B). The upper scale shows the electrode rotation rate corresponding to σ.

TABLE 23 ASV Plating Calculations, Considering the Species $PbX^{(2-m)+}$, Pb^{2+}

Ligand X^{m-}	Speciation		ω= 2400 rpm δ= 1.02x10−3cm		ω = 240 rpm δ =3.22x10−3cm		Type of ASV behaviour	log σ_X	log(σ_X[X]$^{\frac{1}{2}}$)	log ((1+σ_X^{-1})[X])
	Pb^{2+}	$PbX^{(2-m)+}$	Ik/Id	$\varepsilon^{\frac{1}{2}}$(mV)	Ik/Id	$\varepsilon^{\frac{1}{2}}$(mV)				
SO_4^{2-}	0.73	0.27	1.00	-136	1.00	-121	labile	-0.4	0.5	-1.3
Cl^-	0.18	0.82	1.00	-154	1.00	-139	labile	0.6	0.8	-0.2
citrate [a]	0.998	0.002	1.00	-132	1.00	-117	labile	-2.7	1.8	-6.3
OH^-	0.28	0.72	0.72	-144	0.89	-132	partially labile	0.4	3.3	-5.7
humic acid [b]	0.29	0.71	0.41	-136	0.66	-128	partially labile	0.4	3.9	-7.0
CO_3^{2-}	0.03	0.97	0.41	-166	0.68	-158	partially labile [c]	1.6	3.9	-4.6
NTA[a]	0.90	0.10	0.90	-132	0.90	-117	non-labile	-1.0	5.3	-11.6
EDTA [a]	0.002	0.998	0.002	-132	0.002	-117	non-labile	2.7	10.4	-15.3

a Total ligand concentration 10^{-7}M.

b Total ligand concentration 10^{-5}M. Other ligand concentrations as in Table 18.

c If the stability constants determined by Sipos et al. are correct the calculated lability of this complex is greatly enhanced (Fig. 7).
Sipos et al. (Ref. 30).

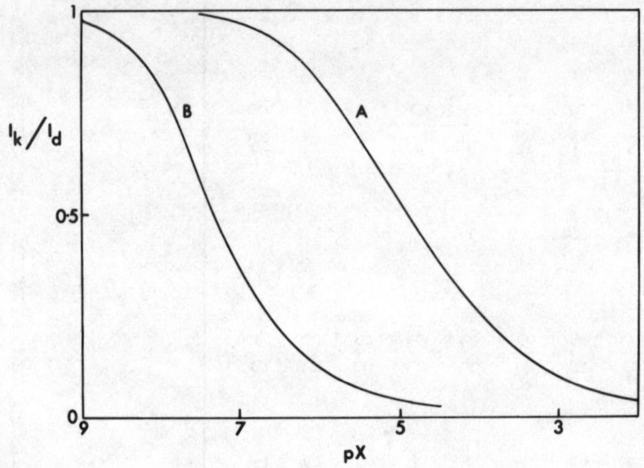

Fig. 6. Calculated dependence of I_k/I_d on ligand concentration (expressed as pX = $-\log_{10}[X]$), illustrated for 1:1 complexation of lead with carbonate (A) and humic acid (B). $\delta = 1.02 \times 10^{-3}$cm in each case.

I_k/I_d is twice as sensitive to variations in log $\beta_1^*(X)$ as to variations in log k_- (Fig. 7), a result which can be deduced from eq. 46 (Table 22) by assuming that $\sigma_X \gg 1$. This is encouraging since we have so few data on which to base our estimates of k_-. However, it does suggest also that calculations of I_k/I_d must reflect uncertainties in β^* values (Fig. 7). This is particularly important in the case of $PbCO_3^{\,o}$ which is predicted to be the major lead species at pH 8. The various log β_1^* (CO_3) values listed in Table 8 can cause this complex to vary from labile to almost non-labile (Fig. 7A). Similarly the various values for log $\beta_1^*(OH)$ affect I_k/I_d but the influence is not so marked in this case (Fig. 7B).

Although the equilibrium calculations showed that σ-values control the influence of ligands on the activity of lead in solution no such direct correlation is observed in electrochemical availability because of the complex form of eq. 47 and 48 (Table 22). However, it is possible to deduce some diagnostic criteria relating to ASV behaviour of a particular lead complex to values of σ_X and X .

Labile behaviour requires that $I_k/I_d \simeq 1$. For $I_k/I_d > 0.99$ eq. 47 gives

$$(\sigma_X/\delta)(D\,\beta^*/k_-(1+\sigma_X))^{\frac{1}{2}} < 10^{-2} \tag{41}$$

Substituting in values for the constants (Table 21) and remembering that $\sigma_X/(1+\sigma_X)$ is always less than unity we find that for <u>labile</u> behaviour

$$\log(\sigma_X[X]^{-\frac{1}{2}}) < 2 \tag{42}$$

This criterion will not be reliable if σ_X <<1, but in this case PbX will not make a significant contribution to the metal speciation.

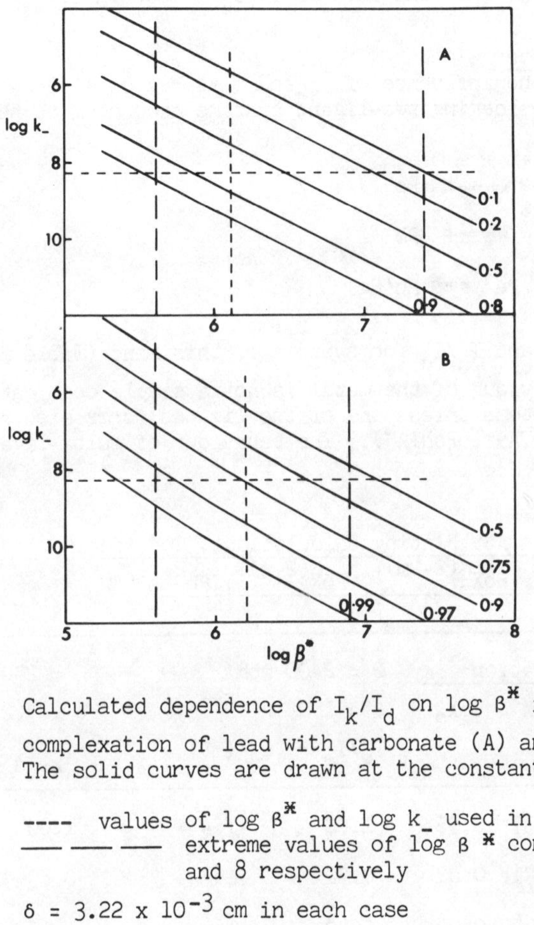

Fig. 7. Calculated dependence of I_k/I_d on log β^* for 1:1 complexation of lead with carbonate (A) and hydroxide (B). The solid curves are drawn at the constant I_k/I_d values shown.

---- values of log β^* and log k_ used in the calculations
—— —— —— extreme values of log β^* considered in Tables 6 and 8 respectively

$\delta = 3.22 \times 10^{-3}$ cm in each case

Non-labile behaviour demands that $V^o = 1$ (eq. 48). This requires that

$$\delta(k_-(1 + \sigma_X)/D\,\beta^*)^{\frac{1}{2}} < 0.1 \tag{43}$$

Substituting for the constants again we find that for non-labile complexes

$$\log\ ((1 + \sigma_X^{-1})\ [X]) < -10 \tag{44}$$

Partially labile complexes will satisfy neither of the above conditions (see Table 23). These predictions rest of course on the assumption that k_- is approximately the same for all the complexes.

By the use of matrix techniques (Ref. 48) the equations in Table 22 can be extended to cover any network of equilibria. A consideration of PbX_2 species (Table 24) indicates that these complexes have a negligible effect on I_k/I_d but, in the case of chloride complexes, the dominance of $PbCl_2^{o}$ causes a significant shift in $\varepsilon_{\frac{1}{2}}$.

(iv) Two-Ligand System

Some indication of the influence of ligand mixtures on electrochemical availability can be gained by considering two-ligand systems according to the scheme

$$PbX^{(2-m)+} \rightleftharpoons Pb^{2+} + X^{m-}$$

$$PbY^{(2-p)+} \rightleftharpoons Pb^{2+} + Y^{p-}$$

$$Pb^{2+} + 2e^- \rightleftharpoons Pb(Hg)$$

Calculations of $\varepsilon_{\frac{1}{2}}$ and I_k/I_d for systems of this kind (Table 25) indicate that the electrochemical behaviour of the metal is not a simple combination of the behaviour of single ligand systems unless one of the ligands forms electrochemically inactive complexes (e.g. $PbEDTA^{2-}$, $PbNTA^-$). The behaviour of multi-ligand systems is now under investigation.

TABLE 24 ASV Plating Calculations, Considering the Species $PbX_2^{(2-2m)+}$, $PbX^{(2-m)+}$, Pb^{2+}

Ligand X^{m-}	Speciation			ω = 2400 rpm, $\delta = 1.02 \times 10^{-3}$ cm		ω = 240 rpm, $\delta = 3.22 \times 10^{-3}$ cm	
	Pb^{2+}	PbX	PbX_2	I_k/I_d	$\varepsilon_{\frac{1}{2}}$ (mV)	I_k/I_d	$\varepsilon_{\frac{1}{2}}$ (mV)
Cl^-	0.08	0.37	0.55	1.00	-164	1.00	-149
OH^-	0.27	0.71	0.02	0.72	-144	0.89	-132
CO_3^{2-}	0.03	0.93	0.04	0.40	-166	0.67	-158

TABLE 25 ASV Calculations, Considering the Species $PbX^{(2-m)+}$, $PbY^{(2-p)+}$, Pb^{2+}

Ligand X^{m-}	Ligand Y^{p-}	Speciation Pb^{2+}	$PbX^{(2-m)+}$	$PbY^{(2-p)+}$	ω = 2400 rpm, $\delta \equiv 1.02 \times 10^{-3}$ cm I_k/I_d	$\varepsilon_{\frac{1}{2}}$ (mV)	ω = 240 rpm, $\delta = 3.22 \times 10^{-3}$ cm I_k/I_d	$\varepsilon_{\frac{1}{2}}$ (mV)
Cl^-	CO_3^{2-}	0.02	0.11	0.87	0.63	-173	0.84	-162
Cl^-	humic acid [b]	0.13	0.56	0.31	0.74	-154	0.87	-141
Cl^-	EDTA [a]	0.002	0.009	0.989	0.011	-154	0.011	-139
Cl^-	NTA [a]	0.18	0.80	0.02	0.98	-154	0.98	-139
OH^-	Cl^-	0.12	0.32	0.56	0.90	-157	0.97	-143
OH^-	humic acid [b]	0.16	0.43	0.41	0.48	-145	0.72	-136
OH^-	EDTA [a]	0.002	0.005	0.993	0.005	-144	0.006	-132
OH^-	NTA [a]	0.27	0.70	0.03	0.70	-144	0.86	-132
CO_3^{2-}	OH^-	0.02	0.91	0.07	0.40	-167	0.67	-159
CO_3^{2-}	humic acid [b]	0.03	0.91	0.06	0.38	-166	0.66	-158
CO_3^{2-}	EDTA [a]	0.00	0.07	0.93	0.03	-166	0.05	-158
CO_3^{2-}	NTA [a]	0.03	0.97	0.00	0.40	-166	0.68	-158

[a] Total ligand concentration 10^{-7}M

[b] Total ligand concentration 10^{-5}M. Other ligand concentrations as in Table 18.

6. SUMMARY AND CONCLUSIONS

The discussion in this paper has been based on the assumption that the various components in the natural system will attain equilibrium and that the biological behaviour of lead, as expressed by a(Pb) will reflect this equilibrium situation. This might be a reasonable assumption for the inorganic speciation of lead since homogeneous ionic interactions tend to equilibrate rather rapidly and the low lead concentration makes it unlikely that heterogeneous solubility reactions will exert a controlling influence on lead chemistry. Furthermore, Schindler (Ref. 27) has calculated that the adsorption of lead onto silica particles will have a negligible effect (<0.5 %) on lead speciation in sea water.

Deviations from equilibrium behaviour are far more likely when organic ligands are considered since (i) the ligands themselves are metastable and will be degraded by organic and inorganic processes in the natural environment (Ref. 64); (ii) the ligand molecules may be effectively adsorbed on to particle surfaces or they may form colloidal aggregates; (iii) the stability constants of many lead-organic ligand complexes are so high that their rates of dissociation may be low compared to their residence time in natural systems and (iv) the species distribution calculated from bulk solution properties might be greatly distorted by high localised concentrations of ligand (e.g. in the gut). In such circumstances a direct measure of lead availability is required to correlate the physiological and chemical characteristics of the medium. At the high lead concentrations used in studies of the physiological effects of lead or human cells (Ref. 65), ion-selective electrodes will be useful and they can be calibrated on the appropriate activity scale using the speciation model given here. At lower concentrations pulse polarography and anodic stripping voltammetry provide direct means of measuring electrochemically available lead in a natural sample. The theory outlined here enables this availability to be related quantitatively to the equilibrium speciation model and provides a first step towards an accurate definition of the metal fraction observed electrochemically. To be generally useful the theory must be extended to include irreversible processes and adsorption processes which might provide pathways for the reduction of complexes considered to be non-labile in the present treatment. The importance of these processes is underlined by the observation that many of the peak current versus plating potential curves observed in natural samples do not correspond to the symmetrical polarographic waves shown in Fig. 4.

Bearing these shortcomings in mind, the speciation model described here and the general treatment described for metal-ligand interactions, will hopefully provide a useful starting point for interpreting the biological impact of lead in the marine environment.

APPENDIX 1 - GLOSSARY OF SYMBOLS

A	Surface area of thin film electrode
a^o_X	Conventional fitting parameter in eq. 10
a(X)	Thermodynamic activity of component X
D	Diffusion coefficient
E	Electrode potential
$E_{\frac{1}{2}}$	Half-wave potential
E^o	Standard electrode potential
F	Faraday
f	= F/RT

$g(X)$	Free activity coefficient of component X
I	Stripping peak current
I_d	Diffusion limited maximum value of I
I_k	Kinetically limited maximum value of I
$K_j^*(X)$	Stepwise stoichiometric stability constant for formation of PbX_j
k_+	Dissociation rate constant
k_-	Formation rate constant
l	Mercury film thickness
n	Stoichiometric number of electrons transferred in the reduction process
Q	$= Dt/\delta l$. In the absence of kinetic control Q is the maximum value of the ratio (amalgam concentration/bulk solution concentration) for the trace metal
R	Gas constant
T	Absolute temperature (^oK)
t	Plating time
V^o	Dimensionless parameter obtained from differential equation solution
$[X]$	Concentration of the species X
$[X]_T$	Total concentration of X in all forms
$[X]_T^*$	Critical total concentration of X inducing significant complexation of lead
Z_X	Charge on component X
$\bar{\alpha}_X$	$= 1 + \sum_X$ (eq. 14)
$\overset{o}{\beta}_j(X)$	Stability constant for formation of PbX_j on the infinite dilution scale
$\beta_j^*(X)$	Stoichiometric stability constant for formation of PbX_j
$\Gamma_j(X)$	Activity coefficient factor for formation of PbX_j
$\gamma_\pm^{(PbX)}$	Mean ion activity coefficient of PbX
δ	Diffusion layer thickness
ε	$= E - E^o$
$\Delta\varepsilon_{\frac{1}{2}}$	Shift in $E_{\frac{1}{2}}$
μ	Ionic strength
θ	Degree of dissociation of metal complex
$\sum_X$	Sum of all individual σ-values involving X
$\sigma_{x,j}$	$= \beta_j^*(X) \cdot [X]^j$
ω	Rotation speed of rotating disc electrode

REFERENCES

(1) M. Burnett and C. Patterson, Analysis of natural and industrial lead in marine ecosystems, in Proceedings of the International Experts Discussion on "Lead-Occurrence, Fate and Pollution in the Marine Environment", Rovinj, Oct. 1977, M. Branica and Z. Konrad (eds.), Pergamon Press, 1979, 15-30.

(2) H. W. Nürnberg, P. Valenta, L. Mart, B. Raspor and L. Sipos, Applications of polarography and voltammetry to marine and aquatic chemistry. II The polarographic approach to the determination and speciation of toxic trace metals in the marine environment, Z. Anal. Chem. 282, 357-367 (1976).

(3) L. Mart, H. W. Nürnberg, and P. Valenta, Comparative base line studies on Pb-levels in European coastal waters, Proceedings of the International Experts Discussion on "Lead - Occurrence, Fate and Pollution in the Marine Environment", Rovinj, October 1977, M. Branica and Z. Konrad (eds.), Pergamon Press, 1979, 155-179.

(4) W. A. MacCrehan, R. A. Durst, and J. P. Bellama, Electrochemical detection in liquid chromatography: application to organometallic speciation, Anal. Lett. 10, 1175-1188 (1977).

(5) M. Whitfield, The ion association model and the buffer capacity of the carbon dioxide system in sea water at 25 ^{o}C and 1 atmosphere total pressure, Limnol. Oceanogr. 19, 235-248 (1974).

(6) T. M. Florence and G. E. Batley, Trace metals species in sea water. I Removal of trace metals from sea water by a chelating resin, Talanta 23, 179-186 (1976).

(7) J. C. S. Lu and K. Y. Chen, Migration of trace metals in interfaces of sea water and polluted sediments, Env. Sci. Technol. 11, 174-182 (1977).

(8) D. Dyrssen and M. Wedborg, Equilibrium calculations of the speciation of elements in sea water, in The Sea (E. D. Goldberg, ed.), Wiley Inter-science, New York 5 181-195 (1975).

(9) A. Zirino and S. Yamamoto, A pH-dependent model for the chemical speciation of copper, zinc, cadmium and lead in sea water, Limnol Oceanogr. 17, 661-671 (1972).

(10) W. Stumm and P. A. Brauner, Chemical Speciation, in Chemical Oceanography, (J. P. Riley and G. Skirrow, eds.), Academic Press, London, 1975, Second Edition, Vol. 1, pp. 173-239.

(11) D. T. Long and E. E. Angino, Chemical speciation of Cd, Cu, Pb and Zn in mixed freshwater, sea water and brine solutions, Geochim. Cosmoshim. Acta 41, 1183-1191 (1977).

(12) M. Truesdell and B. F. Jones, Ion-association in natural brines, Chem. Geol. 4, 51-62 (1969).

(13) M. Whitfield, Sea water as an electrolyte solution, in Chemical Oceanography (J. P. Riley and G. Skirrow, eds.) Academic Press, London, 1975, Second Edition, Vol. 1, pp. 43-171.

(14) C. W. Davies, Ion association, Butterworth, 1962.

(15) D. A. MacInnes, The activities of ions of strong electrolytes, J. Am. Chem. Soc. 41, 1086-1092 (1919).

(16) R. M. Garrels and M. E. Thompson, A chemical model for sea water at 25 °C and 1 atmosphere total pressure, Am. J. Sci. 260, 57-66 (1962).

(17) R. A. Robinson and R. H. Stokes, Electrolyte solutions, Butterworth, London, 1959.

(18) G. M. Lafon, Some quantitative aspects of the chemical evolution of the oceans, Ph. D. Thesis, Northwestern University, Evanston, Illinois, 1969.

(19) H. C. Helgeson, Thermodynamics of hydrothermal systems at elevated temperatures and pressures, Am. J. Sci. 266, 729-804 (1969).

(20) M. Whitfield, The ion association model and the buffer capacity of the carbon dioxide system in sea water at 25 °C and 1 atmosphere total pressure, Limnol. Oceanogr. 19, 235-248 (1974).

(21) F. J. Millero, Sea water as a multicomponent electrolyte solution, in The Sea (E. D. Goldberg, ed.), Wiley, New York, 1975, Vol. 5, pp. 3-80.

(22) A. Ringbom, Complexation in analytical chemistry, Interscience, New York, 1963.

(23) J. F. Elder, Complexation side reactions involving trace metals in natural water systems, Limnol. Oceanogr. 20, 96-102 (1975).

(24) L. G. Sillen and A. E. Martell, Stability constants of metal-ion complexes, Chemical Society, London, 1964, Special Publication No. 17.

(25) L. G. Sillen and A. E. Martell, Supplement No. 1 to Stability constants of metal-ion complexes, Chemical Society, London, 1971, Special Publication No. 25.

(26) C. F. Baes and R. E. Mesmer, The hydrolysis of cations, A critical review of hydrolytic species and their stability constants in aqueous solution, Wiley Interscience, New York, 1976.

(27) P. W. Schindler, Removal of trace metals from the oceans: a źero order model, Thalassia Jugosl. 11, 101-111 (1975).

(28) H. Bilinski, R. Huston and W. Stumm, Determination of the stability constants of some hydroxo and carbonato complexes of Pb(II), Cu(II), Cd(II) and Zn(II) in dilute solutions by anodic stripping voltammetry and differential pulse polarography, Anal. Chim. Acta 84, 157-164 (1976).

(29) I. Hannson, A new set of acidity constants for carbonic and boric acid in sea water, Deep-Sea Res. 20, 461-478 (1973).

(30) L. Sipos, P. Valenta, H. W. Nürnberg and M. Branica, Voltametric determination of the stability constants of the predominant labile lead complexes in sea water, Proceedings of the International Experts Discussion on "Lead - Occurrence, Fate and Pollution in the Marine Environment", Rovinj October 1977, M. Branica and Z.Konrad(eds.), Pergamon Press, 1979, 61-76.

(31) R. Ernst, H. E. Allen and K. H. Many, Characterisation of trace metal species and measurement of trace metal stability constants by electro-

chemical techniques, Water Research 9, 969-979 (1975).

(32) N. N. Baranova, Investigations of the carbonato-complexes of lead at 25 °C and 200 °C, Russ. J. Inorg. Chem. 14, 1717-1720 (1969).

(33) J. J. Lingane, Interpretation of the polarographic waves of complex metal ions, Chem. Rev. 29, 1-35 (1941).

(34) J. J. Morgan, Applications and limitations of chemical thermodynamics in water systems, in Equilibrium concepts in natural water systems (W. Stumm, ed), Am. Chem. Soc., Washington, D. C., Adv. Chem. Ser. 67, 1-29 (1967).

(35) N. Eyring and E. M. Eyring, Reaction rates in solution, in Principles and Applications of Water Chemistry (S. D. Faust and J. V. Hunter, eds.), Wiley, New York, 1967, pp. 1-21.

(36) J. O'M Bockris and A. K. N. Reddy, Modern Electrochemistry, Macdonald, London, 1970, Vol. 1, 297-299.

(37) W. Sunda and R. R. L. Guillard, The relationship between cupric ion activity and the toxicity of copper to phytoplankton, J. Mar. Res. 34, 511-529 (1976).

(38) D. J. Crisp, A. O. Christie and A. F. A. Ghobashy, Narcotic and toxic action of organic compounds on barnacle larvae, Comp-Biochem. Physiol. 22, 629-649 (1967).

(39) T. B. Warner, Ion-selective electrodes in thermodynamic studies, in The Nature of Sea Water, (E. D. Goldberg, ed.), Dahlem Konferenzen, Berlin 1975, 191-217.

(40) A. Siegel, Metal-organic interactions in the marine environment, in Organic compounds in aquatic environments, (S. D. Faust and J. V. Hunter, eds.), Marcel Dekker, New York 1971, 265-296.

(41) E. W. Davey, M. J. Morgan and S. J. Erickson, A biological measurement of copper complexation capacity of sea water, Limnol. Oceanogr. 18, 993-997 (1973).

(42) A. Dickson, R. F. C. Mantoura and J. P. Riley, The complexation of metals with humic materials in natural waters, Estuarine & Coastal Marine Science 6 (1977) (in press).

(43) F. J. Stevenson, Stability constants of Cu^{2+}, Pb^{2+} and Cd^{2+} complexes with humic acids, Soil Sci. Soc. Am. J. 40, 665-672 (1976).

(44) B. Raspor, P. Valenta, H. W. Nürnberg and M. Branica, The chelation of cadmium with NTA in sea water as a model for the typical behaviour of trace heavy metals in natural waters, Sci. Tot. Envir. 9, 87-109 (1978).

(45) J. H. Reuter and E. M. Perdue, Importance of heavy metal-organic matter interactions in natural waters, Geochim. Cosmochim. Acta 41, 325-334 (1977).

(46) J. Gardiner, Complexation of trace metals by ethylenediamine tetra acetic acid (EDTA) in natural waters, Water Research 10, 507-514 (1976).

(47) B. Raspor, H. W. Nürnberg, P. Valenta and M. Branica, The chelation of lead by organic ligands in seawater, Proceedings of the International Experts

Discussion on "Lead - Occurrence, Fate and Pollution in the Marine Environment", Rovinj October 1977, M. Branica and Z. Konrad (eds.), Pergamon Press, 1979, 181-195.

(48) D. R. Turner and M. Whitfield, in preparation.

(49) D. R. Crow, Polarography of metal complexes, Academic Press, London, 1969.

(50) J. Koryta, Electrochemical kinetics of metal complexes, in Advances in Electrochemistry and Electrochemical Engineering (ed. P. Delahay), Interscience, New York, 1967, Vol. 6, 289-327.

(51) J. Heyrovsky and J. Kuta, Principles of Polarography, Academic Press, New York, 1966.

(52) J. Koutecky and J. Koryta, The general theory of polarographic kinetic currents, Electrochimica Acta 3, 318-338 (1961).

(53) S. Bubić and M. Branica, Voltammetric characterisation of the ionic state of cadmium present in sea water, Thalassia Jugosl. 9, 47-53 (1973).

(54) A. Zirino and S. P. Kounaves, Anodic stripping peak currents: electrolysis potential relationships for reversible systems, Anal. Chem. 49, 56-59 (1977).

(55) V. G. Levich, Physicochemical Hydrodynamics, Prentice Hall, Englewood Cliffs, New Jersey, 1962, 2nd Edition.

(56) W. Davison, Defining the electroanalytically measured species in a natural water sample, J. Electroanal. Chem. (1977) (in press).

(57) R. G. Wilkins, The kinetics of complex formation in aqueous solution, in The Nature of Sea water (E. D. Goldberg, ed.), Dahlem Konferenzen, Berlin, 1975, 397-415.

(58) M. Eigen and R. G. Wilkins, The kinetics and mechanism of formation of metal complexes, Am. Chem. Soc., Adv. Chem. Ser. 49, 55-67 (1964).

(59) C. H. Langford, Complex formation and solvent-exchange kinetics: A dynamic approach to electrolyte solutions, in Ionic interactions from dilute solutions to fused salts (S. Petrucci, ed.), Academic Press, London, 1971, Vol. 2, 1-37.

(60) D. J. Hewkin and R. H. Prince, The mechanism of octahedral complex formation by labile metal ions, Coord. Chem. Rev. 5, 45-73 (1970).

(61) J. M. Hale, Transients in convective systems. II Limiting current and kinetically complicated galvanostatic transients at rotating disc electrodes, J. Electroanal. Chem. 8, 332-349 (1964).

(62) D. D. DeFord and D. N. Hume, The determination of consecutive formation constants of complex ions from polarographic data, J. Am. Chem. Soc. 73, 5321-5322 (1951).

(63) M. Branica, D. M. Novak and S. Bubić, Application of anodic stripping voltammetry to determination of the state of complexation of traces of metal ions at low concentration levels, Croat. Chim. Acta 49, 539-547 (1977).

(64) A. Lerman and C. W. Childs, Metal-organic complexes in natural waters: control of distribution by thermodynamic, kinetic and physical factors, in *Trace metals and metal-organic interactions in natural waters*, (P. C. Singer, ed.), Ann Arbor Science, Michigan, 1973, 201-236.

(65) A. B. Fischer, Acute and chronic effects of lead in mammalian cells *in vitro*, *Proceedings of the International Experts Discussion on "Lead - Occurrence, Fate and Pollution in the Marine Environment"*, Rovinj, October 1977, M. Branica and Z. Konrad (eds.), Pergamon Press, 1979, 279-297.

Input of Lead into the Mediterranean Through Rivers

R. Fukai

International Laboratory of Marine Radioactivity, IAEA, Musée Océanographique, Principality of Monaco

ABSTRACT

A series of measurements of lead in water and suspended matter were made on two Mediterranean rivers, one less polluted and the other more polluted. Based on the results of these measurements the annual input of lead into the Mediterranean through rivers was estimated to be around 2.8×10^3 tons Pb/year.

INTRODUCTION

Within the framework of the Coordinated Mediterranean Pollution Monitoring and Research Programme sponsored by UNEP, several aspects of pollution in the Mediterranean are being studied (Ref. 1). One of the major problems is to estimate the input of pollutants from various sources into the Mediterranean. Rivers are known to contribute substantially to the total input of pollutants into the sea. To date studies on the amounts of pollutants carried by rivers have been inadequate for quantitative work. A series of trace element measurements on both river water and suspended matter made during 1977 gave data needed to estimate the river input of various trace elements into the Mediterranean. This paper presents the results of the lead measurements. With these data, an attempt was made to obtain an order-of-magnitude estimate of annual lead input to the Mediterranean through rivers.

MATERIAL AND METHODS

Water was collected from two Mediterranean rivers in February, May and August 1977. Collections were made both at the mouth of the two rivers and at a station located between 30 and 50 km upstream. No river names are given since the purpose of the measurements is not for river pollution control, but for obtaining basic data for an estimation of the overall input. Both rivers flow through calcareous terrain, typical of the northern shore of the Mediterranean, carrying water of pH 8; one river flows mainly through a sparsely inhabited mountain valley without much apparent pollution except near the mouth (river X); the other, one of the major rivers pouring into the Mediterranean, is considerably more polluted and flows through densely populated areas, receiving urban and industrial wastes (river Y). Water samples ($\sim$ 10 liters) were collected from flowing water by a polyethylene bucket connected to a nylon rope and were immediately transferred to polyethylene containers. These were filtered through 0.45 µm membrane filters as soon as they were brought to the laboratory (normally within 23 hours). Suspended matter on the filters was dried at 105-110 °C, weighed and leached at room temperature with 1M HCl for 15 hours. The leachates were filtered and lead in the filtrates was determined with a Perkin-Elmer 403 atomic absorption spectrophotometer in conjunction with a graphite furnace HGA 72 and a deutrium background corrector (Ref. 2). The chemical nature of lead in suspended matter thus measured is hard to define, since the leaching procedure adopted is expected not only to bring anthropogenic lead into solution but also to dissolve natural lead in the crystal lattice of silt particles. Further studies are required to understand the partition of lead between river water and its suspended matter. The recent studies (Ref. 3 and 4) indicate, however, that the leaching condition adopted dissolves practically all anthropogenic lead as well as at least 50 %, or probably much more, of natural particulate lead.

Lead in filtered river waters was measured by differential pulse anodic stripping voltammetry with a composite mercury-graphite electrode (Ref. 5). The apparatus used was a Tacussel pulse polarograph Type CAP4 in conjunction with a programmer Polaromate. Although precautions were taken during collection and analysis to eliminate contamination, some contamination was inevitable without an available "clean laboratory" (Ref. 6). Therefore, the values for soluble lead should be considered to represent the upper limit of the concentration.

RESULTS AND DISCUSSION

The results of lead measurements on suspended matter in Rivers X and Y are given in Table 1 with the data on total suspended load, plus iron and silver. The stations X-1 and Y-1 represent the upstream stations for the respective rivers and the stations X-2 and Y-2 are those at the river mouths. The data show that the levels of lead leachable from a unit mass of suspended matter in River Y are considerably higher than those in River X; comparing the data for the same month, the lead levels in River Y are, on an average, 7 times higher (range: 4-14 times). The difference is considered to result from the different anthropogenic input of lead into the river environment, since mineralogical characteristics of suspended matter are similar for both rivers. This is supported by the data on leachable iron and silver, the levels of which are likely to be much less affected by anthropogenic input than those of lead; iron is higher in River Y, but only by a factor of approximately 2, while silver levels are similar for both rivers. These observations indicate that suspended matter in an apparently polluted river, River Y, is several times more polluted through anthropogenic input with respect to lead, than that in an apparently less polluted river, River X. This does not, however, necessarily imply that the lead load carried by a unit volume of River Y is higher than that of River X, since, as shown in Table 1, the suspended load in River X is generally higher than that in River Y.

Partition of lead between particulate and soluble phases is shown in Table 2, assuming that the leachable lead fraction is approximately equal to the total lead content in suspended matter. It is clear that the major portion of lead is in the particulate phase; more than 90 % of the lead is associated with suspended matter except for the summer season, when the discharge and the suspended load of the rivers are low. In reality, the particulate fraction of lead is considered still greater since, as has been stated, the concentration of soluble lead given represents the upper limit and also since there may exist some unleachable lead in suspended matter. In any case, it should be noted that the lead load of River Y water was much higher than that of River X in February when intermittent heavy precipitation resulted in a high discharge and a high suspended load, while the lead load for both rivers was similar in May, during which snow-melting in the mountain areas caused high discharge and high sediment load. The high lead level of suspended matter in River Y in February is believed to be caused by the wash-out effects of heavy rain over the polluted terrain.

TABLE 1 Lead and Other Metals in Suspended Matter of Mediterranean Rivers

River Station No.		River X* X-1	X-2	River Y** Y-1	Y-2
Suspended load (mg/l)	February '77	434	505	343	208
	May "	673	409	187	53
	August "	13	197	21	31
Pb (μg/g-dry)	February '77	12	10	78	70
	May "	6	10	34	46
	August "	10	4	64	57
Fe (mg/g-dry)	February '77	11.9	10.4	18.5	16.6
	May "	2.3	3.8	7.8	8.6
	August "	5.4	4.1	8.4	9.4
Ag (μg/g-dry)	February '77	1.4	1.2	1.3	1.1
	May "	1.4	1.8	1.4	3.5
	August "	1.5	2.3	1.3	1.7

* River X: apparently less polluted

** River Y: apparently more polluted

TABLE 2 Partition of Lead Between Suspended Matter and Solution in Mediterranean Rivers

Station No.	February		May		August	
	in susp. (μg/l)	in sol. (μg/l)	in susp. (μg/l)	in sol. (μg/l)	in susp. (μg/l)	in sol. (μg/l)
X - 1	5.2	0.5	4.0	0.1	0.1	0.2
X - 2	5.1	0.3	4.1	0.2	0.8	0.1
Average	5.2	0.4	4.1	0.2	0.5	0.2
Total	5.6		4.3		0.7	
Y - 1	27	0.3	6.4	0.1	1.3	0.3
Y - 2	15	0.1	2.4	0.1	1.8	0.2
Average	21	0.2	4.4	0.1	1.6	0.3
Total	21.2		4.5		1.9	

The estimation of the annual input of lead through Mediterranean rivers was derived from the annual average lead load of Rivers X and Y, assuming that these rivers represent respectively "less polluted rivers" and "more polluted rivers" in the Mediterranean region. The annual average lead load of the two rivers was computed on the basis of the data presented in Table 2, taking the seasonal variation of river discharge into account. In order to complete year-round seasonal data, the data missing for autumn in Table 2 were estimated by taking the average of February and May data, since the conditions of the precipitation and wash-out in autumn are considered to be in between February and May. The following seasonal variation of river discharge, derived from the data on the Var River (Ref. 7), was applied for both rivers:

Months	% of annual discharge
January - March	17%
April - June	32%
July - September	19%
October - December	32%

A weighted average of the seasonal data of lead load, computed by taking seasonal discharges as weight, was considered to represent an annual average lead load for each river. Thus, 4.1 μg Pb/l (3.8 particulate plus 0.3 soluble) and 9.7 μg Pb/l (9.5 particulate plus 0.2 soluble) were obtained as the annual average lead load for River X and River Y respectively.

If all the rivers in the Mediterranean region were as little polluted as River X, the annual input of lead into the Mediterranean would be 1.8×10^3 tons/year, 4.1 μg multiplied by 4.4×10^{14} l/year, the total annual discharge of Mediterranean rivers (Ref. 8). If all rivers were polluted like River Y, the annual input would be 4.3×10^3 tons/year. The actual input should be between these two values. On the basis of general pollution observations made along several Mediterranean rivers, we assume simply that 60 % of the total discharge is less polluted and 40 % is more polluted; then the total lead input through rivers into the Mediterranean is estimated to be 2.8×10^3 tons/year.

The world lead input through rivers into the sea has been estimated to be 7.4×10^5 tons Pb/year (Ref. 9). Based on this estimate, the Mediterranean input represents 0.4 % of the world input. Considering the size of the Mediterranean water-shed, the order of magnitude of the Mediterranean contribution to the world input seems to be reasonable.

ACKNOWLEDGEMENT

The author wishes to thank Messrs. B. Oregioni and L. Huynh-Ngoc of the Monaco Laboratory who conducted the lead measurements and also Dr. C. L. Osterberg of the Monaco Laboratory for his critical reading of the manuscript. The International Laboratory of Marine Radioactivity operates under a tripartite agreement between the International Atomic Energy Agency, the Government of the Principality of Monaco and the Oceanographic Institute at Monaco. This work was also supported by the UNESCO-UNEP fund. These supports are gratefully acknowledged.

REFERENCES

(1) UNEP, Administrative Report on the mid-term review meeting on the progress of the co-ordinated Mediterranean pollution monitoring and research programme (MED POL) and related projects of the Mediterranean Action Plan, UNEP/IG. 11/INF. 3, Oct. 1977.

(2) R. Fukai and B. Oregioni, A note on the sensitivity and accuracy of atomic absorption spectrophotometry for trace metal measurements on marine biological samples, Thalassia Jugosl. 13,(1977) (1977).

(3) E. J. Catanzaro, Some relationships between exchangeable copper and lead and particulate matter in a sample of Hudson River water, Environm. Sci. Tech. 10, 386-388 (1976).

(4) J. C. Duinker and R. F. Nolting, Dissolved and particulate trace metals in the Rhine Estuary and Southern Bight, Mar. Poll. Bull.8, 65-71 (1977).

(5) R. Fukai and L. Huynh-Ngoc, Chemical forms of zinc in sea water - Problems and experimental methods, Journ. Oceanogr. Soc. Japan 31, 179-191 (1975).

(6) Meeting Report, Interlaboratory lead analysis of standardized samples of sea water, Mar. Chem. 2, 69-84 (1974).

(7) R. Laurent, Charge solide en suspension et geochimie dans un fleuve côtier mèditerranèen Le Var (Alpes-Maritimes), Thèse Lab. de Geol. et de Sedimentol., Univ. Nice, 1971, p. 249.

(8) UNESCO-UNEP, Role of sedimentation in the pollution of the Mediterranean Sea: Assessment of knowledge and development of guidelines for environmental impact assessment, 1976, p. 50.

(9) J. Murozumi, T. J. Chow, and C. Patterson, Chemical concentrations of pollutant lead aerosols, terrestrial dusts and sea salts in Greenland and Antarctic snow strata, Geochim. Cosmochim. Acta 33, 1247-1294 (1969).

ADDENDUM

After this paper was presented at the meeting in October 1977 at Rovinj, Yugoslavia, a report of the ECE/UNIDO/FAO/UNESCO/WHO/IEAE working group on "Pollutants from Land-based Sources in the Mediterranean" was presented at the UNEP meeting in January 1978 at Monaco.* The group estimated the lead input through rivers into the Mediterranean to be 3.2×10^3 tons/year, of which 2.2×10^3 tons/year was due to pollution.

---

* UNEP, Pollutants from land-based sources in the Mediterranean, UNEP/IG.11/INF. 5 (1977) 46 p.

Comparative Base Line Studies on Pb-Levels in European Coastal Waters*

L. Mart, H. W. Nürnberg and P. Valenta

Institute of Chemistry, Institute 4 Applied Physical Chemistry, Nuclear Research Centre (KFA), Juelich, F.R.G.

ABSTRACT

The contours of the Pb-inventory in waters of the Ligurian and Tyrrhenian coast, the Belgian and Dutch Southern Bight coast and the German North Sea coast have been established from the voltammetric analysis of samples taken with a rather dense network of more than 400 stations. After some general methodological remarks on sampling, sample pretreatment, aspects for selection of differential pulse anodic stripping voltammetry as trace analytical determination method and low temperature ashing by microwave induced oxygen plasma as digestion procedure for suspended particulate matter, the data for the different coastal regions are presented and discussed particularly with respect to the influence of hydrographic parameters and pollution sources on the respective Pb-levels. For the studied coastal zones in the Western Mediterranean and the Belgian/Dutch Southern Bight, data for the dissolved and particulate matter bound Pb-levels and the resulting total Pb-freight in the waters have been measured, while for the German North Sea Coast the discussion has to be based at present only on the already determined dissolved Pb-level and is therefore for this region somewhat preliminary. General characteristics of the different coastal areas are connected with the differences in the particulate matter content, usually very low in the studied Western Mediterranean coastal zone, and the substantially different hydrographic conditions in the various coastal regions studied. While tidal movements and currents are virtually absent in the Western Mediterranean they are an important and complicating parameter at the North Sea coast. Here they exert, due to different directions, pronounced and differing influences on the Pb-level along the Belgian/Dutch Southern Bight coast and the German North Sea Coast, respectively. In a final section some examples for specific local influences (big ports and estuaries) in the elsewhere rather stationary behaving Ligurian and Tyrrhenian coastal waters are described.

INTRODUCTION

The determination of the overall level of Pb present in coastal waters provides a first general information on the occurrence and fate of this toxic metal in this

---

* This work is part of the Doctor Thesis (Dr. rer. nat.) of L. M. submitted to the Rheno-Westfalian Technical University, Aachen, F. R. Germany.

part of the marine environment and on the magnitude of the corresponding pollution burden (Ref. 1 and 2). Such base line studies are also one necessary prerequisite for the identification and characterization of the various chemical forms in which Pb exists in the dissolved state in sea water and in estuaries and for the determination of the distribution of the overall Pb-concentration over those chemical species (Ref. 3 and 4).

Extensive studies have been carried out by us in the last two years in the Western Mediterranean (Ref. 5) (Ligurian and Tyrrhenian coast, 225 stations from Ventimiglia to Cape Anzio, viz. Fig. 1) and along the Belgian, Dutch and German North Sea coast (200 stations from Oostende to the Isle of Sylt, viz. Fig. 2). One aim was to compare the situation in both European coastal regions belonging to different European seas. The rather dense network of over 400 stations and sampling in the various areas within relatively short times yielded a characteristic pattern of the actual Pb-inventory in the studied areas. Based on this general survey, studies of the long term behaviour and of seasonal influences on the Pb-level at selected particularly interesting locations are now in progress.

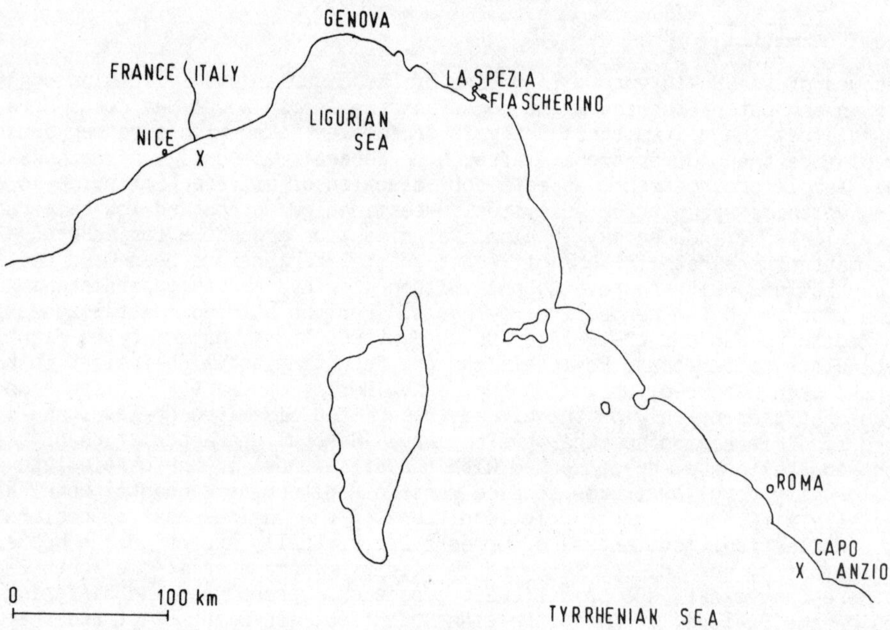

Fig. 1. Sampling area at Ligurian and Tyrrhenian coast, 225 stations, April-June 1976, 60 stations repeated September 1977; fine weather, calm sea.

The amount of Pb present in the dissolved state (Ref. 5) as well as the Pb-amount bound to suspended inorganic and organic particulate matter or incorporated by suspended microorganisms has been determined for each station. The sum of both amounts gives the total Pb-freight at the respective station which can be mobilized under certain conditions and may reach man to a certain extent via the marine food chains.

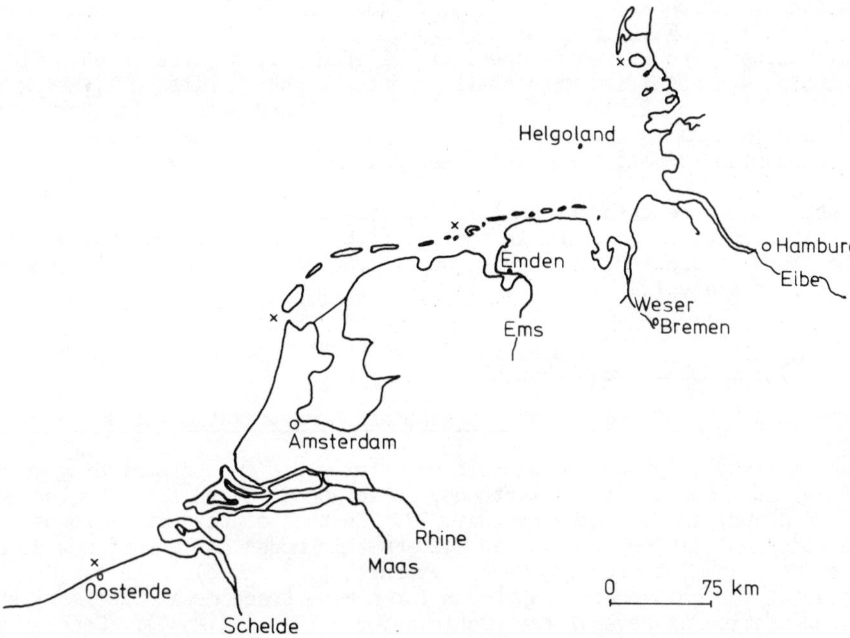

Fig. 2. Sampling areas at North Sea Coast. 1) Oostende to Den Helder (Southern Bight coast), 70 stations, August-September 1976; fine weather, calm sea. 2) Emden to the Isle of Sylt (German coast), 100 stations, June 1977; fine weather, calm sea.

Particular efforts were made to ensure accuracy and reliability of the data by excluding accuracy deficiencies due to contamination or losses in all stages of the investigation from sampling over storage of samples to their measurement (Ref. 5).

SAMPLING AND SAMPLE PRETREATMENT

Samples were taken only from surface waters (0.5 to 1 m below the surface). Avoidance of contamination by the boat is a crucial point. Therefore in Ligurian and Tyrrhenian coastal waters sampling was performed manually in front of the bow of a slowly moving small rubber boat. In this manner any contamination risk by the boat itself and the outboard motor is prohibited. The sampling containers were polyethylene bottles with a volume of 2 l, specially cleaned and subsequently sea water conditioned.

Strength of waves and wind at the North Sea prohibited sampling from a rubber boat. Thus, for these missions a small lowboard wooden fishing cutter was employed. Sampling was performed also with the polyethylene bottles mounted to a 3.5 m long telescope gear (Ref. 6).

The sea water samples were filtrated through precleaned and then conditioned 0.45 μ-filters within 5 h after sampling to remove suspended particulate matter. Extended previous tests had confirmed that no alteration in the level of dissolved

Pb and other trace metals (Cu, Cd, Bi) occurs during this time period. To avoid contamination by airborne dust particles, filtration was performed in a closed device, Sartorius SM16511, under nitrogen pressure. The filters with the suspended particulate matter were stored in specially cleaned polyethylene boxes. The filtrated solution was acidified with small amounts of concentrated HCl, Merck suprapur, (1 ml per 1) to pH 2. In this manner bacterial and microbiological activity is stopped and in samples from a number of stations all of the present rather stable metal complexes decompose within several hours.

If these solutions were then not analysed immediately in our mobile laboratory or our marine station at Fiascherino in the Bay of La Spezia, but had to be shipped to the Institute at Juelich, they were deep frozen at -20°C and gently thawed overnight before analysis.

TRACE ANALYTICAL PROCEDURE

Aspects for selection of the method and general characteristics of the applied procedure

The applied analytical procedure was differential pulse anodic stripping voltammetry (DPASV) at a mercury film electrode, an advanced method from the large arsenal of polarography and voltammetry (Ref. 7). In this context it is to be emphasized that the application of advanced voltammetric modes has opened new dimensions for marine trace chemistry during recent years (Ref. 2 and 8). The striking potentialities of the voltammetric approach for marine trace metal chemistry stem from the accumulation of several favourable aspects (Ref. 1, 2, 9). Generally voltammetric methods combine extraordinary determination sensitivity with good precision and a particular inherent accuracy, as they are based on Faraday's law (n·96500 Cb are equivalent to 1 mole, where the number of electrons n transferred in the elementary step of electrochemical reduction or oxidation equals for metals to 1 to 3). It just so happens that all important trace metals and metalloids, among them the most toxic ones, (e.g. Cd, Pb, Hg, Cu, As, Se, Zn etc.) are well accessible to voltammetry. Several trace metals can be determined simultaneously (e.g. Bi, Cu, Cd, Pb) in one run making voltammetry an oligo-substance approach. Reliable multi-mode instruments of high performance produced by several companies (AMEL, Bruker, Metrohm, PAR and Solea Tacussel) are commercially available. Investments are with less than 10^4 US-dollars for a voltammetric measuring place rather moderate and render this electroanalytical approach to the group with the lowest costs among the methods in modern instrumental trace analysis (Ref. 1, 2, 9, 10). The favourable cost/benefit ratio is an important practical aspect with respect to the application of a battery of voltammetric devices - in the future microprocessor controlled - for the analysis of large sample numbers, as they occur in extended base line studies or monitoring programs, with a satisfactory determination rate. Moreover, sea water is a rather ideal matrix for voltammetry as it contains already the necessary supporting electrolyte due to its salinity.

On the other hand, particularly the alkaline earth ion content of salinity causes severe signal suppression problems for the rather popular alternative in trace metal analysis which is electrothermal atomic absorption analysis (AAS) with the graphite tube (Ref. 11). Except for metals and metalloids (e.g. Hg, As, Se) determinable by the cold vapor mode (Ref. 12), AAS requires prior chemical extraction of the trace metals from the sea water matrix with the involved accuracy risks due to increased contamination possibilities. Moreover, electrothermal AAS is only a sequential single element method and suffers generally from higher inherent accuracy risks due to its specific sensitivity to a variety of sources of systematic errors (Ref. 1 and 11). Cost requirements for electrothermal AAS are also a factor 2 to 3 higher than for a voltammetric multi-mode device (Ref. 10). Thus, suitable modes of voltammetry have become the method of choice for trace metal analysis with respect to determination and speciation of dissolved trace metals for all types of natural waters. Here the voltammetric approach can easily fully display all its favourable potentialities requiring only a minimum of prior sample treatment, which always involves a certain contamination risk, if one has to cope with problems in ultra trace analysis.

Nevertheless, particularly due to its inherent high accuracy degree voltammetry provides also a very reliable route for trace metal analysis of organic and biological matrices if it is combined with a reliable method for the necessary prior sample digestion (Ref. 1, 9, 13). A very satisfactory and also selective approach for the digestion of the suspended particulate matter (together with the filters) is low temperature ashing (LTA) in an oxygen plasma induced by microwave excitation. In this manner volatilisation losses of trace metals are prohibited and selectively biological and organic matter (microorganisms, organic detritus and the organic films on the surface of inorganic particles) as well as easily decomposable inorganic matter (carbonates) are digested. These are the parts of the suspended particulate matter which contain the further amount of trace metals in the sea which may become remobilised under certain conditions, being either converted again into a soluble form or metabolised by filter feeders and fish and gaining thus the potentiality to enter either via the dissolved state or immediately the marine food chains. The ash of the filters with the suspended particulate matter has been subsequently dissolved with HCl, Merck suprapur, and was after adjustment of about pH 2 also subjected to voltammetric analysis (viz. Table 1).

It turned out that at a certain number of sampling stations (about 50% of all stations) the sea water contained a noticeable amount of dissolved organic matter. In those cases the filtrated solutions of the samples require a further pretreatment, because a certain trace metal amount may be bound very strongly, e.g. as metalloproteins and metallothionines or in inert complexes with biopolymers and organophosphates. On the other hand huminates are expected to play an unsignificant role with respect to the required amount and their moderate stability as chelating ligands for Pb in the dissolved state (Ref. 14 and 15). The strongly bound Pb-amount, which exceeded on average not more than 20% of the dissolved trace metal level, will not be released just by acidification of the filtrate to pH 2. It remains, however, of relevance for the marine food chains as it might be resorbed due to enzymatic attack of the organometallic compounds in filter feeders and also fishes. The decomposition of these inert dissolved organometallics requires simultaneously UV-photolysis (150 W, 200-600 nm, 2-3 h) and acid hydrolysis (0.2 M HNO_3) at elevated temperature (95 ^{o}C). A problem were hitherto the contamination risks embodied in the UV-irradiation procedures for natural water treatment described in the literature. These accuracy risks could be eliminated by a new UV-irradiation device and procedure described in detail elsewhere (Ref. 6).

The whole pretreatment sequence applied to our samples is schematically summarized in Table 1.

<u>The applied voltammetric determination method</u> was DPASV at the MFE. Details of the method and the production, polishing and treatment of the test electrode, which are crucial for its in principle high performance, have been described elsewhere (Ref. 6, 7, 16). For the formation of the mercury film the solution is spiked before deaeration with $Hg(NO_3)_2$ adjusting a concentration of about 3×10^{-5}M. A rotatable specially polished glassy carbon electrode (spinning with 1500 rpm during the cathodic deposition stage) on which in situ simultaneously with the cathodic deposition of Pb, Cd, Cu and, if present, Bi as amalgams a thin mercury film (thickness 200 to 1000 Å) was deposited, constitutes the MFE. The principle of DPASV is shown in Fig. 3. After the cathodic deposition time t_d of usually 6 min the rotation is stopped and during a time interval t_r of 30 s the solution comes to rest. Subsequently an aliquot of the trace metals deposited as amalgams are stripped by

TABLE 1 Analytical Procedure for Sea Water Samples

Sampling

surface water from rubber boat or wooden cutter; 0.5-1 l precleaned and with sea water conditioned polyethylene flasks

0.45 μ Filtration

Closed Sartorius device SM 11306 under N_2-pressure

Filter with Suspended Particulate Matter Digestion by low temperature ashing in micro wave induced oxygen plasma, 140 °C, 4 h

Dissolution in aqueous HCl, Merck suprapur, pH 2

Filtrate

UV-irradiation 150 W, 2h, 95 °C, 50 ml, +1ml 10M HNO_3; subsequently pH1.6-2 with 1.1 ml ammonia, Merck suprapur

Acidification with HCl, Merck suprapur, 1ml/l to pH2, **Standing** for several hours to decompose chelates

Voltammetry by DPASV at MFE

spiking with $Hg(NO_3)_2$, 3×10^{-5}M; Deaeration (10-15 min) N_2 99.999%; Cathodic deposition at in situ formed Hg-film, E_d -1.0V(SCE), t_d 6 to 10 min, 1500 rpm; Simultaneous determination of Cd, Pb, Cu (Bi), Evaluation by 2 standard additions

Results

Trace metal amount bound to particulate matter + Dissolved Trace Metal Concentration = Total Amount of Trace Metal in Sample

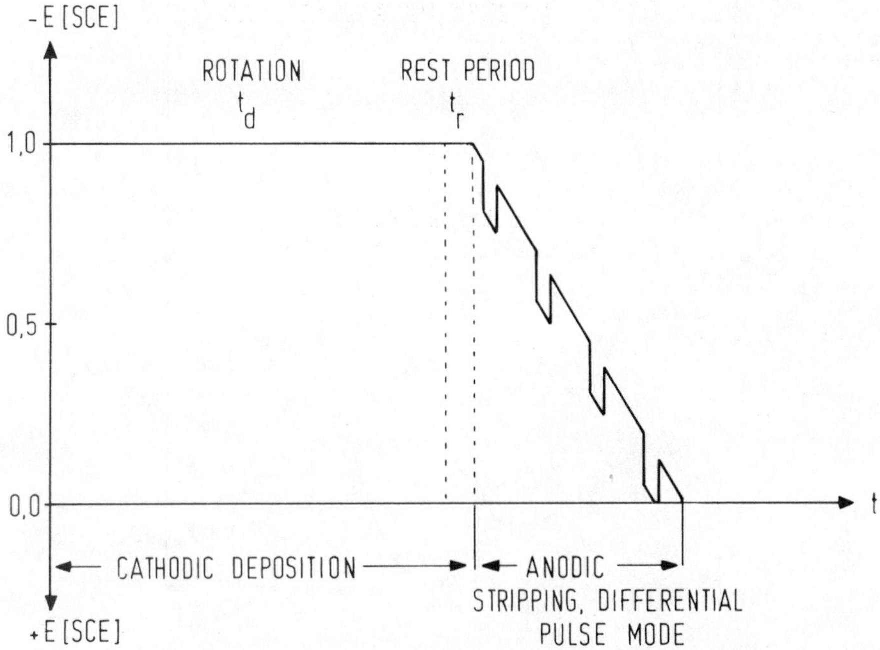

Fig. 3. Principle of Differential Pulse Anodic Stripping Voltammetry (DPASV).

reoxidation in the course of the applied pulse train (Ref. 16). The height of the resulting peaks (viz. Fig. 4) is proportional to the bulk concentration of the trace metals. For the evaluation of the concentration values the standard addition approach is mandatory in connection with the use of a MFE. Usually two standard additions are sufficient. The deposition time t_d is lowered corresponding to the increase in bulk concentration resulting from the standard additions. In this manner the amounts of the trace metal amalgams are kept approximately constant. Thus, all possible interferences resulting from overloading of the mercury film with trace metal are avoided. A further advantage of this standard addition procedure is that the overall time for the voltammetric determination is minimized, an important aspect for the analysis of large sample numbers (Ref. 7, 16, 17). In the present study simultaneously 2 polarographs (PAR 174A) with 4 cells were operated manually to speed up the determination rate. While in two cells voltammetry was going on, in the other two cells deaeration or rinsing was performed, respectively. In this way up to 20 samples have been analysed within 12 hours.

Further automation of all procedures including sample dosage and standard additions - already practised now for voltammetric on-line monitoring of toxic trace metals in drinking water from water works mains (Ref. 18) - will accelerate the analysis considerably. The determination limit of this highly sensitive voltammetric method goes down to the ng/kg-level. Precision and accuracy remain remarkably good even at this ultra trace level. In principle a voltammetric RSD of only $\pm$ 2.5% is attainable down to 50 ng/kg (Ref. 7). The overall precision in the present study including all steps from sampling to voltammetric determination was better than $\pm$ 8%. At the determination limit of 1 ng/kg RSD remains still $<$ 15%.

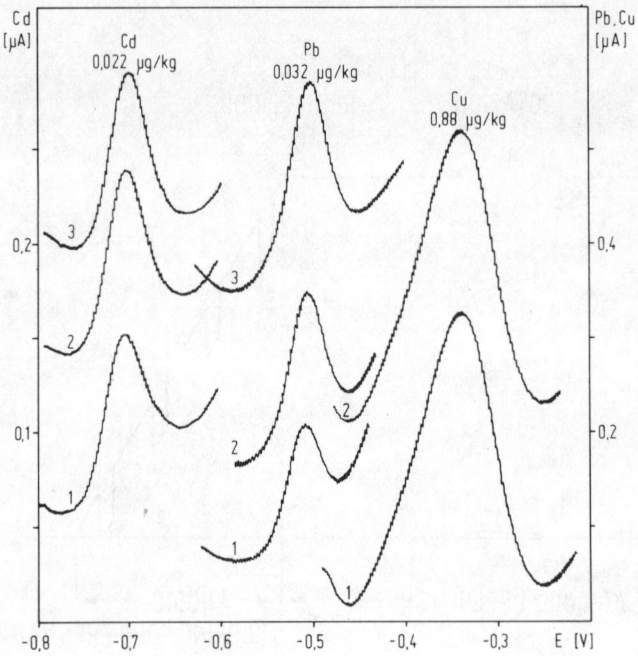

Fig. 4. Voltammograms of simultaneous determination of Cd, Pb and Cu by DPASV in 0.45 µ filtrated sea water (sample weight 56.4 g).
1 Original sample, t_d 6 min (Cd, Pb), 3 min (Cu)
2 First standard addition, t_d 3 min (Cd, Pb), 1.5 min (Cu)
3 Second standard addition, t_d 2 min (Cd, Pb)
Standard additions from concentrated stock solutions by µ l-pipets were equivalent to 1.6 ng Cd, 4 ng Pb and 50 ng Cu.
For details of procedure and evaluation see Ref. 6, 7, and 17.

RESULTS

Total, Dissolved and Particulate Matter Bound Pb-Levels

In this way the contours of the Pb-inventory as well as that of Cd and Cu in coastal waters from the studied regions of the Western Mediterranean, the Belgian and Dutch North Sea Coast (Southern Bight), the German North Sea Coast and the German Bight have been established by a large number of reliable data (Ref. 6). The results for the Ligurian and Tyrrhenian coast and the Belgian and Dutch Southern Bight coast are compared in Fig. 5. For the other studied areas of the North Sea coast analysis of the Pb-amount bound to suspended particulate matter is still in progress and thus only preliminary conclusions based on the already analysed level of dissolved Pb can be communicated for the German North Sea coastal region and the German Bight. The situation of the Pb-inventory and the fate of this trace metal is quite different along the Ligurian and Tyrrhenian coast, the Belgian and Dutch Southern Bight and the German North Sea coastal region. The hydrographic situation is much more complicated in the North Sea regions and exerts significant influences on the respective Pb-level particularly in Southern Bight coastal waters. The kind of tide (low or high) and the respective phase of the tide during sampling is a parameter influencing the actual levels of the Pb-levels. Also the less stable weather situation in the North Sea coastal areas will have influences, due to significant differences in wave strength affecting the distribution degree of polluted water plumes. However, despite this more complicated situation also the contours and ranges of the Pb-inventory in this region seem to emerge from the obtained data.

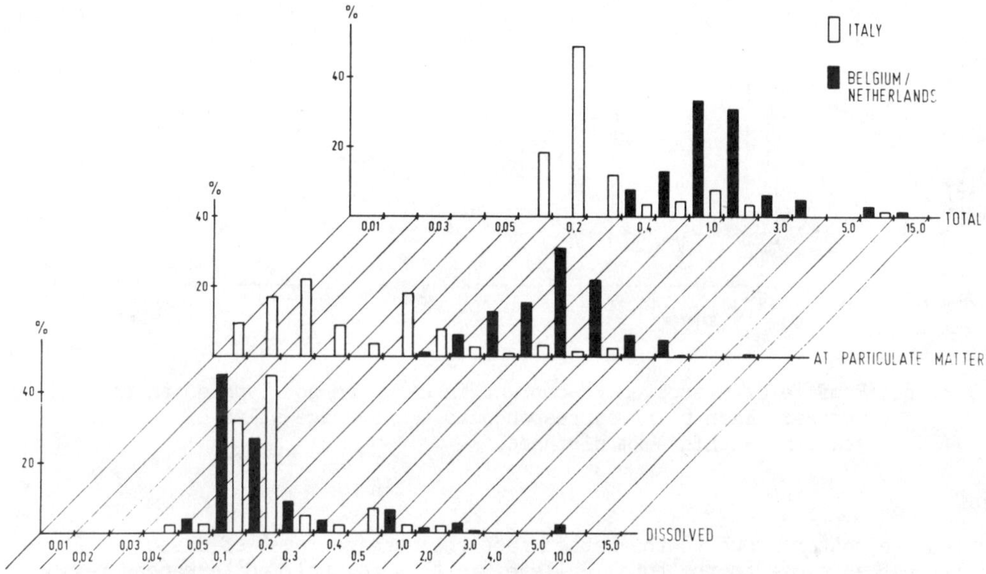

Fig. 5. Percentage distribution of Pb-levels over sampling stations in waters of Ligurian and Tyrrhenian and Southern Bight North Sea coast. The Pb-content of particulate matter is related to µg per kg sea water, as the whole content of suspended particulate matter in 1 kg sampled sea water was collected on the filters. The bars indicate the respective percentage of sampling stations yielding Pb-levels corresponding to the respective interval on the abscissa. Thus, no interpolation within the intervals is possible, as the position of the bars inside their corresponding intervals is arbitrary.

Ligurian and Tyrrhenian Coast

In the studied Western Mediterranean coastal section Pb occurs predominantly in the dissolved state. The reason is that here the amount of suspended particulate matter is usually very low. Sampling stations were 100 to 500 m off the coast. Sampling profiles up to 3 km off the coast at a number of locations yielded no significant differences in the level of dissolved Pb. As Fig. 6 reflects the dissolved Pb-level remains here also rather stationary with the exception of big port areas (Genova and La Spezia) and some estuaries to be discussed later. The increase in the region of Piombino is not of anthropogenic origin but seems to reflect a natural input due to a geological anomaly.

The reason for the observed general behaviour is the absence of significant tidal currents and resulting movements of water mass in the Mediterranean which has the typical oceanographic behaviour of a large inland sea. The resulting overall average value for the dissolved Pb-level of 130 ng/kg is typical for coastal waters with a still rather low Pb-pollution and containing only small amounts of suspended

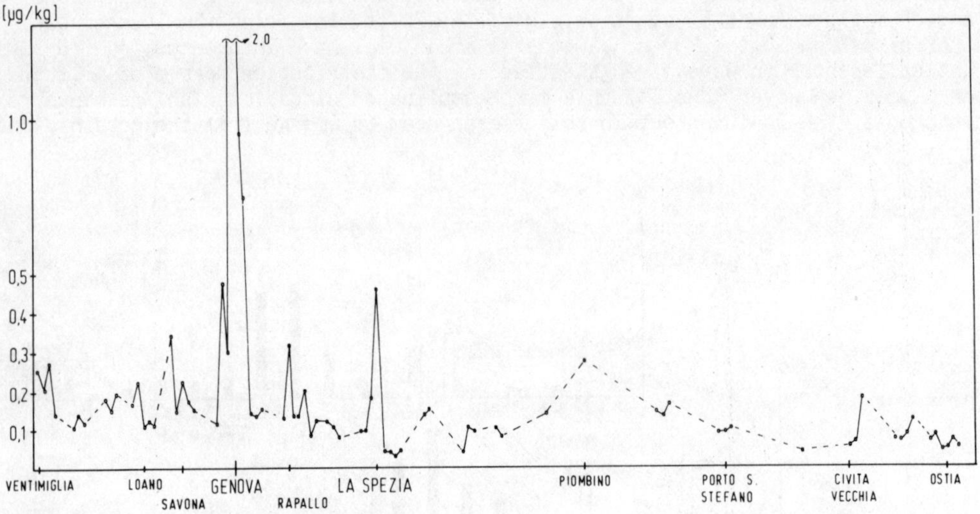

Fig. 6. Profile of dissolved Pb-content in Ligurian and Tyrrhenian coastal waters. Each point corresponds to the average value of sampling stations within 4 km distance.

particulate matter. Due to the latter fact the average level of dissolved Pb is usually rather close to the total Pb-level in this coastal Mediterranean region (viz. Fig. 5). Repetition of the measurements in this coastal section in the subsequent year at a different time yielded, except for big port areas, the same Pb-level confirming the rather stationary situation for this region and this type of coastal waters.

Pb has contrary to Cd in principle a pronounced tendency to be bound to suspended particulate matter. This property of Pb is well reflected by the data from the Belgian and Dutch Southern Bight coastal region (viz. Fig. 5, 7), where the water contains a significant amount of suspended particulate matter. The Pb-content of the particulate matter resulting from the filtration of 1 kg sea water is here usually a factor 8 higher than the Pb-concentration in the dissolved state. The absolute values of the total Pb-level are also significantly higher, i.e. by a factor 5 compared with the Ligurian and Tyrrhenian coastal waters, while the average value of dissolved Pb remains of the same order with 0.11 μg/kg in Belgian and Dutch Southern Bight coastal waters (viz. Fig. 7).

Belgian and Dutch Southern Bight

There are several reasons for the higher Pb-burden in the Belgian and Dutch Southern Bight coastal waters. There is a continuous anthropogenic Pb-input via the polluted rivers Scheldt and Rhine and from industrial activities along the coast. Moreover, the significant tidal currents alternating in direction with low and high tide (Ref. 19) transport clouds of polluted water up and down the coast (viz. Fig. 8). The water remains in continuous movement and this retards sedimentation of Pb-loaded suspended particulate matter and increases the residence time of Pb-pollution in the water in this way. The different behaviour of the suspended particulate matter loaded waters from a river flowing into coastal waters where tidal currents of the discribed kind are absent and present, respectively, is schematically depicted in Fig. 9. This scheme indicates the principles of the estuarian situation in the

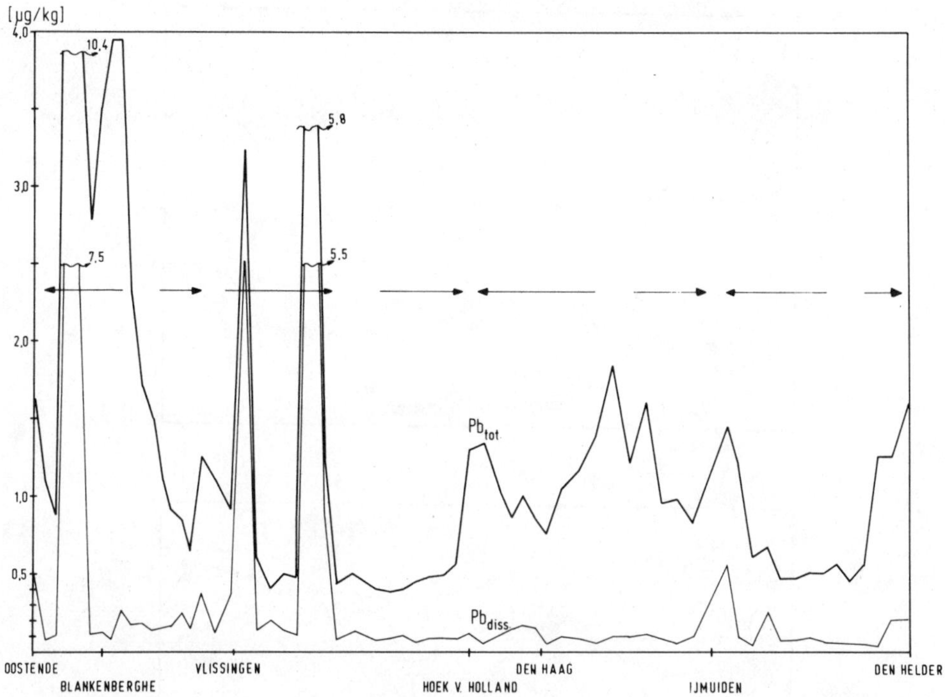

Fig. 7. Profile of dissolved and total Pb-level in Belgian and Dutch Southern Bight coastal waters. Arrows indicate direction of tidal current during sampling (⟵ high tide), (⟶ low tide).

studied Western Mediterranean and Belgian-Dutch Southern Bight coastal zones, respectively. Besides the strong tidal currents periodically altering direction there is a continuous slow residual current going along the Belgian, Dutch and German coast up to Jutland transporting and slowly distributing polluted waters from the Channel and Southern Bight coast into the latter areas.

Figure 7 gives an impression of the typical behaviour of the Pb-level in Southern Bight coastal waters as function of the respective hydrographic situation at the various sampling stations. In this context it has to be emphasized that within a certain range the actual Pb-level will fluctuate according to the given situation during sampling. Usually the sampling stations were between 0.3 and 0.5 km off the coast.

Off Oostende samples were taken during the beginning phase of high tide. As is seen from Fig. 8 one has then a with time progressively increasing tidal current down the coast. This current drives the polluted water plume which had been released from Oostende area during the preceding low tide period down the coast. Samples were taken obviously in the tail region of this plume (viz. Fig. 9) as the progressive decrease in the total Pb-level for stations up the coast reflects.

However, off Blankenberghe influences of a special situation predominate. Here prevails circulation of rather turbid waters containing a relatively high amount of particulate matter (viz. Fig. 10). This water circulation causes the built up of a

a)

b)

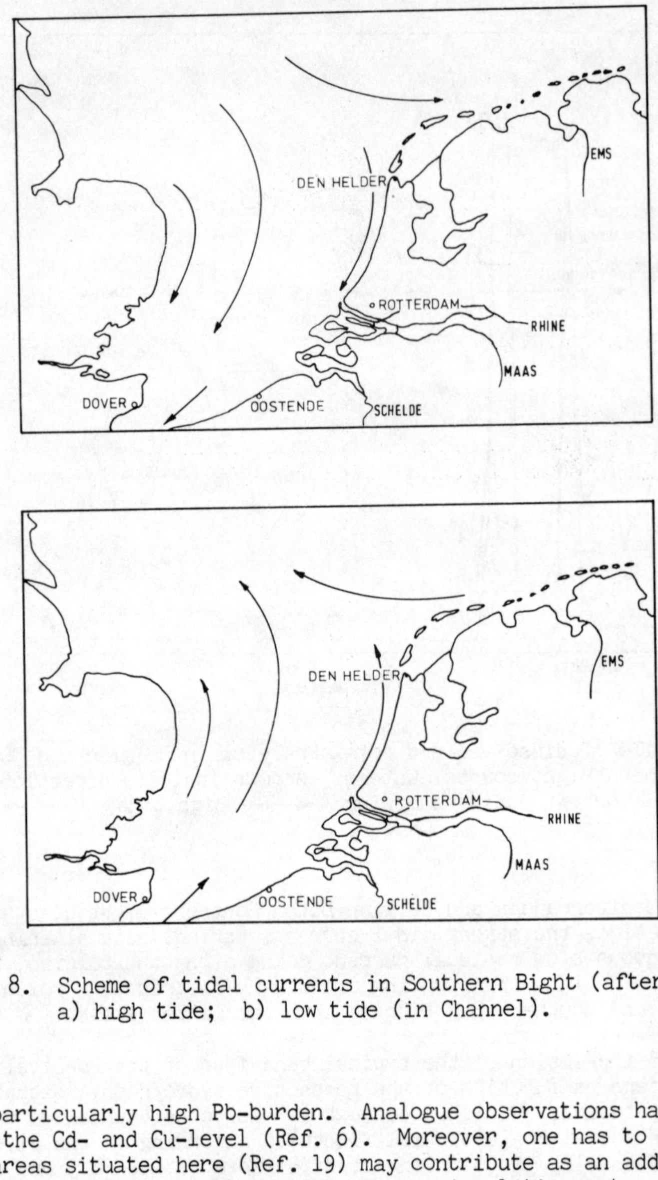

Fig. 8. Scheme of tidal currents in Southern Bight (after Ref. 19) a) high tide; b) low tide (in Channel).

patch with a particularly high Pb-burden. Analogue observations have been made in this area for the Cd- and Cu-level (Ref. 6). Moreover, one has to take into account that dumping areas situated here (Ref. 19) may contribute as an additional continuous source of toxic trace metal input to these circulating waters. Proceeding up the coast the Pb-level decreases again as the area of circulating waters is left and sampling is again performed in the tail region of the beforementioned plume of polluted waters transported during high tide down the coast.

The peak values observed off Walcheren after Vlissingen refer to sampling during low tide. Here the effect of a cloud of polluted water originating from the Wester Scheldt estuary and transported with the low tide current around Walcheren is reflected. It is to be mentioned that in this zone the sampling stations were be-between 3 and 10 km off the coast.

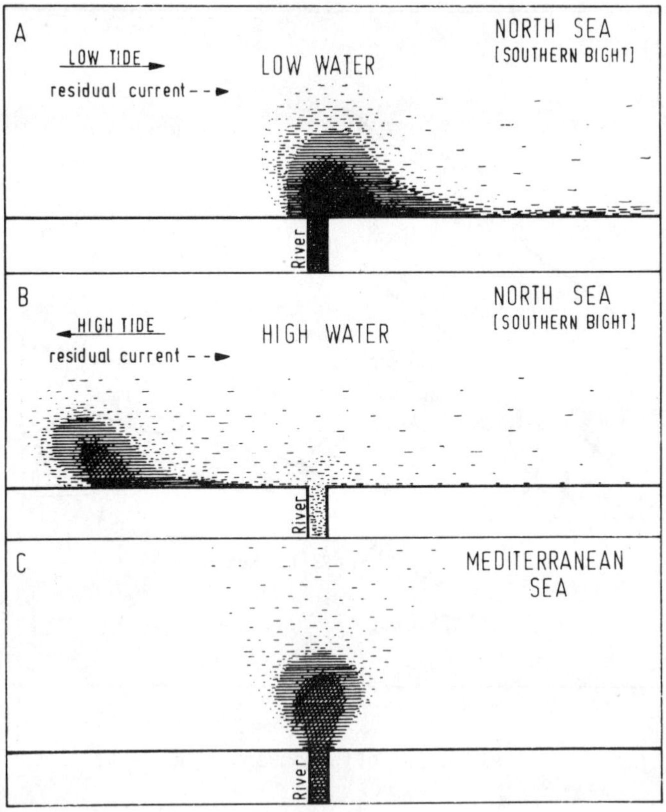

Fig. 9. Fate of waters from river mouth (schematic representation). The large plume released at this tidal phase is depicted. The tail corresponds to the smaller output released during preceding low tide and transported up the coast (viz. Fig. 8).

a) Situation at Belgian-Dutch Southern Bight coast for low water.
b) Situation in same region at high water. During preceding high tide the plume released before at low water had been transported down the coast, changes direction at high water and is during then progressing low tide again transported back up the coast.
c) Mediterranean Sea, predominantly stationary coastal waters.

The increased Pb-level between Hoek v. Holland and Den Haag (sampling stations again as usual 0.3 to 0.5 km off the coast) is to be attributed to the pollution from the Rhine estuary. Sampling was performed during begin of high tide. Thus, plumes of polluted water are driven down the coast. They originate from the output of the Rhine estuary during the preceding low tide period and are therefore already somewhat diluted. This tendency is also reflected by the progressive decrease of the Pb-level up the coast, as one proceeds with sampling in the decreasing part of the plume tail (viz. Fig. 9).

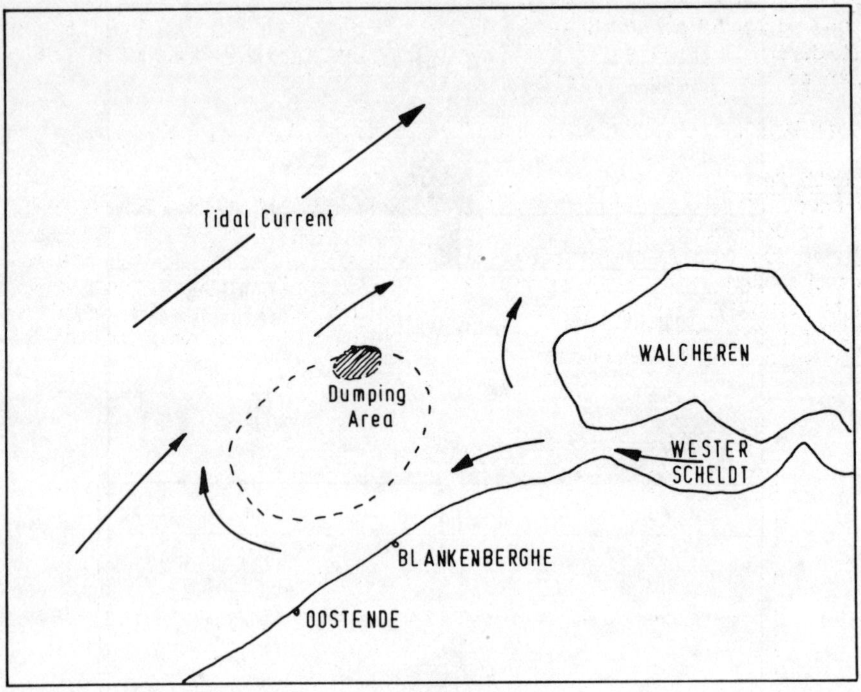

Fig. 10. Hydrographic situation off Blankenberghe (after Ref. 19).

Sampling in the zone off Noordwijk to Ijmuiden has started at the turning point from high to low tide. Thus, one has during progressing low tide an increasing tidal current up the coast (viz. Fig. 8). The source of substantial pollution is the industrial area around Ijmuiden. But the polluted water leaving Ijmuiden during the present low tide is transported up the coast and will thus not affect the sampling stations in the studied area. The polluted water actually sampled there originates, however, from the pollution plume moved down the coast from Ijmuiden during the preceding high tide period (viz. Fig. 8 and 9). When the tide turned (begin the sampling) the front of the polluted water patch released at Ijmuiden at the begin of the prior high tide period had reached a location off Noordwijk and starts now to travel back. Obviously one has under these circumstances to expect a trend for the total Pb-level as is observed. The relative highest value is obtained for the station corresponding to the turning time of the tide. The values for stations closer to Ijmuiden and corresponding to later parts of the low tide period have progressively decreasing total Pb-values, as they are situated in less polluted waters behind the front of the polluted water patch, which had moved down the coast during the prior high tide and travels now more slowly than the sampling vessel. In this context it is to be emphasized that these changes in the total Pb-level, connected with the influences of the tidal currents during the sampling period and the preceding high tide period, are manifested mainly in the alterations of the Pb-amount bound to particulate matter (viz. Fig. 7). This is due to the pronounced general tendency of Pb to be bound to particulate matter, which is present in significant amounts in these coastal regions.

Off Ijmuiden started a new sampling section with different tidal current conditions.

Sampling began in the initial phase of high tide. Consequently the vessel moves against the plume of polluted waters which had been transported from Ijmuiden up the coast during the preceding low tide period and the behaviour of the Pb-profile is similar to that described before.

Finally off Den Halder the plume from the previous high tide period is reached.

German North Sea Coast

The occurence, level and fate of Pb is again different along the German North Sea Coast and in the German Bight. Also here strong tidal currents are operative (viz. Fig. 11) but due to their direction (Ref. 19) they carry out at low tide polluted waters to the open North Sea and at high tide they drive large amounts of unpolluted water to the coastal area which cause a significant dilution of the polluted water. It should be noted that for this area discussion can be based only on the already determined values of dissolved Pb.

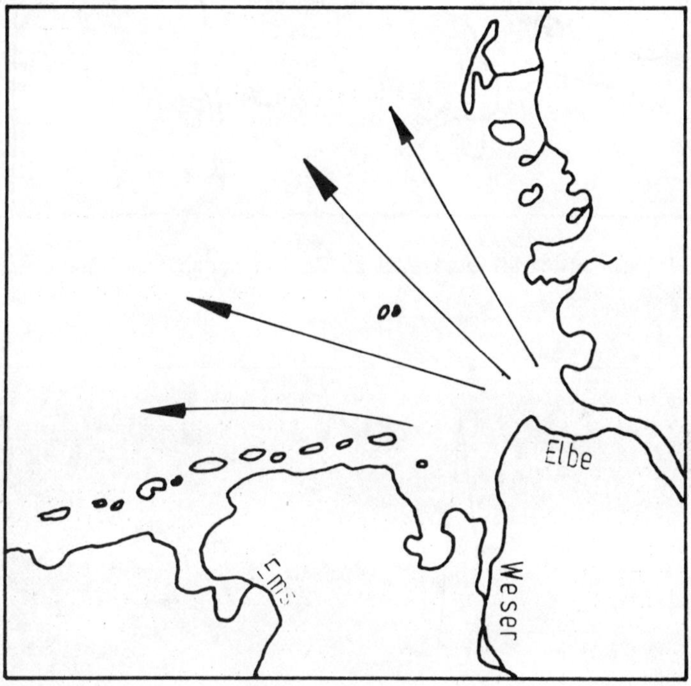

Fig. 11a. Scheme of tidal currents at German North Sea Coast (after Ref. 19); low tide.

In Fig. 12 the Pb-profile resulting for sampling stations from Emden along the East Frisian islands until Cuxhaven is shown. Stations were usually about 3 km off the coast or the isles, respectively. The dissolved Pb-level is generally rather low around 40 ng/kg until Spiekeroog. Then it drops progressively even to values below 20 ng/kg with a minimum value of 13 ng/kg at Scharhörn, Reef (NW of the island Scharhörn). These values are about a factor 6 to 10 lower than the average

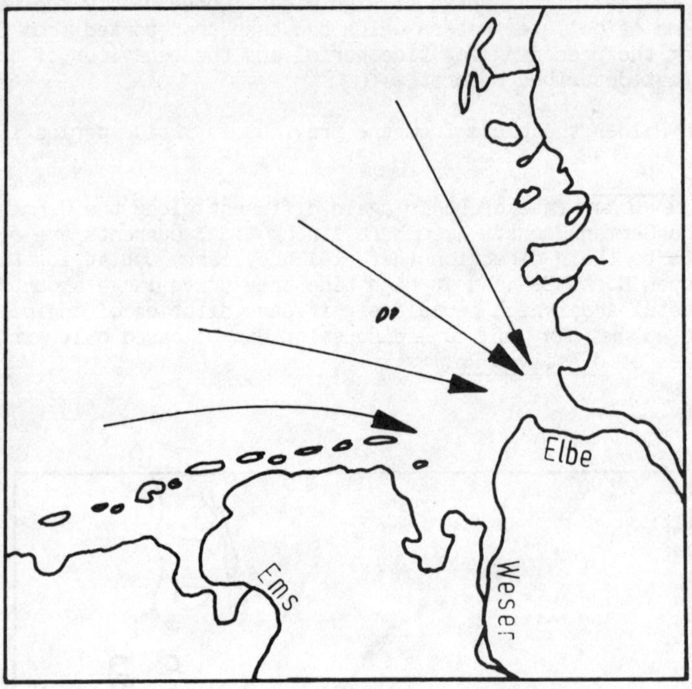

Fig. 11 b. Scheme of tidal currents at German North Sea Coast (after Ref. 19); high tide.

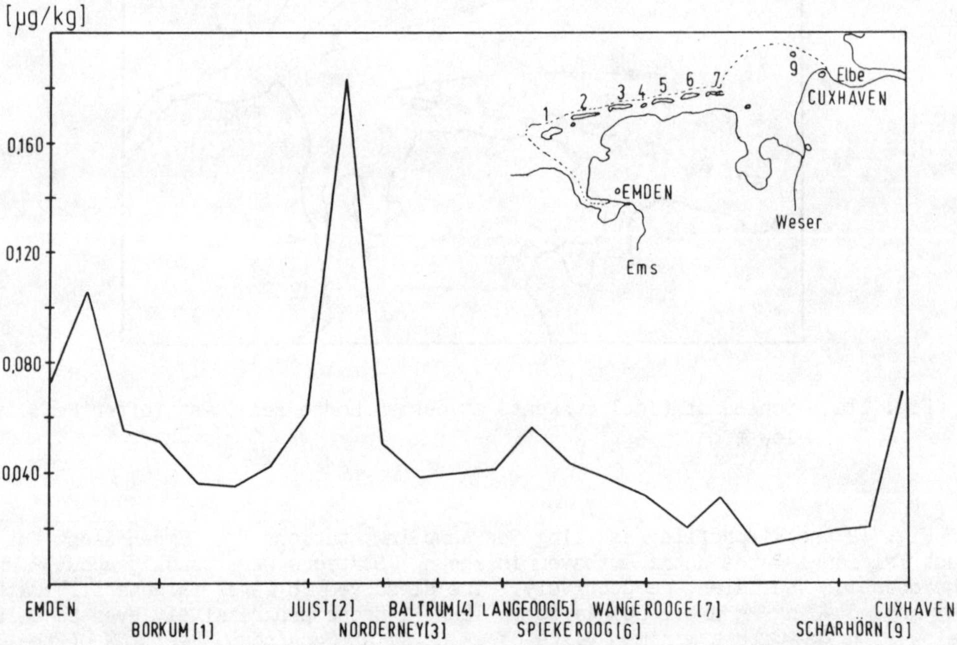

Fig. 12. Profile of dissolved Pb-level for sampling mission outside East Frisian Isles from Emden to Cuxhaven.

level for dissolved Pb observed in the Belgian and Dutch Southern Bight coastal waters (viz. Fig. 5 and 7). No effect from the certainly existing pollution of the rivers Weser and Elbe is noticeable at the sampling stations indicated by the dotted line in the map in Fig. 12.

Although the level of dissolved Pb is significantly higher in the rivers, e.g. 100 to 260 ng/kg in the Weser (Ref. 17), about one order of magnitude lower values are obtained in the other part of their estuaries. Obviously there exists a rather efficient and rapid binding to the substantial amount of particulate matter brought into the estuaries by the rivers. This to be assumed efficient scavenging of Pb is further supported by the algae population observed in these waters, which certainly contain sufficient nutrients to stimulate algae growth. The increases in the Pb-level near Cuxhaven and near Emden reflect the growing influence of pollution in the inner part of the respective estuaries. However, it should be noted that, due to sampling during high tide at these locations, the increase of the Pb-level is moderated by the input of water from the open sea. The described magnitude of the level of dissolved Pb has been confirmed by results of several further missions shown in Fig. 13. The increase off Hooksiel is connected with the shipping route to the industrial area at Wilhelmshaven.

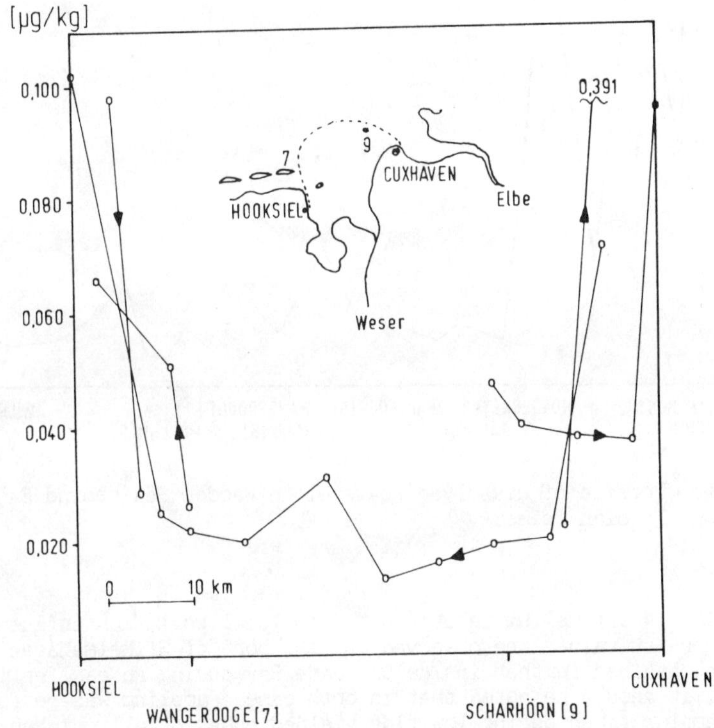

Fig. 13. Profiles of dissolved Pb-levels from several missions from Jade Bay via Weser Estuary to Cuxhaven (Elbe Estuary).

A special pollution situation is manifested by the peak value of 180 ng/kg between the East Frisian Isles, Juist and Norderney in Fig. 12. This pollution is also observed in the Wadden Sea behind these islands to an even more pronounced extent (viz. Fig. 14). Here the results from stations (1-2 km from the isles) in the Wadden Sea are shown. The highest Pb-pollution is observed in the vicinity of the small island Memmert close to the Isle of Juist. Due to the course of the tidal currents directed mainly NW passing thus the West coast of Borkum this significant pollution can not be connected with the certainly existing input from the Ems estuary (Dollart). In eastern direction from this polluted Wadden Sea area the values drop significantly to an average of about 40 ng/kg. The peak near the island Spiekeroog is connected with the shipping route to a fishing harbour at the coast (Neuharlingersiel). These influences are also visible at the corresponding seaward sided station (viz. Fig. 12).

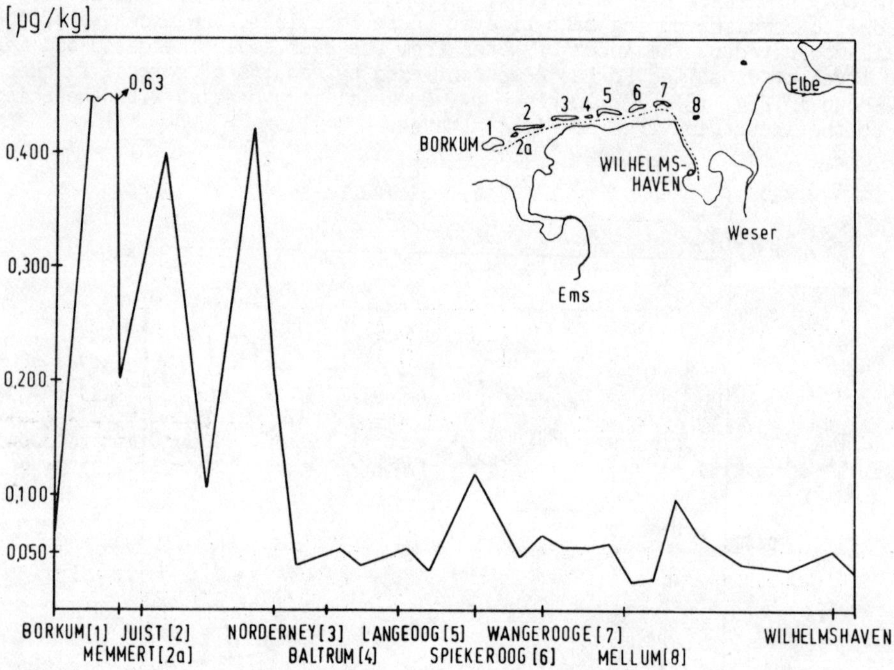

Fig. 14. Profile of dissolved Pb-level in Wadden Sea behind East Frisian Isles.

The peak in Fig. 14 off Mellum is attributed to local pollution influences by ship traffic. Rather low values are observed off the port of Wilhelmshaven. The same level has been obtained further inside the Jade Bay during an earlier mission (Ref. 17). However, it should be noted that in both cases sampling was performed during high tide. Samples taken during low tide yielded substantially higher values for dissolved Pb around 1000 ng/kg reflecting impressively the pollution carried out at low tide from the industrial zone of Wilhelmshaven.

Figure 15 reports the results for the dissolved Pb-level off the coast (10 to 20 km) of Schleswig-Holstein from the area of the North Frisian Isles to Cuxhaven. The peaks off St. Peter-Ording and in the outer Eider Estuary indicate a certain pollution. The increase off Cuxhaven reaches again the same range as in Fig. 12.

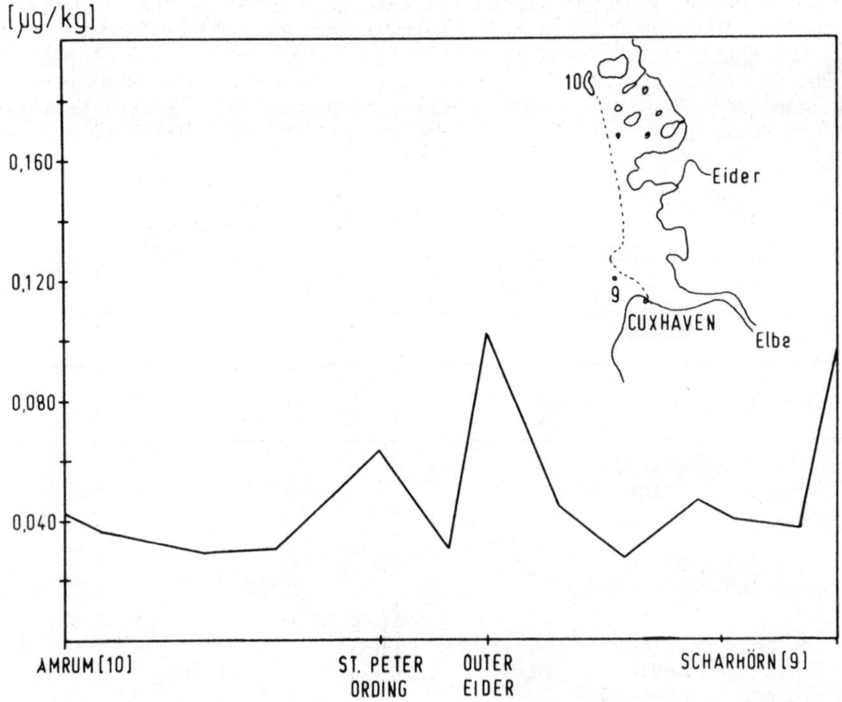

Fig. 15. Profile of dissolved Pb-level along coast of Schleswig-Holstein to Cuxhaven.

In the German Bight around Helgoland the dissolved Pb-level had a remarkably low value $\leqslant 25$ ng/kg, except one station (north of the passage between the island and the Dune) where the dissolved Pb-concentration considerably increases to 166 ng/kg.

GENERAL REMARKS

It is emphasized that, compared with older data in the literature on Pb-levels in coastal waters, the values determined by us are usually lower. This reflects not a decrease of the Pb-level in the sea over recent years but has to be ascribed to the achieved significant progress in trace analytical procedures eliminating the particularly for Pb high contamination risks during sampling, sample storage and pretreatment and the determination stage. This tendency to obtain more accurate Pb-data in sea water by careful and rigorously all contamination sources eliminating work has been recently also demonstrated by other laboratories with particular expertise and long experience in ultra trace analysis, as for instance Patterson et al. (Ref. 20, 21).

The demands with respect to rigorous contamination elimination become even significantly more stringent if deep water samples from the open sea have to be analysed as there the level of dissolved Pb decreases to only several ng/l. An example is the Pb-level of only 3 ng/l in sea water from 1000 m depth, 120 km off the central Californian coast. This value, first determined by Patterson et al. (Ref. 21) with isotope dilution mass spectrometry after prior dithizone extraction, could be confirmed by us recently in direct measurements with DPASV at the MFE (viz. Table 2). The precision of DPASV is comparable to the method used by Patterson et al. (Ref. 21), but the application of DPASV has the advantage that simultaneously the data for Pb, Cd and Cu are obtained and that the whole analysis is much more rapid requiring less than 2 h. This is a further impressive example for the high potentialities and reliability of suitable modes of the voltammetric approach for marine trace metal analysis.

TABLE 2 Comparison of Results Obtained with Different Trace Analytical Methods for Dissolved Trace Metals in Sea Water Sample from 1000 m Depth 100 km off Central Californian Coast (37°7'N, 123°30'W)

Method	Cd	Pb ng/kg	Cu	Investigator
Extraction with CH_3Cl-dithizone and isotopic dilution mass spectrometry	-	3.3 ± 0.2	-	a)
APDC-DDDC-Extraction and electrothermal AAS	105	-	115	b)
DPASV at MFE, pH 2 (HCl), t_d 9 min, 2 standard additions, simultaneous determination	101 ± 4	2.9 ± 0.2	110 ± 8	c)

a) B. Schaule and C. C. Patterson, Div. Geol. Plant. Sci., California Institute of Technology, Pasadena, Cf., USA

b) K. W. Bruland, UC Santa Cruz, California, USA

c) L. Mart, our institute

The sea water sample has been made available to us by C. C. Patterson. It had been removed from a sample taken in April 1977 after storage at 4 °C, unacidified, for 3 1/2 months.

With respect to our investigations on surface coastal waters the average value of 112 ng/kg for the dissolved Pb-level of 80% of our sampling stations (viz. Fig. 5 and 7) along the Belgian and Dutch Southern Bight coast is still lower than the average value of 400 ng/kg reported for the waters off the respective Dutch coast line by Duinker and Nolting (Ref. 22). It is open to what extent this difference is to be related to fluctuations in Pb-pollution in this coastal region as consequence of different weather conditions and of source strength of Pb-pollution. However, no confirmation of the rather high levels of dissolved Pb after acidification reported later by Duinker and Kramer (Ref. 23) has been obtained. On the contrary their average value of dissolved Pb determined in non-acidified samples agrees rather well with our average of 112 ng/kg measured after acidification to pH 2.

Of course, significantly higher Pb-levels may be encountered in parts of the sea where a high permanent heavy metal pollution is maintained, due to continuous significant anthropogenic metal input and absence of purification effects with respect to the given hydrographic situation. A wellknown example are coastal waters in the Irish Sea, where depending on the sampling points Pb-levels between 0.3 and more than 5 μg/l have been reported (Ref. 24) (viz. also our high values at some sampling stations in Fig. 5).

Details of Various Typical Local Situations

The foregoing general considerations on the determined Pb-level in the studied sections of European coastal waters have omitted various particular local situations studied in detail. From the material assembled during our extended study, 3 different typical examples shall be briefly described to show how sensitively and precisely the applied voltammetric method monitors relative differences in the trace level of dissolved Pb, which becomes in this way a good indicator of local pollution and the influences of local ecological parameters. Figure 16 shows the profiles of dissolved Pb and Cd across the main shipping routes into the big port of Genova. The peak-shaped increase is connected with the ship traffic and remains localized to the shipping routes and their immediate vicinity decreasing rapidly to the common average value for the Ligurian coast. In this context, incidentally, also the small peak in the Pb-level across the entrance route to the yacht harbour of Portofino is significant. The strong correlation of the heavy metal pollution to the density of ship traffic to and from Genova is well reflected by the dotted lines which refer to the rather restricted ship traffic on sundays.

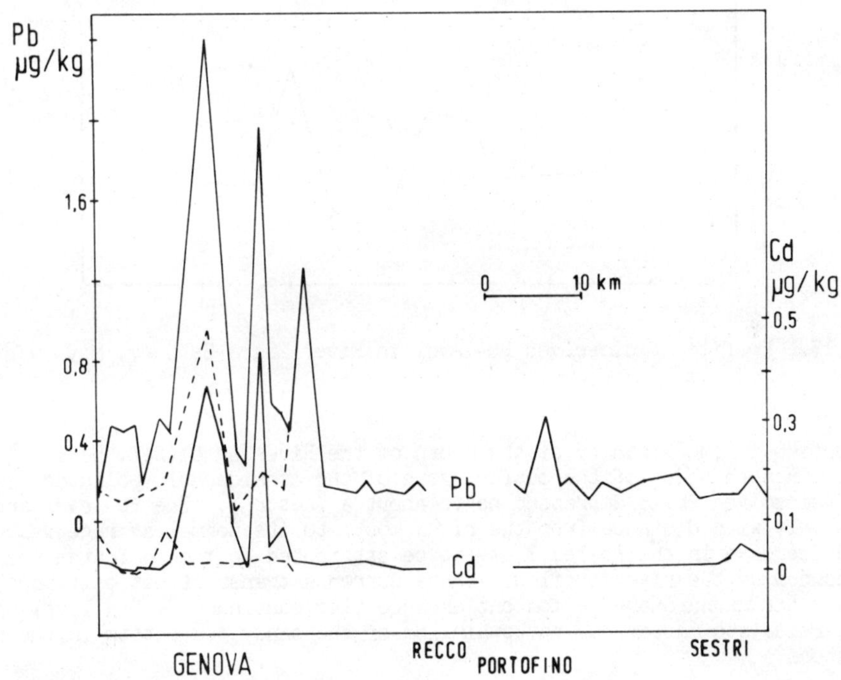

Fig. 16. Profile of dissolved Pb- and Cd-level across shipping routes in port region of Genova, full lines working day, dotted lines sunday.

The course of the level of dissolved Pb in Fig. 17 reflects the fate of this trace metal in the River Magra estuary (Ref. 5). The reason for the by about a factor 3 decreased concentration of Pb, compared with the normal average level, is pronounced algae growth in the estuary due to the continuous input of phosphate rich water by the river. The algae act as efficient scavengers for dissolved Pb. Nevertheless, the shipping route into the port of Marina di Carrara remains indicated by the peaks in the depressed Pb-profile.

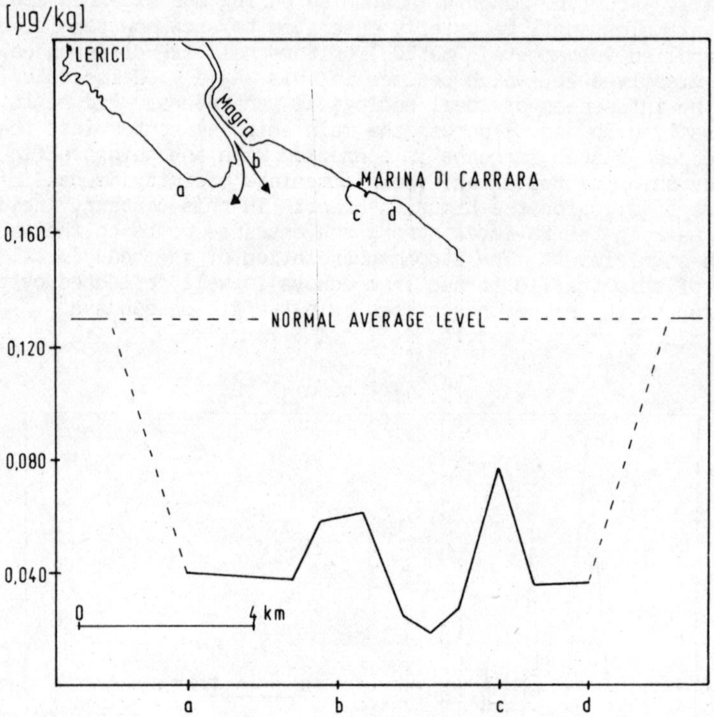

Fig. 17. Profile of dissolved Pb-level in River Magra Estuary, May 1976.

The situation in the vicinity of the mouth of the River Arno (Ref. 5) is shown in Fig. 18. On both sides of the outflow zone of the considerably polluted river the level of dissolved Pb is depressed up to about a factor 3. The Pb-level increases progressively with distance from the river mouth to its normal average value. The observed decrease in the Pb-level has to be attributed to the indicated local currents induced by the river outflow. These currents transport not only unpolluted offshore water to the shore region but enhance also continuously the level of suspended particulate matter due to upwhirling of the sandy sea bottom in the shallow shore waters.

CONCLUDING REMARKS

The reported extended studies emphasize the necessity of these investigations, based on a rather dense network of sampling stations, to elucidate global trends and local details on the occurrence and fate of toxic trace metals, to establish the pattern of their inventory in European coastal waters as function of the geographical, oceanographic and ecological parameters and to monitor the pollution level. The trace level of the toxic metals to be determined and the necessity to analyse large numbers of samples require powerful and reliable trace analytical methods, as suitable modes of voltammetry, which combine the basic prerequisite of high sensitivity with good accuracy and precision, which have a satisfactory and by future automation easily increasable rapidity and which have a favourable cost/benefit ratio.

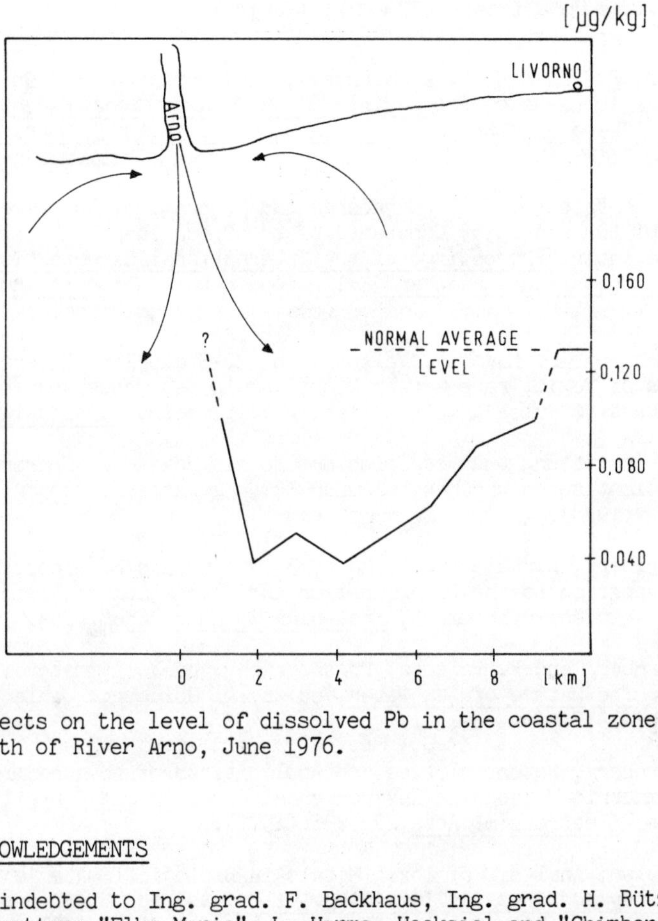

Fig. 18. Effects on the level of dissolved Pb in the coastal zone around mouth of River Arno, June 1976.

ACKNOWLEDGEMENTS

The authors are indebted to Ing. grad. F. Backhaus, Ing. grad. H. Rützel and the captains of the cutters "Elke-Marie", L. Harms, Hooksiel and "Chimborazzo", K. H. Salgert, Veere, for manifold and valuable technical assistance during the various missions.

These studies are in part contributions to the UNEP/FAO program "Pollution in the Mediterranean" and these parts have been performed, based on our institute's Stazione Marina at Fiascherino, Italy. Financial support contributed to other parts of this study by the European Commission, Bruxelles, in the frame of the project "Heavy Metal Pollution in European Seas", contract No. EC-173-77-1 ENVD, is gratefully acknowledged.

REFERENCES

(1) H. W. Nürnberg, Polarography and Voltammetry in Studies of Toxic Metals in Man and His Environment, Pure Appl. Chem. (in press).

(2) H. W. Nürnberg, Potentialities and Applications of Advanced Polarographic and Voltammetric Methods in Aquatic and Marine Trace Metal Chemistry, Proc. Int. Symp. "Structure and Dynamics in Chemistry" Uppsala 1977, Acta Universitatis Upsaliensis 12, 270-307 (1977).

(3) B. Raspor, P. Valenta, H. W. Nürnberg,and M. Branica, The Chelation of Lead by Organic Ligands, Proceedings of the International Experts Discussion on the "Lead - Occurrence, Fate and Pollution in the Marine Environment", Rovinj 1977, M. Branica and Z. Konrad (eds.), Pergamon Press, 1979,181-195.

(4) L. Sipos, P. Valenta, H. W. Nürnberg, and M. Branica, Voltammetric Determination of the Stability Constants of the Predominant Labile Lead Complexes in Sea Water, Proceedings of the International Experts Discussion on the "Lead -Occurrence, Fate and Pollution in the Marine Environment", Rovinj 1977, M. Branica and Z. Konrad (eds.), Pergamon Press, 1979, 61-76.

(5) H. W. Nürnberg, L. Mart, P. Valenta, and M. Stoeppler, Determination of Levels of Toxic Trace Metals Dissolved in Sea Water and Inland Waters by Differential Pulse Anodic Stripping Voltammetry, Thalassia jugosl.14, 171 (1978).

(6) L. Mart, Ermittlung und Vergleich des Pegels Toxischer Spurenmetalle in Nordatlantischen und Mediterranen Küstengewässern, Doctor Thesis, RWTH Aachen (1978).

(7) H. W. Nürnberg, P. Valenta, L. Mart, B. Raspor and L. Sipos, The Polarographic Approach to the Determination and Speciation of Toxic Trace Metals in the Marine Environment, Fresenius Z. Anal. Chem. 282, 357 (1976).

(8) H. W. Nürnberg and P. Valenta, Polarography and Voltammetry in Marine Chemistry in The Nature of Sea Water, ed. E. D. Goldberg, Dahlem-Konferenzen, Berlin, 1975, pp. 89-136.

(9) H. W. Nürnberg, Potentialities and Applications of Advanced Polarographic and Voltammetric Methods in Environmental Research and Surveillance of Toxic Metals, Electrochim. Acta 22, 935 (1977).

(10) M. Stoeppler, Analysis of Toxic Metals in Biological Materials, Clinical Chemistry and Chemical Toxicology of Metals, ed. S. S. Brown, Elsevier/North Holland Biomedical Press, Amsterdam, 1977, p. 307.

(11) M. Stoeppler, Present Potentialities and Limitations of Atomic Absorption Spectroscopy in Environmental and Biological Materials with Particular Reference to Lead Determinations, Proceedings of the International Experts Discussion on the "Lead - Occurrence, Fate and Pollution in the Marine Environment", Rovinj 1977, M. Branica and Z. Konrad (eds.), Pergamon Press 1979, 207-224.

(12) M. Stoeppler and W. Matthes, Storage Behaviour of Inorganic Mercury and Methylmercury Chloride in Sea Water, Anal. Chim. Acta (in press).

(13) P. Valenta, H. Rützel, H. W. Nürnberg, and M. Stoeppler, Trace Chemistry of Toxic Metals in Biomatrices, Fresenius Z. Anal. Chem. 282, 25 (1977).

(14) L. Musani, P. Valenta, H. W. Nürnberg, Z. Konrad, and M. Branica, Interaction of Pb-21o and some Natural Organic Materials in Sea Water, Proceedings of the International Experts Discussion on the "Lead - Occurrence, Fate and Pollution in the Marine Environment", Rovinj 1977, M. Branica and Z. Konrad (eds.), Pergamon Press 1979, 197-206.

(15) M. Whitfield and D. R. Turner, Theoretical Studies of the Chemical Speciation of Lead in Sea Water, Proceedings of the International Experts Discussion on the "Lead - Occurrence, Fate and Pollution in the Marine Environment", Rovinj 1977, M. Branica and Z. Konrad (eds.), Pergamon Press, 1979,109-148.

(16) P. Valenta, L. Mart, and H. Rützel, New Potentialities in Ultra Trace Analysis With Differential Pulse Anodic Stripping Voltammetry, J. Electroanal. Chem. 82, 327 (1977).

(17) P. Valenta, L. Mart, H. W. Nürnberg, and M. Stoeppler, Voltammetrische simultane Spurenanalyse toxischer Metalle in Meerwasser, Binnengewässern, Trink - und Brauchwasser, Vom Wasser 48, 98 (1977).

(18) P. Valenta, H. Rützel, P. Krumpen, K. H. Salgert, and P. Klahre, Device for the Automated Simultaneous Voltammetric on Line-Determination of Toxic Trace Metals in Drinking Water, Fresenius Z. Anal. Chem. (in press).

(19) J. C. J. Nihoul, Modèlisation des Systèmes Marins, Programme National de Recherche et de Developpement, Environnement, Eau, Projet Met, Rapport final, 1977, Vol. 1, Services du Premier Ministre, Programmation de la Politique Scientifique, Bruxelles.

(20) C. Patterson, D. Settle, and B. Glover, Analysis of Lead in Polluted Coastal Sea Water, Marine Chemistry 4, 305 (1976).

(21) M. Burnett, and C. Patterson, Analysis of Natural and Industrial Lead in Marine Ecosystems, Proceedings of the International Experts Discussion on the "Lead - Occurrence, Fate and Pollution in the Marine Environment", Rovinj 1977, M. Branica and Z. Konrad (eds.), Pergamon Press, 1979,15-30.

(22) J. C. Duinker and R. F. Nolting, Dissolved and Particulate Trace Metals in the Rhine Estuary and the Southern Bight, Marine Poll. Bull. 8, 65 (1977).

(23) J. C. Duinker and C. J. M. Kramer, An Experimental Study on the Speciation of Dissolved Zink, Cadmium, Lead and Copper in River Rhine and North Sea Water by Differential Pulse Anodic Stripping Voltammetry, Marine Chemistry 5, 2o7 (1977).

(24) M. I. Abdullah and L. G. Royle, The Occurrence of Lead in Natural Waters, Comm. Eur. Communities (1973), p. 113, Cent. Inf. Doc., Luxembourg.

The Chelation of Lead by Organic Ligands in Sea Water

B. Raspor*,**, H. W. Nürnberg*, P. Valenta* and M. Branica**

*Institute of Chemistry, Institute 4 Applied Physical Chemistry, Nuclear Research Centre (KFA), Juelich, F.R.G.

**Center for Marine Research, "Rudjer Bošković" Institute, Zagreb, Croatia, Yugoslavia

ABSTRACT

The specific effects of the major salinity components of sea water on the chelation of Pb(II) by organic ligands have been studied. Application of a suitable voltammetric method (differential anodic stripping voltammetry, DPASV) enabled measurements at the realistic ultra trace level of 10^{-8}M for the dissolved Pb(II) concentration due to the excellent sensitivity and the particular general potentialities of the voltammetric approach for this kind of speciation studies. The results obtained for Pb(II)-chelation with the well defined organic ligand NTA confirm and support the general conclusions for trace metal chelation in sea water emerging from our previous analogous studies with Cd(II). The generally operative specific effects are in the first place the pronounced competition of the Ca(II) and Mg(II) excess for chelating organic ligands and furthermore the side effects due to the respective particular labile complex formation of the respective trace metal. A third influence to be taken into account stems from comparable levels of other trace metals forming rather stable chelates, e.g. Zn(II). The general aspects of trace metal chelation by dissolved organic matter are discussed in terms of the two intercorrelated parameters, stability constant and ligand concentration.

INTRODUCTION

The chemical species and the binding state in which dissolved toxic metals exist in the sea is of particular significance for their uptake by organisms and thus their entrance into the marine food webs. Toxic trace metals dissolved as stable chelates are regarded according to the present state of knowledge as less hazardous than ionic or labile complex forms, e.g. chloro-, carbonato-, hydroxo-complexes, as the uptake of chelates seems to be difficult, while uptake of unchelated Pb and Cd and its accumulation, particularly by plankton and molluscs, but also by crustacea and fish, have been observed (Ref. 1). It is to be expected that as consequence of biological activity (detritus) and/or anthropogenic pollution, particularaly in coastal waters, in various regions of the sea different amounts of organic matter with strong chelating properties can occur. However, for humic acids theoretical considerations (Ref. 2) and recent experimental findings (Ref. 3) have shown an unsignificant chelation power for most trace metals. This has been for Pb also experimentally confirmed by recent electrophoretic studies (Ref. 4). Ultimately the natural strongly chelating organic substances have to be isolated from sampled sea water, identified and studied in detail. However, as we have previously shown

(Ref. 5-7) for the toxic trace metal Cd the common typical specific influences of chemical and physicochemical properties of sea water, as pH, chemical nature and quantity of the major salinity components and of certain trace metals present in excess, e.g. Zn, on the stability of trace metal chelates and the kinetics and mechanism of their formation can be elucidated applying suitable voltammetric methods and strongly chelating organic ligands with defined and well known properties in terms of coordination chemistry. In our previous and several recent further studies on Cd and Zn (Ref. 5-10) we have used as ligands NTA and EDTA. Both ligands have not only well known properties and strong chelating power for heavy metals, but may also occur in coastal waters already as pollutants, possibly to an increasing extent in the future. The latter statement refers particularly to NTA, which is in some countries already used as substitute for the eutrophicating phosphates in detergents. Also EDTA, originating from anthropogenic activities, enters via rivers and waste water outfalls coastal waters. Thus, these ligands are no more without immediate relevance to the marine environment.

In the present study we shall report our findings on the chemical and physicochemical aspects of the chelation of Pb by organic ligands in sea water. NTA was used for the outlined measurements on chelate stability. Into the general considerations will be also included the results of investigations on the kinetics and mechanism of Pb-chelate formation using EDTA as a ligand of which the details are described elsewhere (Ref.10). The measurements were made using the excellent sensitivity potentialities of the applied voltammetric method (Ref. 11 and 12) at a total Pb-Concentration of 4 μg/l. This concentration is a factor 10 to 30 higher than the average dissolved level in unsignificantly to moderately Pb-polluted coastal waters in the Western Mediterranean and the Southern Bight of the North Sea (Ref. 13-15). However, Pb-levels in the order of μg/l may be actually encountered in more heavily Pb-polluted waters, e.g. certain coastal areas of the Irish Sea (Ref. 16).

The strategy of our investigations was similar to that adopted in previous studies on the chelation of Cd and Zn (Ref. 5, 6, 8). At first the individual influence of the respective major salinity components was studied at an ionic strength and pH corresponding to sea water in model solutions of the respective single component. Then Pb-chelation was studied in artificial sea water, prepared according to Lyman and Fleming (Ref. 17), where the specific effects of the salinity components act in an accumulative manner and finally the measurements were performed in sampled natural sea water. The applied voltammetric method was, with respect to the sensitivity requirements at an adjusted total Pb-level of μg/l or about 2×10^{-8}M differential pulse anodic stripping voltammetry (DSASV) at the hanging mercury drop electrode (HMDE). For basic details on DPASV see (Ref. 5 and 12).

EXPERIMENTALS AND METHODOLOGICAL ASPECTS

Chemicals

Model solutions were prepared from Merck Suprapur chemicals, except $MgCl_2$, which was p.a. grade. In the model solutions usually a chloride concentration of 0.55 M was adjusted. The sea water analogous ionic strength of 0.7 M was established by respective additions of $NaClO_4$ if necessary. The total concentration of dissolved Pb(II) was always 4 μg/l.

The pH was adjusted usually as in the sea by a carbonate buffer system. The solution contained 2×10^{-3}M $NaHCO_3$ and a constant CO_2 partial pressure was maintained above it in the following way (Ref. 18). Argon was passed before entering the voltammetric cell through a washing bottle, containing a suspension of magnesium hydroxycarbonate in a borate buffer of pH 8. In this manner the appropriate small CO_2 partial pressure could be adjusted in the argon gas stream so that in the model solutions a pH of about 8 was established. The CO_2 loaded Ar (99.999% purity) served also for 30 min deaeration of the solutions prior to DPASV. In NaCl-solutions for comparison also measurements have been made, where a pH of 7.24 has been adjusted by a borate buffer ($5x10^{-3}M\ H_3BO_3 + 5x10^{-4}M$ NaOH). But the majority of the measurements have been performed with the more realistic adjustement of pH by the carbonate buffer system. All solutions were prepared in specially purified quartz flasks. Aliquots of the standard stock solutions of Pb(II), Zn(II) and NTA were added with Eppendorf microliter pipettes. The largest volume was 100 µl.

Apparatus

The voltammetric measurements were carried out under potentiostatic potential control with the PAR-polarograph, model 174A, in conjunction with the automatic PAR-unit 315, in a 3-electrode equipped quartz cell thermostated via a double jacket to 25 °C. The test electrode was HMDE, type E410, Metrohm. The auxiliary electrode was an Ag/AgCl-electrode in saturated KCl connected to the cell via a salt bridge filled with 0.7 M NaCl. The principle of DPASV is depicted in Fig. 1.

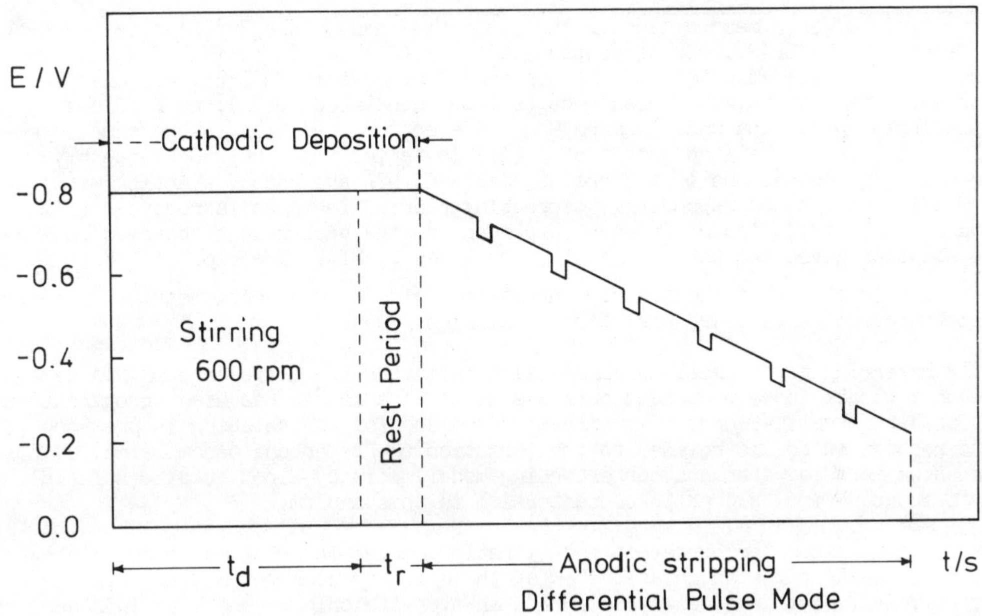

Fig. 1. Principle of DPASV.

At the automatic PAR-unit 315 a deposition potential t_d of 60 s was selected and a cathodic deposition potential E_d of -0.8 V was adjusted. During the cathodic deposition time the solution was stirred with a magnetic bar with 600 rpm to speed up mass transfer towards the HMDE. After t_d had elapsed, stirring was stopped and a time interval of 15 s was spent as resting period t_r of the solution. Subsequently an aliquot of the amount of Pb deposited as amalgam at the HMDE during t_d and t_r was reoxidized (stripped) in the differential pulse mode employing a train of rectangular voltage pulses of 50 mV height and 57 ms duration with a clock time of 0.5 s and a scan rate of 2 mV x s^{-1}. The resulting voltammetric Pb-peak is proportional to the bulk concentration of unchelated Pb(II), which is reversibly reduced and reoxidized. It has been experimentally established in an analogous manner as in Ref. 6 that the response is diffusion controlled and is not affected by noticeable contributions from additional Pb(II) produced by dissociation of the trace metal chelate. During the rather short total cathodic deposition period (t_d+t_r) of 75 s the dissociation rate of the small [PbNTA]$^-$-concentration ($\leq$ 1x10^{-8}M) remains negligible. The trace metal chelates are also reducible at a several hundred mV more negative electrode potential in an irreversible electrode reaction. On this fortunate potential difference between the electrode reaction of the unchelated trace metals being mainly present as labile complexes [MeX_j], where X equals e.g. Cl^-, CO_3^{2-}, OH etc., or to a small extent also as hydrated cations Me^{2+}, the applicability of the chosen voltammetric procedure is based.

pH was controlled with an Orion pH-meter, Mod. 701, using an Ingold glass electrode, type 405 NS. For standardisation a borate buffer of pH 8.0, Merck Titrisol, was used.

RESULTS AND DISCUSSION

Stability of [PbNTA]$^-$ in Various Media and in Sea Water

At the bigin of each measurement in the respective medium the Zn(II)-concentration was determined by DPASV. Subsequently a total Pb(II)-concentration of 4 μg/l (about 2x10^{-8}M) was adjusted and the corresponding reversible DPASV-response was recorded. The peak height corresponds to 100% unchelated Pb(II), as Pb^{2+} and labile [PbX_j]-complexes undergo a reversible electrode reaction in the same potential range. The unchelated amount of Pb(II) is termed ionic Pb. Subsequently increasing concentrations of the chelating ligand NTA are added, starting with 10^{-8}M NTA, and the correspondingly decreasing peak of ionic Pb is recorded (viz. example in Fig. 2). The difference in percent to the peak height observed in NTA-free solution gives the percentage of Pb chelated by NTA. The plot of % Pb_{ionic} versus the logarithm of the NTA concentration yields thus a voltammetrically determined titration curve of Pb-chelation (viz. Figs. 3-6).

Fig. 3 referring to the NaCl-model solution indicates the difference in NTA demand between a borate and a carbonate buffer system. The latter was used throughout in the further investigations. The higher NTA-demand for Pb-chelation in presence of the carbonate buffer is related to the formation of Pb-carbonato-complexes, as has been shown for sea water and corresponding model media by Sipos et al. (Ref. 18) and to about 4 times higher Zn-concentration in this medium.

Figs. 4 and 5 show the corresponding titration curves in model media containing similar concentrations as in sea water of the salinity components Ca(II) and Mg(II), respectively. Of course, always a pH and an overall ionic strength as in sea water was adjusted. Compared with the titration curve for the carbonate buffer in 0.55 M NaCl-solution (viz. Fig. 3) the titration curves in the alkaline earth model solutions (Fig. 4 and 5) are shifted to higher NTA-concentrations indicating that the Pb-chelation is less favoured in these media. The reason is the efficient competition of the excess of the alkaline earth ions for the chelating ligand. The titration curves for Ca(II)-and Mg(II)-model solutions virtually coincide (see Fig. 4 and 5). This indicates that Ca(II) is more efficient in competing for NTA than Mg(II) (Ref. 6).

Figure 6 shows the voltammetric titration curves corresponding to [PbNTA]$^-$-formation in artificial and natural sea water. For comparison also the titration curves in 0.55 M NaCl and in 10^{-2}M Ca(II) are drawn. Compared with the alkaline earth model solutions the demand in NTA-concentration for 50% Pb-chelation has in artificial sea water further increased about 2 times towards 10^{-5}M NTA. This is due to the accumulative action of all competitive effects of the salinity components on Pb-chelation. For natural sea water the demand of NTA for 50% Pb-chelation has again

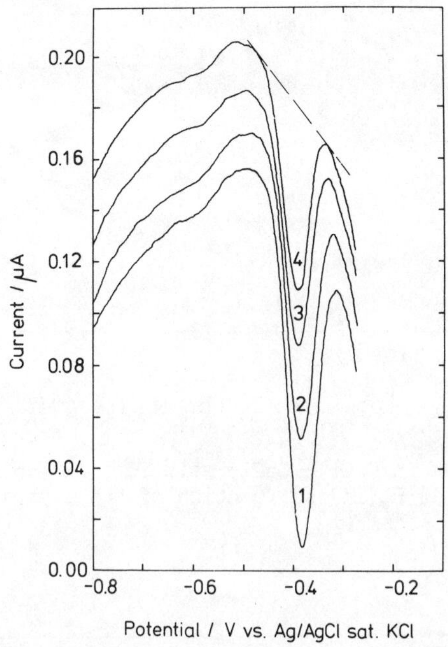

Fig. 2. Differential Pulse Anodic Stripping Voltammograms (DPASV) in 0.45μ filtrated sample of Ligurian sea water, pH 7.8, spiked with 4 μg/l Pb(II). (1) 4 μg/l Pb(II); decreased peak height of ionic Pb(II) after addition of (2) $1x10^{-6}$M NTA; (3) $4x10^{-6}$M NTA; (4) $8x10^{-6}$M NTA. The evaluation of the peak height versus the respective base line is shown on voltammogram (4).

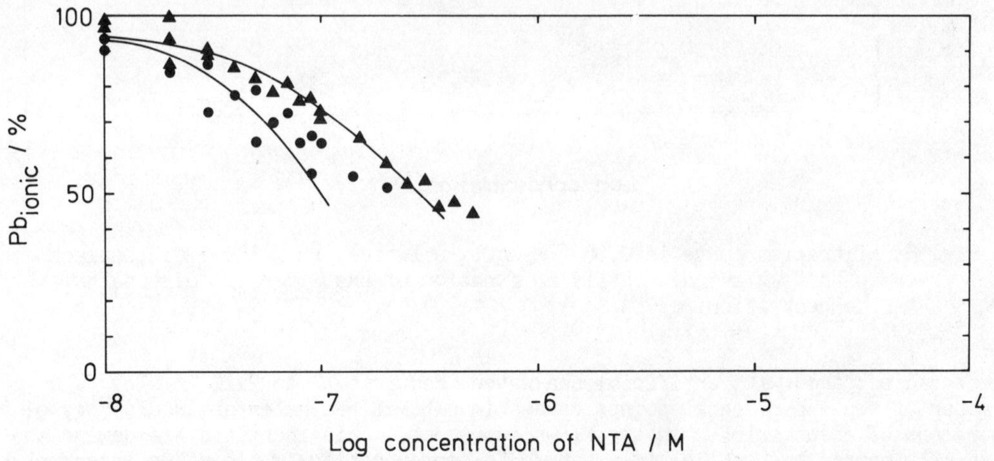

Fig. 3. Titration curves in (●) 0.67 M NaCl, borate buffer, pH 7.24, and (▲) 0.55 M NaCl, carbonate buffer, pH 7.70, spiked with 4 μg/l Pb(II), as function of the logarithm of total NTA concentration. In both cases the ionic strength was 0.7.

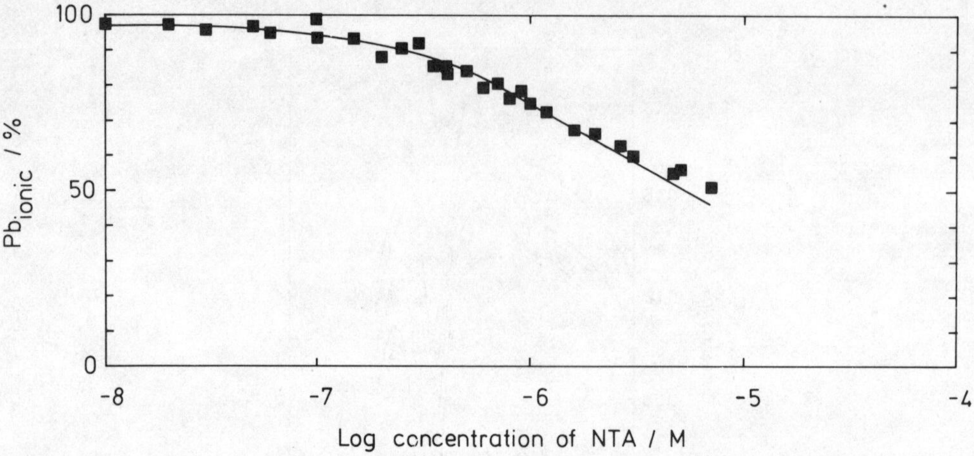

Fig. 4. Titration curve in 0.01 M $CaCl_2$ solution, pH 7.55, I 0.7, spiked with μ g/l ionic Pb(II) as function of the logarithm of total NTA concentration.

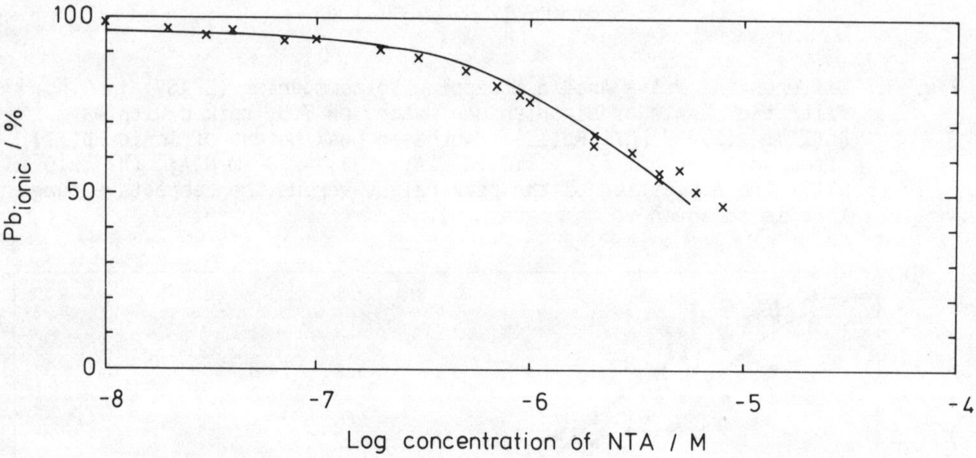

Fig. 5. Titration curve in 0.0536 M $MgCl_2$ solution, pH 7.55, I 0.7, spiked with 4 μg/l ionic Pb(II) as function of the logarithm of total NTA concentration.

increased compared with artificial sea water, requiring 2 to 3×10^{-5}M NTA. The scatter of the experimental points causes in natural sea water an uncertainty of an increment of about 1×10^{-5}M in the NTA-requirement. This increased NTA demand is partially caused by the 1.8 times larger Zn-concentration in the natural sea water sample. A further part of the increase in NTA-demand is to be ascribed to the competition for the ligand by various other trace metals dissolved in natural sea water. Compared with a carbonate buffered 0.55 M NaCl-model solution (viz. Fig. 3) which contains approximately the same amount of Zn as the studied sea water sample a 50% chelation of Pb requires in natural sea water (viz. Fig. 6) an about a factor 100 higher NTA-concentration, i.e. 2 to $3x10^{-5}$M instead of $3x10^{-7}$M. For Cd-chelation the NTA demands are even larger (Ref. 6 and 8). This is connected with the somewhat lower stability of $[CdNTA]^-$ compared with $[PbNTA]^-$.

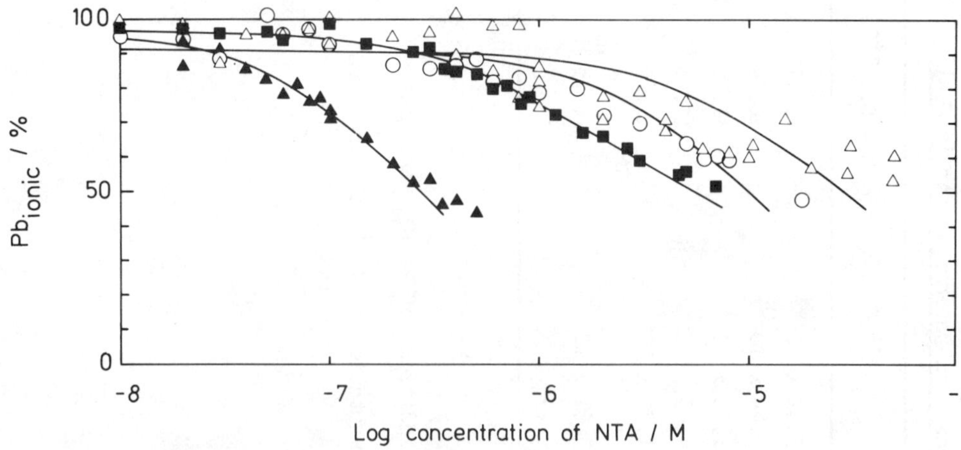

Fig. 6. Sequence of titration curves of 4 μg/l ionic Pb(II) as function of logarithm of total NTA concentration in (▲) 0.55 M NaCl; (■) 0.01 M $CaCl_2$; (o) artificial sea water, pH 7.63 and (△) Ligurian sea water sample, pH 7.8.

The titration curves in Figs. 3-6 were evaluated from the experimental points by polynomial fitting with a computer program. The reciprocal value of the NTA concentration corresponding to 50% Pb-chelation is the operational conditional stability constant K' of $[PbNTA]^-$ in the respective medium. K' reflects the operative specific influences of pH, salinity components and other relevant trace metals, e.g. Zn, on the formation of $[PbNTA]^-$, due to competition for the chelating ligand NTA and the tendencies to form labile complexes $[PbX_j]$. The lower the value of K' for the various media studied the higher the effects of these side reactions are. The K'-values for the various media studied are given in Table 1.

In general it turns out that a rather substantial concentration of a strongly chelating ligand, as NTA, in the order of $1x10^{-5}$M to $3x10^{-5}$M has to be dissolved in sea water to achieve a 50%-chelation of the dissolved Pb(II)-concentration (compare Fig. 6).

Determination of the Common Conditional Stability Constant

Provided all relevant complex and protonation equilibria, determining in the respective medium the concentrations of Pb^{2+}, i.e. hydrated Pb-ions, and the actually chelating ligand NTA^{3-}, are taken into account the common conditional stability constant K corresponding to an ionic strength of 0.7M can be computed from eq. (1).

$$K_{PbNTA} = \frac{[PbNTA]^-}{[Pb^{2+}]\,[NTA^{3-}]} \qquad (1)$$

TABLE 1 Operational and Common Conditional Stability Constants Determined for $[PbNTA]^-$

Electrolyte*) /M	pH**)	K'/M^{-1}	log K'	K/M^{-1}	log K $\pm$ 0.25***)
0.67 NaCl	7.24	1.0×10^7	7.00	5.92×10^{10}	10.77
0.55 NaCl	7.70	3.7×10^6	6.57	5.44×10^{10}	10.74
0.01 $CaCl_2$	7.55	1.8×10^5	5.26	1.60×10^{11}	11.20
0.0536 $MgCl_2$	7.55	2.0×10^5	5.30	9.58×10^{10}	10.98
Artificial Sea Water	7.63	1.0×10^5	5.00	1.59×10^{11}	11.19
Natural Sea Water (Ligurian Coast)	7.80	3.5×10^4	4.54	5.61×10^{10}	10.75

*) The ionic strength was always 0.7, if necessary appropriate additions of $NaClO_4$ were made.

**) The differences in pH have been taken into account in the computation of K via the involved dissociation equilibrium of $HNTA^{2-}$.

***) This is the maximal uncertainty in lg K.

The same K-value should be obtained from any point of the experimental titration curves (Fig. 3-6) in all studied media. Thus, this evaluation will be an important prove of the general consistency of the applied approach and the emerging conclusions on chelation behaviour in the various studied media. Figure 7 shows the equilibria, which affect in sea water to a noticeable extent the chelation equilibrium of $[PbNTA]^-$. From the trace metals only Zn(II) has to be taken into account, as it is present at a similar level as Pb(II) and both metals from NTA-chelates of about the same stability. All equilibrium constants of complex and protonation equilibria required for the computation of $[Pb^{2+}]$ and $[NTA^{3-}]$ are listed in Table 2.

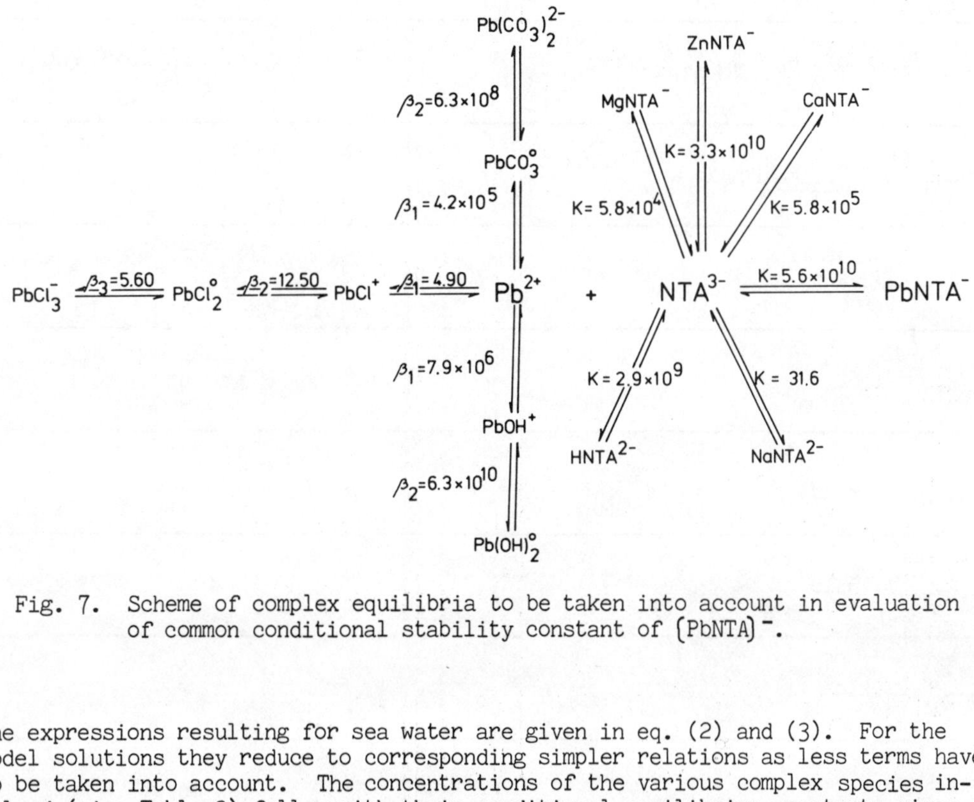

Fig. 7. Scheme of complex equilibria to be taken into account in evaluation of common conditional stability constant of $[PbNTA]^-$.

The expressions resulting for sea water are given in eq. (2) and (3). For the model solutions they reduce to corresponding simpler relations as less terms have to be taken into account. The concentrations of the various complex species involved (viz. Table 3) follow with their conditional equilibrium constants given in Fig. 7 and Table 2.

$$[Pb^{2+}] = [Pb(II)_{ionic}] - \{[PbCl^+] + [PbCl_2^+] + [PbCl_3^-] + [PbOH^+] + [Pb(OH)_2^o] + [PbCO_3^o] + [Pb(CO_3)_2]^{2-}\} \quad (2)$$

$$[NTA^{3-}] = [NTA]_T - [HNTA^{2-}] + [NaNTA^{2-}] + [CaNTA^-] + [MgNTA^-] + [ZnNTA^-] + [PbNTA^-] \quad (3)$$

TABLE 2 Stability Constants of Complexes Involved in Chelation Equilibrium of Pb(II) with NTA in Studies Media

Stability Constants	Ref.	Remarks
β_{PbCl} 4.9 β_{PbCl_2} 12.5 β_{PbCl_3} 5.6	19	
β_{PbOH} 7.94×10^{6} $\beta_{Pb(OH)_2}$ 6.31×10^{10}	2o	in 1.o M KNO_3
β_{PbCO_3} 4.17×10^{5} $\beta_{Pb(CO_3)_2}$ 6.25×10^{8}	18	
K_{ZnNTA} 3.31×10^{10}	8	
K_{NaNTA} 31.62	21	at I = 0.1
K_{CaNTA} 5.75×10^{5}	6	
K_{MgNTA} 5.75×10^{4}	6	
K_{HNTA} 2.9×10^{9}	6	
pK_W 13.78	24	in 0.7 M NaCl
pK_W 13.2o	24	in sea water

All charges and dimensions have been omitted for simplicity. Overall stability constant $\beta_1 = K$.

$Pb(II)_{ionic}$ and NTA_T are known and $[PbNTA^-]$ and $[ZnNTA^-]$ have been determined experimentally. The concentrations of the chloro- and hydroxo-complexes of Pb(II) are evaluated from literature data and for the carbonato-complexes from the results of the recent study of Sipos et al. (Ref. 18). The concentrations of $[CaNTA]^-$ and $[MgNTA]^-$ are calculated using the conditional stability constants with ion sensitive electrodes for the conditions of the systems studied (Ref. 6). The concentrations of $[HNTA]^{2-}$ are computed from the pH and literature data (Ref. 6) of the conditional stability constant. Also the evaluation of $[NaNTA^-]$ is based on literature data (Ref. 21). Although the stability of this chelate is low, its influence is not negligible due to the high concentration (0.47 M) of Na^+ in sea water.

As the data in Table 1 show the evaluation of the experimental titration curves, measured at a realistic trace level of the total Pb(II) concentration, yields in all media studied within the error limits a common value of the conditional stability constant referring to an ionic strength of 0.7 M. This maximal uncertainty in lg K amounts to 0.25 logarithmic units. This uncertainty is encountered in the most complicated medium natural sea water sampled at the Ligurian coast. Here the scatter of the experimental points (viz. Fig. 6) has the relative largest amount. The general coincidence of the lg K-values in Table 1 is a very satisfactory result confirming the consistency of the applied voltammetric approach and showing also that all significant specific influences in the chelation of Pb(II) in sea water by strongly chelating organic ligands have been elucidated in this study with the in terms of coordination chemistry well defined ligand NTA.

A study being at present in the final stage (Ref. 22) has revealed that in a rather highly concentrated multielectrolyte, as sea water, the formation of coulombic ion-pairs between the anions and the alkali- and alkaline earth cations deserves attention to achieve en even more accurate picture of the actual distribution of the total level of dissolved trace metals over the various existing labile complex species, e.g. chloro-, hydroxo- and carbonato-complexes will be affected more considerably, if the mentioned ion-pair formation between the major components is taken into account, no dramatic influences on the amount of Pb(II)-chelate with strong organic ligands, as NTA, are to be expected, because the predominant influence of this ion-pairing will be a redistribution of the relative amounts of the various labile complex species.

In general, the results of the present study on $[PbNTA]^-$ add further confirming and supporting experimentally based evidence on the general conclusions drawn from our previous studies on $[CdNTA]^-$ (Ref. 6 and 7) on the typical and common aspects of chelation trace levels of heavy metals in sea water and other types of natural waters. Substantial excess amounts of rather strongly chelating organic ligands are required to achieve in sea water chelation of a significant fraction of the total dissolved amount of trace metals. Corresponding to the higher stability constant of $[PbNTA]^-$ (lg K 11) with about 10^{-5} M NTA an about 10 times lower ligand concentration is necessary to chelate 50% of the total dissolved trace metal (viz. titration curve for artificial sea water in Fig. 6), than for $[CdNTA]^-$-formation, which requires about 10^{-4}M NTA with lg K 9.5. The main reason is in both cases the pronounced competition of the alkaline earth excess for the organic ligand and in the second place the respective particular speciation of the trace metal in form of labile complexes.

Also the investigations on the kinetics of $[PbEDTA]^{2-}$-formation (Ref. 10) yield in general analogous findings, as have been previously obtained for the rate constant of $[CdEDTA]^{2-}$-formation (Ref. 7). The conditional rate constant in sea water is about a factor 10 larger for the formation of $[PbEDTA]^{2-}$ than for $[CdEDTA]^{2-}$ and $[ZnEDTA]^{2-}$ (Ref. 7 and 10). Experiments with equal ampounts of Pb(II) and Zn(II) at the 10^{-7} level have revealed that the simultaneous formation of the respective EDTA-chelates occurs without mutual interference. The mechanism of $[PbEDTA]^{2-}$-formation is as for $[CdEDTA]^{2-}$ (Ref. 7) and $[ZnEDTA]^{2-}$ again ligand exchange, predominantly with the a priori formed $[CaEDTA]^{2-}$.

TABLE 3 Total Concentrations of Major Components and Zn(II) Involved in Chelation Equilibrium of Pb(II) with NTA in Studied Media

Fig.	Cl^- M	Na^+ M	Ca^{2+} M	Mg^{2+} M	Zn(II) M	Remarks
3	0.55	0.670			9.3×10^{-9}	Borate Buffer
3	0.55	0.697			3.48×10^{-8}	Carbonate Buffer
4	0.55	0.667	0.01		4.41×10^{-8}	Carbonate Buffer
5	0.55	0.537		0.0536	3.24×10^{-8}	Carbonate Buffer
6	0.55	0.472	0.0102	0.0538	2.33×10^{-8}	Artificial Sea Water
6	0.55	0.472	0.0102	0.0538	4.2×10^{-8}	Ligurian Sea Water

GENERAL ASPECTS OF CHELATION OF HEAVY METAL TRACES IN SEA WATER

On the basis of the assembled experimental evidence and available literature data one can put forward the following approximate and preliminary general arguments with respect to the chelation of trace metals in sea water by strongly chelating dissolved organic matter.

In oceanic waters typical values reported for the total dissolved organic matter (DOM) range in the surface zone from 0.8 to 2.0 mg C/l and decrease gradually with depth to a constant level of 0.6 mg C/l below 5oo m (Ref. 23). Presuming that about 50% of the total weight of DOM is carbon one arrives at rounded concentrations for DOM between 2 and 5 mg/l in the surface zone decreasing to about 1.2 mg/l below 5oo m. The values in the surface zone correspond for molecular weights of 10^2, 10^3 and 10^4 daltons to 2 - 5×10^{-6}M and 2 - 5×10^{-7}M, respectively, while in deep waters they are smaller, i.e. about 10^{-5} to 10^{-7}M.

Presuming a level of about 10^{-8}M for the trace metals Cd, Zn and Pb in the dissolved state our chelation studies (see also Ref. 5-1o) with NTA have shown that according to the stability constants of the respective chelates $[MeNTA]^-$, which ranged from lg K 9.5 to 11, NTA-concentrations beteeen 10^{-4} and 10^{-5}M are required to chelate 5o% of the present trace metals in sea water. Thus, one concludes that even if the total DOM would yield significant trace metal chelation, if lg K $\geq$ 1o prevails. In reality, of course, only a fraction of DOM will have chelation potentialities and the requirements on the minimum value of lg K become even more stringent. This will be particularly the case,if the chelating components of DOM belong to the higher molecular weight fraction, an assumption which seems not improbable. On the other hand, as then each molecule of this chelating material of higher molecular weight might have several chelator sites, this property will moderate somewhat the actual demands on the two always intercorrelated parameters, ligand concentration and stability constant.

Extended base line studies in European coastal waters on the level of dissolved trace metals (Ref. 14, 15) emphasize that there are at a number of locations strongly chelating or more generally trace metal binding fractions of DOM requiring UV-photolysis for decomposition and release of the bound amount of trace metals. The nature of these organic compounds has still to be identified. It can be only concluded at present, that contrary to a hitherto widespread believe for most trace metals, expect Cu and Hg (Ref. 3), these strongly chelating organic material cannot belong to the humic acid fraction of DOM. As has been discussed in detail in a further contribution (Ref. 4) for Cd(II) and Zn(II) the stability constants of the humates, ranging according to Mantoura et al. (Ref. 3) from l g K 3 to 6 are far too low in correlation with the available concentration of the dissolved humic material. The same conclusion holds (Ref. 4) for the formation of Pb-humate, for which according to Whitfield and Turner (Ref. 2), a lg K of 7.5 is to be assumed. The negative findings in our recent electrophoretic tests on chelation of Pb-210 with dissolved humic acid (Ref. 4) confirm experimentally these conclusions. For Pb-humate having requirements on ligand excess will become considerably less for inland waters and estuaries, provided the corresponding salinity decrease leads to a proportional decrease of the concentration of the two predominantly for organic ligands competing components, Ca^{2+} and Mg^{2+}. This is, however, not necessarily the case for all types of inland waters, as studies on Zn-chelation with NTA in Lake Ontario have revealed (Ref. 9).

ACKNOWLEDGEMENT

This study is part of the joint research project "Environmental Research in Aquatic Systems" of the Institute of Chemistry, Institute 4 Applied Physical Chemistry, Nuclear Research Center (KFA) Juelich, F. R. Germany and the Center for Marine Research, "Ruđer Bošković" Institute, Zagreb, Yugoslavia. Financial support during the attachment of Dr. B. Raspor to the Institute 4 Applied Physical Chemistry of KFA by the International Bureau of Nuclear Research Center (KFA), Juelich, on the basis of the bilateral scientific and technical cooperation between the Socialist Federative Republic Yugoslavia and the Federal Republic Germany is gratefully acknowledged.

REFERENCES

(1) M. Bernhard and A. Zattera, Major Pollutants in the Marine Environment, in Marine Pollution and Marine Waste Disposal, E. A. Pearsin and E. DeFraja Frangipane, eds., Pergamon Press, Oxford, 1975.

(2) M. Whitfield and D. R. Turner, Theoretical Studies of the Chemical Speciation of Lead in Sea Water, Proceedings of the International Experts Discussion on "Lead - Occurrence, Fate and Pollution in the Marine Environment", Rovinj 1977, M. Branica and Z. Konrad (eds.), Pergamon Press, 1979, 109-148.

(3) R. F. C. Mantoura, A. Dickson, and J. P. Riley, The Complexation of Metals with Humic Materials in Natural Waters, Estuarine and Coastal Marine Science, 6, 387 (1978).

(4) Lj. Musani, P. Valenta, H. W. Nürnberg, Z. Konrad, and M. Branica, Interactions of Pb-210 with Some Natural Organic Materials in Sea Water, Proceedings of the International Experts Discussion on "Lead - Occurrence, Fate and Pollution in the Marine Environment", Rovinj 1977, M. Branica and Z. Konrad (eds.), Pergamon Press, 1979, 197-206.

(5) H. W. Nürnberg, P. Valenta, L. Mart, B. Raspor, and L. Sipos, Application of Polarography and Voltammetry to Marine and Aquatic Chemistry. II. The Po-

larographic Approach to the Determination and Speciation of Toxic Metals in the Marine Environment, Fresenius Z. Anal. Chem. 282, 357 (1976).

(6) B. Raspor, P. Valenta, H. W. Nürnberg, and M. Branica, The Chelation of Cadmium with NTA in Sea Water as a Model for the Typical Behaviour of Trace Heavy Metal Chelates in Natural Waters, The Science of the Total Environment 9, 87 (1978).

(7) B. Raspor, P. Valenta, H. W. Nürnberg, and M. Branica, Polarographic Studies on the Kinetics and Mechanism of Cd(II)-Chelate Formation with EDTA in Sea Water, Thalassia jugosl. 13, 79 (1977)

(8) B. Raspor, P. Valenta, H. W. Nürnberg, and M. Branica, Studies with Differential Pulse Anodic Stripping Voltammetry on the Chelation of the Trace Levels of Cadmium and Zinc with NTA in Sea Water, Marine Chem. Comm. (in press).

(9) B. Raspor, H. W. N. Nürnberg, P. Valenta, and M. Branica, Voltammetric Studies on the Chelation of Zinc with NTA and EDTA in Lake Ontario Water as Example for the Chelation of Trace Metals in Inland Waters, The Science of the Total Environment (to be submitted).

(10) B. Raspor, P. Valenta, H. W. Nürnberg, and M. Branica, Voltammetric Determination of Kinetics and Mechanism of Pb(II)- and Zn(II)-EDTA Chelate Formation in Sea Water, (1978) Proc. V. Int. Symp. Chemistry of the Mediterranean, Rovinj, May 1978.

(11) H. W. Nürnberg and P. Valenta, Polarography and Voltammetry in Marine Chemistry, E. D. Goldberg (ed.), The Nature of Sea Water, Dahlem-Konferenzen, Berlin, 1975, pp. 89-136.

(12) H. W. Nürnberg, Potentialities and Applications of Advanced Polarographic and Voltammetric Methods in Environmental Research and Surveillance of Toxic Metals, Electrochim. Acta 22, 935 (1977).

(13) J. C. Duinker and R. F. Nolting, Dissolved and Particulate Trace Metals in the Rhine Estuary and the Southern Bight, Marine Poll. Bull. 8, 65 (1977).

(14) L. Mart, Ermittlung and Vergleich des Pegels Toxischer Spurenmetalle in Nordatlantischen und Mediterranen Küstengewässern, Doctor Thesis, RWTH Aachen, 1978.

(15) L. Mart, H. W. Nürnberg, and P. Valenta, Comparative Base Line Studies on Pb-Levels in European Coastal Waters, Proceedings of the International Experts Discussion on "Lead - Occurrence, Fate and Pollution in the Marine Environment", Rovinj, October 1977, M. Branica and Z. Konrad (eds.), Pergamon Press, 1979, 155-179.

(16) M. I. Abdullah, L. G. Royle, and A. W. Morris, Heavy Metal Concentration in Coastal Waters, Nature 235, 158 (1972).

(17) R. A. Horne, Marine Chemistry, Wiley, New York, 1969, p. 140.

(18) L. Sipos, P. Valenta, H. W. Nürnberg, and M. Branica, Voltammetric Determination of the Stability Constants of the Predominant Labile Lead Complexes in Sea Water, Proceedings of the International Experts Discussion on "Lead - Occurrence, Fate and Pollution in the Marine Environment", Rovinj, October 1977, M. Branica and Z. Konrad (eds.), Pergamon Press, 1979, 61-76.

(19) M. Lovrić, Trace Metal Characterization by Voltammetric Methods, M. Sc. Thesis, Univ. Zagreb, 1977.

(20) L. G. Sillén and A. E. Martell, Stability Constants of Metal Ion Complexes, Chem. Soc. Spec. Publ. 17, 1971.

(21) R. Pribil, Analytical Applications of EDTA and Related Compounds, Pergamon Press, Oxford, 1972, p. 10, Table 1.

(22) L. Sipos, B. Raspor, H. W. Nürnberg, and R. M. Pytkowicz, Interaction of Metal Complexes with Coulombic Ion Pairs in High Salinity Aqueous Media, Marine Chem. (in press).

(23) N. Ogura, Rate and Extent of Decomposition of Dissolved Organic Matter in Surface Sea Water, Marine Biology 13, 89 (1972).

(24) D. Dryssen and M. Wedborg, Equilibrium Calculations of the Speciation of Elements in Sea Water, in The Sea, E. D. Goldberg (ed.), Wiley-Interscience, New York, Vol. 5, 1973.

Interactions of Pb-210 with Some Natural Organic Materials in Sea Water

Lj. Musani*,**, P. Valenta*, H. W. Nürnberg*,
Z. Konrad** and M. Branica**

*Institute of Chemistry, Institute 4 Applied Physical Chemistry, Nuclear Research Centre (KFA), Juelich, F.R.G

**Center for Marine Research, Institute "Rudjer Bošković", Zagreb, Croatia, Yugoslavia

ABSTRACT

The chelating ability of Pb-210 for two natural organic materials, Na-alginate and a humic acid type of terrestrial origin, has been tested in sea water and other related media by measurements with high voltage paper electrophoresis. For the polysaccharide alginate Pb-210 chelation in sea water is indicated by the observed change of electrophoretic mobility. The chelation becomes favoured with decreasing salinity as measurements in sea water diluted with distilled water reveal. Humic acid has been tested in sea water and 0.55 M NaCl. No significant Pb-210 chelation is indicated up to humic acid concentrations far above the level to be presumed according to available literature data as dissolved in oceanic waters. This stability constant of Pb-chelates with humic acids, the competition of alkaline earth ions and of other inorganic complexing anions affecting significantly the chelation of trace metals in sea water, as has been established previously by comprehensive voltammetric investigations with organic ligands of larger and well defined chelating potentialities.

INTRODUCTION

Pollutants of natural waters are toxic metals, radionuclides, and various organic products. Toxicity and fate of metals in the aquatic environment and their participation in the biogeochemical cycle depend strongly on the chemical and physicochemical properties of the respective chemical species in which metals are present in natural waters. Therefore it is important to determine and characterize those metal species actually dissolved in aquatic systems. Electrophoretic investigations offer one convenient route to test, whether strong interactions between a trace metal or a radionuclide and organic substances exist, which are to be interpreted as chelation taking into account the structure of the respective organic material. The electrophoretic approach is rendered particularly suitable for these kinds of tests on metal chelation, if a radioactive isotope of the trace metal in question is studied. In the course of previous work in that manner physico-chemical forms of various radionuclides entering sea water have been studied, their speciation with time and specific processes occuring in sea water (Ref. 1 and 2). The interactions of Pb-210 with two natural organic materials occuring in sea water in the dissolved state and possibly also bound to the surface of suspended particulate matter were Na-alginate and a humic acid of terrestrial origin. Both substances have a structure with functional groups (carboxyl, carbonyl, OH) positioned suitable to chelate metal ions. Alginates are polysaccharides containing guluronic and mannuronic acid. These polysaccharides occur in marine brown algae. Humic compounds have been hitherto assumed frequently as possibly significant chelators of trace metals in sea water, although for many metals inspection of the known stability constants (Ref. 3) should have thrown severe doubts on this widespread belief, at least with respect to the amount of humic acids dissolved in sea water.

Pb-210 is a naturally occurring radionuclide which belongs to the U-238 family. The concentration of Pb-210 in surface water off the Californian coast has been determined to 0.1 dpm/kg (Ref. 4) and 0.15 dpm/kg for the North Pacific Ocean (Ref. 5). (1 dpm is equivalent to 5.1 x 10^{-15}g) This concentration is more than corresponding to the equilibrium with the probable concentration of Ra-226. This might be attributed to an input of Pb-210 from the atmosphere which is washed out after formation of Rn-222. It seems that the input from nuclear weapons tests is negligible (Ref. 6). The concentration of Pb-210 in waters below 1000 m ranged from 0.09 to 0.27 dpm/kg (Ref. 4). The residence time of Pb-210 is calculated to be 54 years for deep water (Ref. 7) and 0.63 years for water above 100 m (Ref. 8). Pb-210 is concentrated in organisms but less than its daughter product Po-210. It is more accumulated in zoo- and phytoplankton than in pelagic fish (Ref. 4).

It is well known that chelation of metals by various organic and inorganic ligands maintains a reservoir of metals in solution. There is good reason for the common belief that strong chelation will usually prohibit or at least strongly retard the uptake of toxic trace metals and radionuclides by planktonic organisms and will thus weaken this ingestion pathway into the marine food webs.

EXPERIMENTAL

Details for experiments with the Pb-210-alginate systems are given in (Ref. 9). The samples for paper electrophoresis of the Pb-210-humic acid systems were prepared in the following way. Solutions of the radionuclide Pb-210 from Radiochemical Centre Amersham, UK, had a specific radioactivity between 21 to 60 mCi/mg Pb.

An aliquot was evaporated to dryness. The residue was dissolved in sea water or other investigated media adjusting a specific radioactivity of 0.27 μCi/20 μl. This corresponds to a Pb-210 concentration between 3 x 10^{-6}M according to the above given range of specific radioactivity. 20 μl are the amount of solution applied to the starting point of the electrophoretic paper strip wetted previously with the respective solvent containing eventually the respective ligand concentration studied. This causes in the liquid phase on the electrophoretic paper a dilution of the Pb-210 concentration by a factor 6 and renders thus due to the now correspondingly larger excess of the potential chelator the equilibrium conditions even somewhat more favourable for chelation. For experiments in presence of the chelator concentrations given in Table 2-4 the respective chelator concentration was adjusted in the test solution and from the latter was after 30 min an aliquot of 20 μl subjected to electrophoresis. This measurement is termed 0 day. The measurements were extended for each adjusted chelator concentration over a period of 0 to 12 days to detect possible aging effects.

Electrophoretic experiments were performed in a high voltage device, Flat Plate Electrophoresis, Savant Instruments Inc.. Temperature was maintained at 20° ± 0.1 °C with an ultrathermostate. The electrophoretic conditions were for humic acid investigations as follows: Whatman 3 MM filter paper with a free length and width of 85x1.5 cm; voltage and current per strip of paper 1000 V and 13.1 - 25 mA;

current per 1 cm width of the strip 8.7 - 16.7 mA; specific electric effect on the paper strips 0.10 - 0.20 $VAcm^{-2}$; duration of one experimental run 102 min.

In all investigated systems the behaviour of Pb-210 was followed by measuring its electrophoretic mobility u $[cmV^{-1}s^{-1}]$, corrected as described in (Ref. 10), in the cationic and anionic zone and by evaluating the amount of Pb-210 in the three observable zones, i.e. the cationic, anionic and the immobile zone at the starting point.

The electrophoretic zones could be detected due to the radioactivity of Pb-210 conveniently by autoradiography exposing an X-ray film overnight to the electrophoretic strip. For the determination of the Pb-210 amount in the respective radioactive zones these were cut from the paper strips and counted in a β-liquid scintillation counter. The electrophoretic strips had to be stored before β-counting for about one month, because the counted species is the β-radiating Bi-210, which is formed from Pb-210. Within the stated period radioactive equilibrium between Pb-210 and Bi-210 is attained.

The studied chelator Na-alginate was from BDH, Poole, UK, and the investigated humic acid was a product of Aldrich Chemicals, Milwaukee, USA.

The experiments have been performed with the humic acid in natural sea water samples of different origin and in 0.55 M NaCl-solution. The sea water was sampled in the North Sea near Helgoland and in the Mediterranean in the Bay of La Spezia near Fiascherino. The only pretreatment was filtration through 0.45 membrane filter to attain separation from suspended particulate matter. The natural level of dissolved Pb(II) was determined by differential pulse anodic stripping voltammetry according to the procedure outlined elsewhere (Ref. 11). The North Sea sample contained 0.16 μg/l while the Medittrranean sample had 0.49 μg/l. Sea water for the Na-alginate experiments were from the North Adriatic Sea with a Pb-level > 0.05 μg/l. Thus, the natural Pb-levels were negligible compared with the adjusted Pb-210-concentration. Na-alginate was studied in sea water and in sea water diluted with distilled water to 30% and 10% to simulate approximately salinity conditions encountered in estuaries.

RESULTS AND DISCUSSION

Studies with Na-Alginate

Electrophoretic studies on the interactions of various trace metals and radionuclides with this natural chelating ligand have been reported elsewhere (Ref. 9). Here the effects on Pb-210 shall be considered in detail (viz. Table 1). In blank solutions of sea water, as well as in estuarine situations simulating diluted sea water, two electrophoretic zones are observed, an immobile zone at the starting point and a zone moving into cathodic direction. The Pb(II)-species in the immobile zone at the starting point are either uncharged and/or strongly adsorbed to the paper. With decreasing salinity the electrophoretic mobility of Pb-210 increases in the cationic zone up to a factor of about 10 in a medium having a factor 10 lower salinity compared with genuine sea water. This correlation of the mobility increase with the decrease in salinity obviously indicates besides the common mobility increase due to dismantling of the ion clouds also alterations in Pb(II)-speciation with respect to the rather labile complexes forming ligands present in genuine sea water. Probably the formation of free hydrated Pb^{2+}-ions is also favoured by progressive dilution with distilled water while in genuine sea water a significant amount of Pb(II) is present as rather stable carbonato complex (Ref. 12 and 13). No anionic zone was detectable in the blank media.

If the chelator Na-alginate had been added to the test solutions the situation changed significantly. This natural organic ligand has the tendency to move into

TABLE 1 Electrophoretic Mobility of Pb-210 in Sea Water - Na-alginate Systems

Electrolyte	Electrophoretic Mobility of Pb-210 $(cm^2V^{-1}s^{-1}) \times 10^4$ 1st zone	2nd zone
100% sea water	- 0.33	0
100% sea water + 4.6 g Na-alginate	+ 0.60	0
30% sea water	- 2.17	0
30% sea water + 3.3 g Na-alginate	+ 0.90	0
10% sea water	- 3.49	0
10% sea water + 2.1 Na-alginate	+ 1.20	0

anionic direction. At the same time the mobile amount of Pb-210 changes its direction and produces now an anionic zone. The change in direction of electrophoretic mobility is the more pronounced the lower the salinity of the medium despite the fact that the alginate concentration is about a factor 2 higher in 100% sea water. The conclusion from the described electrophoretic behaviour is that Pb-210 moving in the blank media into cationic direction becomes chelated by alginate as is indicated by the drastic change of electrophoretic mobility. This result is in general agreement with the prounounced chelation tendency for Pb of guluronic and mannuronic acid contained in the polysaccharide alginate (Ref. 14). The chalation efficiency of this natural organic ligand grows with decreasing salinity. This is obviously connected with two effects, the corresponding decrease in alkaline earth ion concentration, particularly Ca(II), competing for the chelation sites and the general enhancement of the level of more mobile Pb(II) reflected by the results for the blank media. In general accordance with this conclusion on the alkaline earth ion competition are also previous observations on the chelation by alginate of the even significantly stronger chelates forming Sr-90 (Ref. 9).

No aging effects have been observed with this system at any ligand concentration up to seven days.

Studies with Humic Acid

Measurements have been made in sea water and for comparison also in the model solution 0.55 M NaCl (viz. Table 2-4). The pH was 8.1 in both cases. In sea water

TABLE 2 Distribution of Pb-210 Species in Sea Water - Humic Acid Systems

Concentration of Humic Acid (mg/l)	Distribution of Pb-210 Species (%) C	S	A
0	6.33	93.53	0.14
3	1.91	97.03	1.06
15	1.18	98.44	0.38
65	0.02	98.36	1.62
100	0.05	98.48	1.47

C = cationic zone;
S = zone at the starting point of electrophoretic strip; A = anionic zone (tailing).

TABLE 3 Electrophoretic Mobility of Pb-210 in 0.55 M NaCl - Humic Acid Systems

Concentration of Humic Acid (mg/l)	Electrophoretic Mobility of Pb-210 $(cm^2V^{-1}s^{-1}) \times 10^4$ C	S	A
0	- 0.44	0	tailing
13	- 0.43	0	tailing
17	- 0.52	0	tailing
20 (0-3 days)	-	0	tailing
20 (t=12 days)	- 0.30	0	tailing
25 (0-12 days)	-	0	tailing

C = cationic zone
S = zone at the starting point of electrophorectic strip
A = anionic zone

TABLE 4 Distribution of Pb-210 in 0.55 M NaCl - Humic Acid System

Concentration of Humic Acid (mg/l)	Distribution of Pb-210 over Electrophoretic Zones C	S (%)	A
0	97.07	2.77	0.16
13	94.61	5.24	0.15
17	93.39	6.50	0.11
20 (0-3 days)	0.22	99.18	0.60
20 (t=12 days)	80.05	19.40	0.55
25 (0-3 days)	0.02	97.74	2.24
25 (t=12 days)	0.02	97.74	2.24

C = cationic zone
S = zone at the starting point of electrophoretic strip
A = anionic zone (tailing)

(Table 2) the concentration of humic acid was varied from 3 to 100 mg/l. In this context it is to be remarked that above 15 mg/l the humic acid remains not dissolved but tends to precipitate. Nevertheless, electrophoresis is still possible. Up to a concentration of 15 mg/l no significant effect on the electrophoretic distribution of Pb-210 was noticeable. Above this level of humic acid concentration the cationic zone, at any rate containing only a small Pb-210 amount, begins to disappear. Yet still the majority of Pb-210, i.e. 98%, remains in the immobile zone at the starting point up to the highest adjusted humic acid concentration of 100 mg/l. At the same time some anionic tailing becomes visible at 3 mg/l not exceeding 1.5% even at 100 mg/l. As the humic acid is distributed on the electropheregrams from the starting point into anionic direction it is concluded that the chelation of Pb-210 with the studied humic acid remains unsignificant in sea water. No aging effects up to 12 days were observable.

In the 0.55 M NaCl model solution the electrophoretic distribution is different (viz. Table 4). The pattern of the electrophoretic mobilities in this medium is given in Table 3. In contrast to sea water the majority of Pb-210 (97%) is found in blank 0.55 M NaCl-solution in the cationic zone. Up to 17 mg/l of humic acid Pb-210 remains to more than 90% in the cationic zone. Also the cationic mobility remains approximately the same up to this humic acid level. It has to be noted that in this model medium a pH of 8.1 was adjusted by addition of a corresponding amount of concentrated NaOH. Thus, the complexation of Pb(II) by rather strong carbonato complexes prevailing due to the carbonate buffer system in sea water (Ref. 12 and 13) is not operative in this model medium and thus in the blank and at lower humic acid concentrations the majority of Pb-210 remains distributed in the cationic zone. If the humic acid concentration is elevated to 20 mg/l the cationic zone exhibits a transient behaviour. For electrophoresis of freshly prepared solutions the cationic zone nearly disapperas and the Pb-210 is accumulating in the immobile zone at the starting point. This seems to indicate some interaction with the humic acid, which is distributed from the starting point region into anionic direction. However, this is only a transitory state lasting up to 3 days. Subsequently about 80% of Pb-210 are again in the cationic zone and this is now a stationary state as the observations of aging effects up to the 12th day reveal. If the concentration of humic acid is increased further to 25 mg/l the cationic zone virtually disappears immediately and this state remains stationary. Nearly all Pb-210 remains in the immobile starting point zone while about 2% appear as anionic tailing. This result seems to indicate some permanent chelating interaction between Pb-210 and humic acid at this elevated humic acid concentration level in 0.55 M NaCl while in sea water (viz. Table 2) for this and higher humic acid levels the similar small anionic tailing effect is even less. It should be added that in 0.55 M NaCl humic acid remains in the dissolved state up to the highest tested concentration of 25 mg/l.

The facts, that due to the different Pb(II)-speciation prevailing in sea water, over 90% of Pb-210 are located already in blank sea water at the starting point region (viz. Table 2, zone S) and that on the other hand humic acid distributes from this region into anionic direction, might introduce still some ambiguity in the conclusion that the electrophoretic tests indicate no significant Pb-chelation. However, in 0.55 M NaCl (viz. Table 4), where due to a different Pb-speciation 97% of Pb-210 are located for the blank solution in the cationic zone, up to 17 mg/l humic acid over 90% of Pb-210 remains also in the cationic zone. This clearly settles this point. At least up to this humic acid level Pb-chelation by the studied humic acid type remains unsignificant in both media, the 0.55 M NaCl-solution and also in sea water.

The emerging central experimental result, reflected by the data in Table 2 and 4, that in sea water up to rather high concentrations of the studied humic acid type no significant chelation of Pb-210 occurs, correlates well with the at present available knowledge on the chelating abilities of various humic acid types and of organic chelators with well defined properties for a number of metals. These properties have to be considered in context with the levels of humic material expected to be usually dissolved in the oceans.

Values reported for total dissolved organic matter (DOM) in oceanic waters ranged from 0.8 to 2 mgC/l in surface waters and decreased gradually with depth. Below 500 m depth DOM remained constant around 0.6 mg C/l (Ref. 15). In the Sargasso Sea, for instance, only 5% of the total organic carbon (TOC) have been found to belong to the humic substance fraction in surface waters, while the humic fraction increases to 23% of TOC in deep water (Ref. 16). The differences in humic acid contributions to TOC are connected with the fact that in the euphotic zone nonhumic substances fractions from freshly produced organic material due to the biological activity prevail.

If one assumes according to the aforementioned data that only 5-10% of DOM are humic substances in surface waters one obtains a concentration range $\leqslant$ 0.2 mgC/l. Elemental analysis of the studied terrestrial humic acid yielded for the carbon content 45%. Consequently the order of concentration of humic acids dissolved in oceanic surface and deep waters should not be larger than 0.5 mg/l (viz. TOC in Ref. 16). If one inserts as average molecular weight 3.2×10^4 to 4.6×10^3 daltons for terrestrial humic acids (isolated from peat, Ref. 18) this dissolved level would become only 10^{-8} to 10^{-7}M. Other sources (Ref. 17) report for humic acids originating from peat even smaller average molecular weights of 1.4×10^3 daltons. Due to the polydispersity of humic acids certainly such smaller molecules are present to a certain extent. If they were predominating in sea water they would render the resulting concentration value somewhat more favourable to 3.6×10^{-7}M. One might conclude from the proposed structure of this lower molecular weight humic acid (Ref. 17) that one molecule would have not more than about 10 chelator sites. More probably it has less chelator sites. This would render with respect to maximal chelation potentialities the dissolved concentration of 3.6×10^{-6}M of a ligand with only one chelator function.

These at present only possible and very crude estimates on the possibly existing level of dissolved humic acids (having several chelator sites per molecule) in surface oceanic waters have to be correlated with the known data on stability constants for metals with various types of humic acids (Ref. 18 and 19) and with the general tendencies elucidated for chelation of trace metals by organic ligands in sea water (Ref. 20). Very recently Mantoura et al. (Ref. 18) have presented for a number of metals, but not for Pb, and a variety of humic acid types originating from peat, lakes, rivers, lochs, sediments, sea water and soil stability constants of the formed chelates. Logarithmic K-values for Ca(II) and Mg(II) value from 4.65 to usually 3.4, for Cd(II) the range is 4.57 to 5.08 and for Zn(II) it extends from about 3 to 6. These rather moderate stabilities are also reported for Mn(II), Co(II) and Ni(II). Therefore the authors (Ref. 18) concluded that in sea water the chelation ability of dissolved humic acids for the mentioned trace metals remains unsignificant and over 90% of their chelation sites will be occupied by Ca(II) and Mg(II) being present in sea water in relatively large excess.

An exception are only Cu(II) with lgK between 7.8 and 11.4 and Hg(II) with lgK between 18 and 21.3 forming thus very stable chelates with humic material. It should be noted that the cited stability constant values refer to very low ionic strength (usually I = 0.02). Extrapolations using the Davies equation to sea water ionic strength (I = 0.7) for a number of various types of metal chelates, including humic acid as ligand by Whitfield and Turner (Ref. 19) suggest, however, that the alterations due to the ionic strength effect (common salt effect) are not dramatic and will only have the general tendency to lower the stability constant values somewhat by a few tenth of one logarithmic unit. For the humic acid chelate of Pb(II), not treated in (Ref. 18), Whitfield and Turner (Ref. 19) computed in sea water for lgK a value of 7.49.

Now we will correlate this order of the stability constant for Pb(II)-humate and the previously estimated level of dissolved humic acid, even taking into account that per humic acid molecule up to 10 chelator sites might be operative, with our general findings on the requirements with respect to the two always intercorrelated parameters, stability constant value and ligand concentration, for the achievement of substantial metal chelation in sea water. Our conclusions are based on the generalisable findings of our comprehensive voltammetric studies on this problem close to the usually encountered actual level of dissolved trace metals (Ref. 20, 21, 22). As has been established (Ref. 22) with the well defined ligand NTA for the chelation of about 1×10^{-8}M dissolved Pb(II) an excess of NTA amounting to about 2 to 3×10^{-5}M is required, although the stability constant of Pb(II) has at I = 0.7 a lgK value of about 11. This lgK is more than three orders of magnitude higher than that lgK-value predicted from the computation of Whitfield and Turner (Ref. 19) for Pb(II)-humate. For Cd(II) the demanded NTA-concentration is even about one order of magnitude higher, as the stability constant of $[CdNTA]^-$ in sea water with lgK 9.44 is about one order of magnitude smaller than of $[PbNTA]^-$ (Ref. 20). Again here the chelates of Cd(II) with various types of humic material would have according to (Ref. 18) a 4.5 to 5 orders of magnitude lower stability. The reasons for the required excess of organic ligands to achieve significant chelation of trace metals in sea water are in the first place the always existing competition of the excess alkaline earth ions, Ca^{2+} and Mg^{2+}, for the organic ligands and furthermore the respective side reactions leading to the formation of more labile complexes of the trace metals with inorganic ligands. Sometimes the latter might even form relatively stable complexes as in the case of the in sea water for Pb(II) prevailing carbonato species (Ref. 12 and 13).

The results of our electrophoretic investigations indicating for the studied terrestrial humic acid type virtually no chelation of Pb-210, even up to humic acid concentrations exceeding by about two orders of magnitude the level of humic material estimated, based on the available data, to be dissolved in oceanic surface waters fits well into the general conclusions emerging. At present electrophoretic and voltammetric investigations with humic material isolated from marine sediments are in progress to confirm the foregoing conclusions with a for the marine environment more realistic humic acid type.

Although it is to be expected that, except for Cu(II) and Hg(II), humic material dissolved in sea water is for most metals an unsignificant chelator, the situation might tend for Pb(II) to differ in fresh water systems considerable, as the requirements on the magnitude of organic ligand concentration decrease with lowered salinity, particularly as the concentration of the competing Ca^{2+} and Mg^{2+} becomes lower.

Thus, one concludes from (Ref. 18) that for a lgK-value around 7.5 in fresh water about 50% or more of the dissolved Pb(II) might become chelated by humic material. Consequently the situation in estuaries might differ from that in the open sea particularly for salinities $\leqslant 5^{o}/oo$ where a Pb(II)-chelation by dissolved humic material (lgK $\simeq$ 7.5) of $\geqslant$10% of the dissolved Pb(II) cannot be excluded.

Moreover, even with respect to the open sea (salinity $35^{o}/oo$) one should bear in mind that there remains still a possibility of chelating action on dissolved Pb(II) by humic material bound to the surface of suspended particulate matter, as there the interfacial concentration will be enhanced by several orders of magnitude. A very crude estimate shows for instance that a surface concentration of 10^{-10} moles cm^{-2} assuming a typical thickness of 10 Å for that phase boundary monolayer would be at the surface of the particulate matter particles equivalent to a hypothetical volume concentration of the order of 1 mole/liter. This possibility for formation of trace metal humates with adsorbed humic material layers deserves also particular attention with respect of interstitial water in marine sediments. Thus, surface layers of humic material at suspended particulate matter might even act as chelator for other beforementioned trace metals having lower stability constants, according to (Ref. 18), than to be expected for Pb(II)-humates. Of course, in the sea there always remains the competition of the salinity components, Ca^{2+} and Mg^{2+}, present in excess and this competition will remain of particular significance, if the respective trace metals, e.g. Cd and Zn form humates of not significantly higher stability than the alkaline earth ions.

ACKNOWLEDGEMENT

The authors thank Dr. J. Golimowski and S. Kozar, M. Sc., for the voltammetric determination of natural levels of dissolved Pb(II) in the utilized sea water samples.

This study is part of the joint research project "Environmental Research in Aquatic Systems" of the Institute of Chemistry, Institute 4 Applied Physical Chemistry,

Nuclear Research Centre (KFA) Jülich, F. R. Germany, and the Center for Marine Research of the "Rudjer Bošković" Institute, Zagreb, Yugoslavia. Support of this study, during the attachment of Dr. Lj. Musani to the Institute 4 Applied Physical Chemistry of KFA by the International Bureau of Nuclear Research Centre, Jülich (KFA) on the basis of the bilateral scientific and technical cooperation between Socialist Federative Republic of Yugoslavia and the Federal Republic Germany is gratefully acknowledged.

REFERENCES

(1) Lj. Marazović and Z. Pučar, Two-dimensional electrochromatography of ^{106}ruthenium and some other radiomicroconstituents in sea water, J. Chromatogr. 27, 450 (1967).

(2) M. Ishikawa and Z. Pučar, High-voltage electrophoresis of ^{106}Ru - sea water systems, J. Radioanal. Chem. 11, 197 (1972).

(3) L. G. Sillèn and A. E. Martell (eds.), Stability Constants of Metal Ion Complexes, Spec. Publ. Nr. 17, The Chem. Society, London (1964).

(4) Radioactive nuclides in the marine environment, Chemical Oceanography (J. P. Riley and G. Skirrow, eds.), Academic Press, London-New York-San Francisco, Ch. 18, Vol. 3 (1975).

(5) Y. Nozaki and S. Tsunogai, A simultaneous determination of lead-210 and polonium-210 in sea water, Anal. Chim. Acta 64, 209 (1973).

(6) T. M. Beasley, Lead-210 production by nuclear devices: 1946-1958, Nature 224, 573 (1969).

(7) H. Craig, S. Krishnaswami, and B. L. K. Somayajulu, ^{210}Pb-^{226}Ra: Radioactive disequilibrium in the deep sea, Earth planet Sci. Lett. 17, 295 (1973).

(8) Y. Nozaki and S. Tsunogai. Lead-210 in the North Pacific and the transport of terrestrial material through the atmosphere, Earth planet Sci. Lett. 20, 88 (1973).

(9) Lj. Musani and Z. Konrad, Interaction of ^{90}Sr-^{90}Y and some other radionuclides in sea water and 0.55 M NaCl solutions, Thalassia Jugosl. 13, 93 (1977).

(10) Z. Pučar, Elektrophorese der Halogenkomplexe von Hg(II), Bi(III), Cd(II), Pb(II) and Cu(II). I Chlorokomplexe in Chlorwasserstoffsäure, Anal. Chim. Acta. 17, 476 (1957).

(11) L. Mart, H. W. Nürnberg, and P. Valenta, Comparative base line studies on Pb-levels in European coastal waters, The International Experts Discussion on the "Lead - Occurrence, Fate and Pollution in the Marine Environment", Rovinj, October 1977, M. Branica and Z. Konrad (eds.), Pergamon Press, 1979, 155-179.

(12) L.Sipos, P. Valenta, Voltammetric Determination of the Stability Constants of the Predominant Labile Lead Complexes in Sea Water, The International Experts Discussion on the "Lead - Occurrence, Fate and Pollution in the Marine Environment", Rovinj, October 1977, M. Branica and Z. Konrad (eds.), Pergamon Press, 1979, 61-76.

(13) L. Sipos, B. Raspor, H. W. Nürnberg, and R. M. Pytkowicz, Interaction of metal complexes with coulombic ion-pairs in High salinity aqueous media, Marine Chem. (in press).

(14) A. Haug and C. Smidsrød, Selectivity of some anodic polymers for divalent metal ions, Acta Chem. Scand. 24, 843 (1970).

(15) N. Ogura, Rate and extent of decomposition of dissolved organic matter in surface sea water, Marine Biology 13, 89 (1972).

(16) D. H. Stuermer and G. R. Harvey, The isolation of humic substances and alcohol-soluble organic matter from sea water, Deep-Sea Research 24, 303 (1977).

(17) E. Rammler, Technologie and Chemie der Braunkohlenverwertung, VEB Verlag f. Grundstoffindustrie, Leipzig (1962).

(18) R. F. C. Mantoura, A. Dickson, and J. P. Riley, The complexation of metals with humic materials in natural waters, Estuarine and Coastal Marine Science 6, 387 (1978).

(19) M. Whitfield and D. R. Turner, Theoretical studies of the chemical speciation of lead in sea water, The International Experts Discussion on the "Lead - Occurrence, Fate and Pollution in the Marine Environment",

(20) B. Raspor, P. Valenta, H. W. Nürnberg, and M. Branica, The chelation of cadmium with NTA in sea water as a model for the typical behaviour of trace heavy metal chelates in natural waters, The Science of the Total Environment 9, 87 (1978).

(21) H. W. Nürnberg, P. Valenta, L. Mart, B. Raspor, and L. Sipos, The polarographic approach to the determination and speciation of toxic metals in the marine environment, Fresenius Z. Anal. Chem. 282, 357 (1976).

(22) B. Raspor, P. Valenta, H. W. Nürnberg and M. Branica, The Chelation of lead by organic ligands in sea water, The International Experts Discussion on the "Lead - Occurrence, Fate and Pollution in the Marine Environment", Rovinj, October 1977, M. Branica and Z. Konrad (eds.), Pergamon Press, 1979, 181-195.

Present Potentialities and Limitations of Atomic Absorption Spectroscopy in Environmental and Biological Materials with Particular Reference to Lead Determinations

M. Stoeppler

Institute of Chemistry, Institute 4 Applied Physical Chemistry, Nuclear Research Centre (KFA), Juelich, P.O. Box 1913, D-5170 Juelich, F.R.G.

ABSTRACT

Atomic Absorption Spectroscopy (AAS) despite some disadvantages is a useful tool for trace element determinations due to its sensitivity and easy operation. This is particularly valid for the toxic trace element lead. The absolute detection limit is $\leq 10^{-11}$ g for electrothermal and ≤ 0.02 µg/ml for flame AAS.

The present general potentialities and limitations of AAS with particular reference to lead as well as the necessary sampling, sample storage and pretreatment procedures, are discussed critically. Finally a condensed overview on lead levels which have been determined in selected environmental and biological materials is given and the approach to reliable analysis of these materials is outlined.

1. INTRODUCTION

Although it has been known since around 1860 when Kirchhoff and Bunsen made their fundamental studies, that every substance is able to absorb light at the same wavelength at which it emits light, the practical application of this only began about a century later.

In 1955 Walsh (Ref. 1) proposed to utilize this principle for elemental analysis. This was the beginning of the development of one of the most powerful and wide spread spectroscopic techniques in trace analysis used at present. Figure 1 shows the principle of an atomic absorption spectrometer also including flameless techniques. Compared with emission spectra the advantage of absorption spectra is that they contain far fewer analytical lines. They therefore require significantly less sophisticated monochromator systems and relatively simple light sources made from the elements to be determined. The practical consequence is that for all elements with absorption lines between 190 and 850 nm, the spectral selectivity is high and the occurrence of direct spectral interferences by other elements is very rare (Ref. 2).

Since this method is based upon the atomization of the element to be determined the ratio of the number of vaporized atoms in the ground state to the total number of atoms of the respective element has a very important influence on sensitivity. At first the only technique used to reach efficient atomization was the production of the analyte solution into a very hot flame. Because of its moderate cost, versatility, sensitivity and relatively high precision, despite the limitation to single element analysis, this technique soon played a leading role in the determination of minor and trace constituents with detection limits around or below 1 μg/ml for a considerable number of elements. Under optimal conditions using the flame, lead shows a detection limit of ≤ 0.02 μg/ml (Ref. 2).

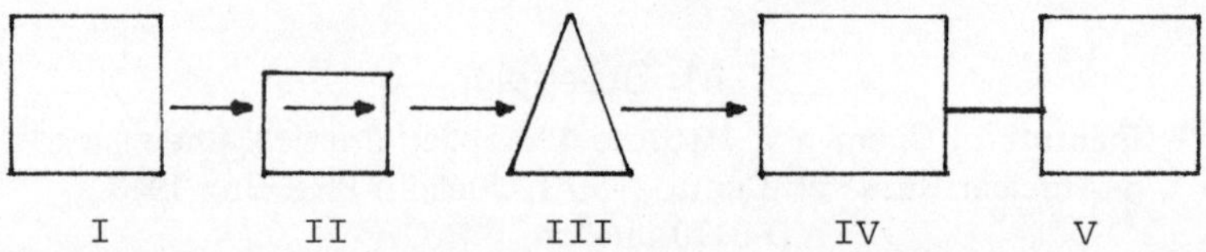

Fig. 1. Principle of an Atomic Absorption Spectrometer.

I Radiation Source; hollow cathode or electrodeless discharge lamp

II Production of Atomic Vapour (Flame, cold vapour, electrothermal)

III Wavelength isolator (Monochromator)

IV Radiation detector (Photomultiplier)

V Readout system (Digital or Recorder)

The widespread use of atomic absorption spectroscopy also introduced AAS in laboratories heretofore without experience in trace analysis. This and the failure of some manufacturers of AAS instruments to ignore related problems and the urgent need for steady and critical improvement in the knowledge of methodological and instrumental limitations have lead to an overly optimistic estimation of the potentialities of this method. Too many laboratories with poor experience, insufficiently trained staff, and often a very little additional instrumentation relied too heavily upon a single, seemingly advanced and error-free apparatus used in a "black-box" manner, although it actually could only be operated successfully by experts and in comparison with the results from other independent, e.g., colorimetric or electrochemical methods. Thus, neglecting of serious errors due to nonspecific absorption and other effects only recognizable by experts caused significant regression with respect to accuracy risks.

The introduction of background compensation, either as the two line technique using a nonabsorbing line influenced only by background effects close to the analytical line, or as the continuum (i.e., deuterium lamp (Ref. 3)) technique, improved this situation markedly. The first instruments with built-in deuterium compensation practically eliminating this source of error for less experienced laboratories and routine use, appeared on the market in 1968 (Ref. 4).

Since flame techniques are limited in sensitivity a more powerful method allowing more complete vaporization and longer residence times of atoms in the light beam was sought. The construction of flameless "electrothermal" atomizers developed by several scientists (Ref. 5-9) and the commercial introduction of graphite furnaces or graphite rods (Ref. 2) based on the prototype built by Massman (Ref. 6) fulfilled these requirements and provided sensitivities down to the absolute level of 10^{-13} g or even less.

Electrothermal atomic absorption spectroscopy reaches for Pb a practically attainable absolute detection limit of $\leq 10^{-11}$ g. That is a good, but not extraordinary sensitivity (Ref. 2).

As a result of the highly enhanced sensitivity, electrothermal atomizers also show an enormous increase in possible source of error. Thus, the need for extended investigations was stressed by their frequently observed failure to determine the desired element when a particular matrix was present. Considering some emphatic forecasts about the possibility of the direct introduction of biological fluids, solutions from different procedures and even of solid samples, this was a rather severe disappointment. Advocated by manufacturers of AAS instruments as a straightforward technique, electrothermal atomization proved in a relatively short time to be as vulnerable to numerous sources of error which are at this time only very poorly understood as it was striking in its sensitivity. The remaining reality was an extremely sensitive technique which needs much more expertise than the flame, and is far from being the ultimate and absolute answer to long-term demands in trace and ultratrace analyses. In the meantime, owing to the contributions of numerous scientists from theoretical and practical standpoints, a much better knowledge of the physical and chemical interactions in electrothermal atomic absorption spectroscopy has been attained. Since these highly sensitive techniques are at present indispensable in biochemical and environmental research, the state-of-the-art of AAS with particular reference to electrothermal devices will be discussed below.

2. STATE-OF-THE-ART OF ATOMIC ABSORPTION SPECTROSCOPY

2.1 Error Sources and Their Elimination

There are some major sources of error at present, which are only partly eliminated by most of the commercially available AAS instruments. They are commented upon in the following without the intention of presenting a comprehensive contribution. For more detailed information, the cited articles may be helpful.

Nonspecific losses of light, i.e., light scattering by solid particles or smoke or broad band molecular absorption due to dissociation phenomena within the band pass (slit width) of the monochromator, may be corrected in conventional instruments with deuterium compensation under optimal conditions up to an absorbance of about 1.0 (Ref. 2, 10). In a routine analytical laboratory, however, the less than optimal conditions are due to a nearly unavoidable optic or electronic mismatching of measurement and reference signals often far from desired values. Positive or negative errors may therefore be generated if the analytical system is operated near the determination limit as is often the case in environmental research and surveillance. Thus, the compensating ability of an AAS system must be continuously controlled by the application of deuterium compensation at a nonabsorbing line. An improvement for routine measurements up to an absorbance of ≥ 1.7 also seems attainable, if the recently developed Zeeman-effect compensated AAS instruments, which utilize the polarization characteristics of the analyte resonance line and are able to compensate at the resonance wavelength exactly are applied (Ref. 11-13). The simple optical system which requires only one element lamp promises, in this case, an end to optical mismatching.

Another source of error are complex effects during the temperature programs of electrothermal atomizers in the graphite tubes or rods, such as recondensation on colder parts which can not be completely overcome at present, using commercial devices (Ref. 14). Thus, the frequently recommended application of signal integration seems for some particular problems, e.g., nonlinear calibration graphs due to the influences of the measuring system or the analyte solution, the best approach (Ref. 15, 16). If a better reproducibility is desired, however, it is not superior to the peak height registration (Ref. 17). Also, belonging to this category are deviations due to signal fluctuations highly dependent on individual tubes, even if taken from the same batch, and on the lifetime of these tubes (Ref. 17-19).

Strong influences are observed by chemical interferences, acting predominantly in flames, e.g., via the formation of stable oxides (refractory elements) and the minimization of the atomization process by the formation of undissociated molecules (Ref. 2). In carbon furnaces various similar but also quite different reactions appear besides the well known carbide formation (Ref. 2) which are at present frequently not yet clearly understood (Ref. 14, 22-24), but in some cases identified as influences of metal halides, which may produce volatile, non-absorbing molecules by direct reaction with the atoms of the respective element (Ref. 19, 25) or cause by its volatization (Ref. 26) strong suppressions. Additionally only partly understood effects - e.g., from different acids (Ref. 27-29) - sometimes lead to a strong signal decrease or even its total disappearance. These effects may be at least partly overcome for some elements and matrices in flame AAS by the use of flames with higher temperatures (e.g., for refractory elements), the addition of dissociating compounds to increase the percentage of atoms of the elements to be determined (Ref. 2) and by prior solvent extraction. The latter, besides the careful evaluation of the respective temperature program for a given matrix and a given element (Ref. 2, 30) is also always recommendable in graphite furnace AAS (Ref. 2, 19). Since it is well known that these effects are highly time-dependent, some sources of error may be eliminated for furnace AAS by direct time-resolved measurements with modified instruments (Ref. 25, 31, 32). Another approved approach is the use of matrix modifying agents to alter or to remove the disturbing species prior to the introduction into commercially available devices (Ref. 25, 33-35).

Particularly but not very frequently in flames as well as in furnaces, spectral interferences caused by a nonlinear background structure due to line-rich electronic excitation molecule spectra occurs. These interferences cannot be properly corrected by deuterium compensation, since the actual background is absolutely dependent on whether or not a rotational line of the molecular spectrum coincides with the respective resonance line. In some cases, the variation of slit width showing changing values for the corrected signal may be a rough indication of such an effect. The measurement of the interfering molecular spectra with high resolution instruments is, however, at present the only certain way to obtain the needed information via a spectrometric method (Ref. 39-42). In analytical practice, and if the obtainable precision of the procedures used is adequate, the application of at least one independent method is by far the best way to recognize systematic errors due to spectral interferences. Concerning the determination of Pb in biological and environmental materials, voltammetry seems the most desirable approach from the authors experience for this purpose at present due to its reliability and the moderate instrumental costs besides isotope dilution mass spectrometry (Ref. 43-47).

Also the use of Zeeman-effect compensation, particularly useful for furnace AAS, may be advantageous if the spectrum of the background is structured (Ref. 11-13), but hitherto no detailed experience has been gained, because the background absorbing species may sometimes exhibit a preferential absorption of one polarization in the magnetic field of such an instrument.

2.2 Improved Accessories

Though at present AAS instruments produced are far from being ideally designed, they, nevertheless, in general represent an acceptable compromise and a few new conceptions, such as Zeeman-effect and very sophisticated computer controlled devices which seem to behave rather satisfactorily. In addition, in the last several years a remarkable improvement in peripheric systems has been noted.

While flame techniques show excellent precision in the optimal range of $\geq$ 0.1 absorbance units - within-run reproducibilities $\leq$ 0.5 % - and pose in principle no problems for automated sample introduction (Ref. 48) this was not initially the case for electrothermal atomizers.

Thus, until recently electrothermal techniques were not known to be very reliable due to the necessary manual sample introduction connected with obvious error sources. To improve this, in the last few years systems for automated sample injection have been developed, checked and subsequently introduced commercially (Ref. 17, 20, 21, 49, 50). These devices circumvent some former disadvantages of graphite furnace AAS. The automated delivery of the analyte solution via automatically cleaned teflon tubes improves the <u>accuracy</u> by avoiding the hitherto frequently observed contamination during injection caused by an undesired contact with the, often highly contaminated, outside near the injection hole or from the trace element contents of the commonly used disposable polyolefine tips. In addition, also the <u>precision</u> could be substantially improved avoiding improper manual dosage, thus resulting in within-run reproducibilities around 1 % for absorbances $\geq$ 0.1 (Ref. 17). An example of the practically obtainable precision in blood Pb determinations (Ref. 38) is shown in Table 1.

This progress renders possible an increased number of analyses per working day and the additional introduction of computer coupling (Ref. 38), since the hitherto relatively low number of firings per day obtainable is now increased by at least a factor of two, therefore now giving electrothermal devices a better chance from the standpoint of speed and costs per analysis compared with flame AAS (Ref. 38, 51).

From this view the useful analytical lifetime of graphite tubes is of paramount importance for the cost-benefit ratio of an automated system with the capability of performing round-the-clock measurements, since the replacement of a tube commonly requires new calibration and a series of checking measurements. Thus, extensive studies on the practically attainable tube-life time were performed. They show, with the exception of the occasionally early breakdown of tubes (Ref. 52), quite satisfactory average lifetimes, well above 200 firings even in routine use and also with comparatively aggressive solutions (Ref. 17, 18, 20, 21).

In order to enhance the sensitivities for some carbide forming and/or less volatile elements and also to improve the tube properties in general, impregnation procedures with tungsten, niob, tantalum, and zirconium salts (Ref. 53-56) or a coating with pyrolytic carbon (Ref. 17, 18, 21, 70-78) have been advocated. In all cited studies, it could be shown that these treatments are very promising for signal enhancement as well as for a marked increase in average tube lifetimes. Thus, it is strongly recommended to use these possibilities whenever practicable since they are applicable without undue expense and easy to perform.

3. SAMPLING, SAMPLE STORAGE AND SAMPLE PRETREATMENT

The initial steps in trace and ultratrace analyses are frequently the source of systematic errors and therefore usually the most important and also the most crucial part of an analytical procedure (Ref. 43, 44, 46, 47, 61, 62).

TABLE 1 Pb in whole blood, HNO_3-procedure, automated electrothermal AAS (HGA) 74 - PE 400 with AS-1 unit

Calculation of standard deviations (S.D.), relative standard deviations (R.S.D.) and confidence intervals from pairs of subsamples arranged in 7 groups with increasing lead concentration; total: 282 samples, values in µg/l, Ref. 37 .

Number of pairs	Mean	Range	S.D.	R.S.D. %	Confidence intervals (%) p = 0.1	p = 0.05
24	52.4	36- 60	4.88	9.3	15.9	19.2
45	73.2	61- 80	6.55	8.9	14.9	17.8
45	100.8	81-120	8.50	8.4	14.1	16.8
40	136.6	121-160	7.68	5.6	9.4	11.2
40	174.5	161.200	8.72	5.0	8.4	10.0
40	232.7	201-300	10.38	4.47	7.5	9.0
48	479.0	301-700	16.63	3.47	5.8	7.0

Remarks: The above values are mean values from 2-3 firings evaluated against a matrix matched average calibration graph (at least 2 matrix additions).

The standard deviation was computed from the difference of two parallel determinations at different times, the number of pairs indicated above for each group following the equation:

$$S.D. = \sqrt{\frac{\Sigma(\Delta x)^2}{2n}}$$

where Δx = difference between the two single values; n = number of choosen pairs.

3.1 Sampling and Sample Storage

The sampling, of particular importance for an element such as Pb which is spread in the environment in excessive amounts by human activities (Ref. 63, 64), has to be performed with all possible precautions with respect to contamination and also to the conservation of the original sample properties. This includes the use of clean benches and the scrupulous cleaning of all labware prior to use to ensure the lowest attainable blank level.

To obtain reliable information about contamination levels and limits a series of measurements of rinsing solutions has to be performed to develop the best suitable cleaning procedures for labware and storage containers. Since, on the other hand, absorption losses may also occur, if, e.g., liquids with very low trace element levels have to be analyzed, the use of practically carrier-free solutions of radionuclides, e.g., ^{210}Pb and ^{212}Pb for Pb is also an excellent way to check this behaviour (Ref. 43).

The storage of environmental and biological samples should be preferably performed in carefully cleaned containers made from teflon, polyolefines or quartz, and at lower temperatures to prevent a microbial degradation. With the exception of whole blood stabilized with heparine, sodium citrate or EDTA, stored at a temperature of 278-283 K, a temperature blow 225 K is advisable for biological and environmental materials.

In some cases, lyophilization followed by sterilization is also applicable if long-term storage is desired and relatively stable compounds without extremely low trace element levels have to be considered, as is the case for some standard reference materials (Ref. 65). In principle, knowledge about storage behaviour is limited at present and that is very urgent to learn more about this subject.

For reliable data, normalization to a meaningful weight basis is necessary and may be realized at present by the simultaneous use of an actual weight for analytical purposes and a dry weight after freeze or oven drying as the reference weight (Ref. 66).

3.2 Sample Pretreatment

The pretreatment steps prior to AAS analysis are manifold and may be subject to contamination as well as to elemental losses in digestion and concentration (e.g. extraction) procedures. These procedures may introduce a broad variety of errors if performed by a nonexperienced staff. Thus, for the standardization and relevant application of a method all these steps must be checked very carefully to obtain routine procedures which are as simple as possible but sufficiently accurate and precise.

To circumvent contamination and cumbersome methodological developments, direct measurements, e.g., in liquid samples, are sometimes advantageous as they may in addition increase the determination rate (Ref. 43, 47, 67, 68). In this case, rather frequently encountered disadvantages are rather specific but inaccurate results can be due to matrix and background influences as mentioned above. Thus, accuracy checks must be performed as often as possible by the continuous application of suitable (i.e., carefully pre-analyzed) working standards or standard reference materials. If such a material is not at hand, analysis with at least one independent method is necessary. An excellent check for Pb (and also Cd) in body fluids is an additional analysis by differential pulse anodic stripping voltammetry (Ref. 43, 44, 46, 47) as demonstrated for whole blood in Table 2.

Direct AAS measurements with milligram amounts of solid samples are also advocated in the literature. This may be only useful in particular cases (Ref. 69) and for preliminary rough information about trace element levels in an unknown material at the present technical state of AAS due to statistical reasons.

Precise and statistically accurate data for solid materials are only to be expected if a sufficiently high sample amount is digested prior to analysis. In the last decade, numerous papers have appeared on this subject dealing with the optimization of classical, i.e., wet and dry ashing and combustion methods, with respect to a decrease of contamination and time and the achievement of a complete dissolution of "difficult" matrices and of mechanization (Ref. 43, 47, 67, 70-74). The growing use of nitric acid pressure decomposition in teflon crucibles according to slightly differing principles, with and without springs (Ref. 75-81) also with multisample units for routine analysis (Ref. 80), as shown in Fig. 2, provides an excellent way for rather contamination free solutions and is particularly suitable for volatile elements and, in general, for atomic absorption spectroscopy and radiochemical methods. A disadvantage is the incomplete destruction of organic matter (Ref. 43, 80, 81), probably due to the temperature limit of the teflon vessels used (about

TABLE 2 Blood Pb, Intercomparison between Differential Pulse Anodic Stripping Voltammetry (DPASV) at the Hanging Mercury Drop Electrode after Microwave induced ashing with an oxygen plasma (Ref. 46) and automated Electrothermal Atomic Absorption Spectrometry after deproteinization and matrix modification (Ref. 38), n = 15 samples from unexposed adults, males and females (values in µg/l)

	DPASV	AAS
	108	91
	93	90
	71	95
	211	180
	85	86
	40	39
	35	32
	36	36
	87	97
	137	110
	96	115
	69	80
	188	224
	103	105
	97	80
mean	97.1	97.3
stand. dev.	50.4	50.5
stand. dev. of the mean	13.0	13.0

430 K) which makes these techniques somewhat unsuitable for electroanalytical methods (Ref. 43, 81). However, low temperature ashing in a microwave excited oxygen plasma, if properly operated, has proven to yield excellent ashes for numerous biological and environmental materials with extremely low blanks (Ref. 43, 46, 82-84) very useful for all relevant trace analytical methods.

The solutions from the described digestion procedures were introduced either into the flame or the graphite furnace if necessary after appropriate dilution.

Flame atomic absorption spectrometry offers, in comparison to the graphite furnace, a considerably higher speed, about a factor of two better precision and a much less vulnerability against matrix effects. Hence the application of the flame is preferable, if the analyte concentration allows its use. Further, if an automated injection of 100 µl sample aliquots into the flame (Ref. 48) is applied, the required total amount of the analyte solution can be minimized.

Fig. 2. Exploded view of a multisample decomposition unit made from an aluminum alloy for the simultaneous treatment of 9 teflon crucibles with 35 ml volume, allowing the decomposition of sample weights expressed as dry weight up to 1000 mg per crucible (Ref. 80).

If the concentration in the resulting solution is very low, or if salts interfere, Pb and also other trace elements, have to be extracted by Ammonium Pyrrolidine Dithiocarbamate (APDC), other chelating agents or suitable mixtures of chelates (Ref. 47, 67, 85, 86).

Solvent extraction is necessary, if liquid samples with a high salt content and very low trace element levels (urine, tap, sea water, etc.) have to be analyzed.

4. TYPICAL LEAD VALUES AND ANALYTICAL DATA FOR BIOLOGICAL AND ENVIRONMENTAL MATERIAL

Table 3 shows typical ranges and, if useful, average levels of Pb in several selected materials together with references to possible and/or recommendable procedures in a condensed overview, exclusively based on studies performed in our laboratory (Ref. 38, 43, 46, 47, 66, 87-99). Since Pb determinations are frequently subject to contamination due to the ubiquitous distribution of Pb and the continuous introduction of this metal into the atmosphere (dust) via anti-knock agents, the

Table 3 Typical Pb Content of Some Environmental Materials with Remarks on Recommendable Procedures and Feasibility, values in ng/g

Sample type	range	average	pretreatment	AAS-technique	feasibility	Remarks on sample and sample origin	Ref.
I. Human Samples							
washed hair, near root ($\leq$5 cm) unexposed	$50- 15 \cdot 10^3$	$1.1 \cdot 10^3$	Dissol.with HNO_3 or pressure digestion	flame or electrothermal	easy		44,88,91
whole blood occupational exposed	$80-1.2 \cdot 10^3$	270	Deproteinisation + Matrix modification	electrothermal	"	n >300, Western Europe	43,44,87, 88
whole blood total, unexposed	35-425	120	"	"	"	n > 800, Western Europe	87,88
whole blood males, unexposed	35-425	140	"	"	"		87,88
whole blood females, unexposed	35-200	90	"	"	"		87,88
urine	5-100	20	pH adjustment, liquid extraction	"	feasible	Fed.Rep.Germany	47,94
II. Environmental Samples							
Coal, soil,	$100-300 \cdot 10^3$		dry, wet or pressure decomp.	flame or electrothermal	easy	Fed.Rep.Germany	88,95
river sediments	$40 \cdot 10^3-800 \cdot 10^3$		pressure decomp. (HNO_3 + HCl)	flame	"	Rhine and tributaries, Fed.Rep. Germany	88,95
grass etc.	$1.10^3-10.10^3$		wet or pressure decomp. (HNO_3)	flame or electrothermal	"	without traffic	44,88
sea water, overall	0.003-5.0		liquid extraction	electrothermal	very difficult	not recommendable for AAS, best method DPASV/MFE	89,90
tap and inland water	0.02-30.0		direct introduction or liquid extraction	electrothermal	difficult	DPASV by far more advantageous	88,97

III. Biological Materials and Food

Sample type	range	average	pretreatment	AAS-techniques	feasibility	Remarks on sample and sample origin	Ref.
fish fillet marine fish		<20	pressure decomp. or low temp.ashing direct, or liquid extraction	electrothermal	difficult	Mediterranean Sea, North and Baltic Sea, values possibly below 5 ng/g or less, limited number of species	66,88
sepia		<20	"	"	"	"	66,88
mussels	220-2.5 . 10^3	1.1 . 10^3	pressure decomp. or wet	flame or electrothermal	easy	"	66,88
crabs	150-420	230	"	electrothermal	"	"	66,88
flour		<20	pressure decomp. or low emp.ashing direct, or extraction	"	difficult		83,98
sugar		<10	"	"	"		88,98
starch products		<10	"	"	"		88,98
wines, red wines, white	80-700 90-700		direct after acidifying or pressure decomp.	"	easy	now under detailed study, may sometimes contain volatilisable lead compounds	44,88,99
beverages	5-100	50	direct or liquid extraction	"	feasible		44,88,98
cereals	5-100	30	pressure decomp. or low temp.ashing	"	"		88,98
baby meals	10- 50	20	"	"	"		88,98
canned food	10-400		"	"	"	lead increase possibly due to solders, limited number of samples	44,88,98

analysis of Pb at the ng/g or ng/ml level is extremely difficult and requires rigid contamination precautions (Ref. 92, 93). Thus, for the reliable determination of very low Pb levels, advanced laboratory equipment and a skillful and well-trained staff is mandatory.

5. ORGANOLEAD COMPOUNDS

Hitherto it was supposed that Pb occurs in the environment predominantly in the ionic form. Thus, only a few studies have been devoted to the question of a possible biotransformation into the much more toxic organolead compounds and their contribution to environmental pollution (Ref. 100-102).

GC-AAS coupling is very useful for separating and detecting organically bound lead (Ref. 103). The recently performed identification of compounds of this type in seafood using an extraction/digestion/electrothermal AAS method was reported to yield appreciable amounts of organolead compounds (Ref. 104).

The possible occurrence of organolead compounds in environmental materials should thus stimulate careful investigations on this subject, including specific food samples of marine and terrestrial origin.

REFERENCES

(1) A. Walsh, The Application of Atomic Absorption Spectra to Chemical Analysis, Spectrochim. Acta 7a, lo8 (1955).

(2) B. Welz, Atomic Absorption Spectroscopy, Verlag Chemie, Weinheim, 1976.

(3) R. A. Newstead and P. I. Whiteside, Background Correction for Atomic Absorption Spectrophotometers, European Spectroscopy News 11, 8 (1977).

(4) H. L. Kahn, A Background Compensation System for Atomic Absorption, At. Absorption Newslett. 7, 4o (1968).

(5) B. V. L`vov, The Analytical Use of Atomic Absorption Spectra, Spectrochim. Acta 17, 761 (1961).

(6) H. Massmann, Bestimmung von As mittels Atomabsorption, Fresenius Z. Anal. Chem. 225, 2o3 (1967).

(7) H. Massmann, Vergleich von Atomabsorption und Atomfluoreszenz in der Graphitküvette, Spectrochim. Acta 23B, 215 (1968).

(8) T. S. West and X. K. Williams, Atomic Absorption and Fluorescence Spectroscopy with a Carbon Filament Atom Reservoir, Anal. Chim. Acta 45, 27 (1969).

(9) R. Woodriff, Atomization Chambers for Atomic Absorption Spectrochemical Analysis: A Review, Applied Spectroscopy 28, 413 (1974).

(10) M. Dujmović, P. Adrian and E. Knözinger, Eine neuartige Methode der Kompensation unspezifischer Lichtverluste in der Atomabsorptions-Spektroskopie, Fresenius Z. Anal. Chem. 272, 177 (1974).

(11) H. Koizumi and Yasuda, An Application of the Zeeman Effect to Atomic Absorption Spectrometry: A New Method for Background Correction, Spectrochim. Acta 313, 237 (1976).

(12) E. Grassam, I. B. Dawson and D. I. Ellis, Application of the Inverse Zeeman Effect to Background Correction in Electrothermal Atomic-absorption Analysis, Analyst lo2, 8o4 (1977).

(13) S. D. Brown, Zeeman Effect-Based Background Correction in Atomic Absorption Spectrometry, Anal. Chem. 49, 1269A (1977).

(14) B. V. L'vov, Elektrothermal Atomization - The Way Towards Absolute Methods of Atomic Absorption Analysis, 3rd Annual Meeting FACSS, Nov. 15-19, Philadelphia (1976).

(15) P. Schramel, The Application of Peak Integration in Flameless Atomic-absorption Spectrometry, Anal. Chim. Acta 72, 414 (1974).

(16) H. J. Issaq and R. M. Young, Peak Area versus Peak Height in Flameless Atomic Absorption Measurements, Appl. Spectroscopy 31, 171 (1977).

(17) M. Stoeppler, M. Kampel, and B. Welz, Contributions to Automated Trace Analysis, I. Studies on Automated Flameless Atomic Absorption Spectrometry, Fresenius Z. Anal. Chem. 282, 369 (1976).

(18) K. R. Sperling, Determination of Heavy Metals in Seawater and in Marine Organisms by Flameless Atomic Absorption Spectrophotometry. V. Methane Pyrolysis Treatment of the Graphite Tube Atomizer, Fresenius Z. Anal. Chem. 283, 3o (1977).

(19) G. Volland, G. Kölblin, P. Tschöpel and G. Tölg, Einige systematische Fehlerquellen bei der Bestimmung von Elementen im ng- und pg-Bereich mit der Atomabsorptionsspektrometrie mit flammenloser Anregung im Graphitofen, Fresenius Z. Anal. Chem. 284, 1 (1977).

(2o) R. Dahl and M. Stoeppler, Studien zur automatischen Probenaufgabe bei der flammenlosen AAS mit der Perkin-Elmer Graphitküvette HGA 72, Ber. KFA Jülich, Jül- 1234 (1975).

(21) M. Stoeppler and M. Kampel, Langzeitversuche mit dem Autosampler AS-1 zur automatischen Probeneingabe in die Graphitküvette HGA 74/76, Ber. KFA Jülich, Jül-136o (1976).

(22) R. B. Cruz and J. C. van Loon, A Critical Study of the Application of Graphite Furnace Non-flame Atomic Absorption Spectrometry to the Determination of Trace Base Metals in Complex Heavy Matrix Sample Solutions, Anal. Chim. Acta 72, 231 (1974).

(23) J. Smeyers-Verbeke, Y. Michote, P. van den Winkel, and D. L. Massart, Matrix Effects in the Determination of Copper and Manganese in Biological Materials Using Carbon Furnace Atomic Absorption Spectrometry, Anal. Chem. 48 125 (1976).

(24) T. C. Woodis, Jr., G. B. Hunter, and F. J. Johnson, Statistical Studies of Matrix Effects on the Determination of Cadmium and Lead in Fertilizer Materials and Plant Tissue by Flameless Atomic Absorption Spectrometry, Anal. Chim. Acta 9o, 127 (1977).

(25) E. J. Czobik and J. P. Matousek, Interference Effects in Furnace Atomic Absorption Spectrometry, Anal. Chem. 5o, 2 (1978).

(26) W. Oelschläger and W. Lautenschläger, Interferenzeinflüsse in der Graphitküvette am Beispiel der Bestimmung von Blei und Cadmium in biologischen und anderen Materialien, Fresenius Z. Anal. Chem. 287, 28 (1977).

(27) A. Andersson, Interferences from Sulfur in the Determination of Lead with the Graphite Furnace, At. Absorption Newslett. 15, 71 (1976).

(28) W. Wegscheider, G. Knapp, and H. Spitzy, Statistical Investigation of Interferences in Graphite Furnace Atomic Abosorption Spectroscopy, III. Lead, Fresenius Z. Anal. Chem. 283, 183 (1977).

(29) K. Julshamn, Inhibition of Response by Perchloric Acid in Flameless Atomic Absorption, At. Absorption Newslett. 16, 149 (1977).

(30) C. Hendriks-Jongerius and L. de Galan, Practical Approach to Background Correction and Temperature Programming in Graphite Furnace Atomic Absorption Spectrometry, Anal. Chim. Acta 87, 259 (1976).

(31) F. D. Posma, J. Balke, R. F. M. Herber, and E. J. Stuik, Microdetermination of Cadmium and Lead in Whole Blood by Flameless Atomic Absorption Spectrometry Using Carbon-Cup as Sample Cell and Comparison with Flame Studies, Anal. Chem. 47, 834 (1975).

(32) P. del Castilho and R. F. M. Herber, The Rapid Determination of Cadmium, Lead, Copper and Zinc in Whole Blood by Atomic Absorption Spectrometry with Electrothermal Atomization. Improvements in Precision with a Peak-Shape Monitoring Device, Anal Chim. Acta 94, 269 (1977).

(33) R. D. Ediger, Atomic Absorption Analysis with the Graphite Furnace Using Matrix Modification, At. Absorption Newslett. 14, 127 (1975).

(34) J. G. T. Regan and J. Warren, A Novel Approach to the Elimination of Matrix Interferences in Flameless Atomic Absorption Spectroscopy Using a Graphite Furnace, Analyst 101, 220 (1976).

(35) D. J. Hodges, Observations on the Direct Determination of Lead in Complex Matrices by Carbon Furnace Atomic Absorption Spectrophotometry, Analyst 102, 66 (1977).

(36) H. T. Delves, A Simple Matrix Modification Procedure to Allow the Direct Determination of Cadmium in Blood by Flame Microsampling Atomic Absorption Spectrophotometry, Analyst 102, 4o3 (1977).

(37) F. Alt, Eine einfache, schnelle und zuverlässige Bleibestimmung in Blut mittels Atomabsorptionsspektrometrie, Fresenius Z. Anal. Chem. 29o, 1o8 (1978).

(38) M. Stoeppler, K. Brandt, and T. C. Rains, Contributions to Automated Trace Analysis. Part II. Rapid Method for the Automated Determination of Lead in Whole Blood by Electrothermal Atomic Absorption Spectrophotometry, Analyst 1o3, 714 (1978).

(39) H. Massmann and S. Güçer, Physikalische und chemische Prozesse bei der Atomabsorptionsanalyse mit Graphitrohröfen, Spectrochim. Acta 29B, 283 (1974).

(40) R. Höhn and E. Jackwerth, Nicht kompensierbarer Untergrund als systematischer Fehler bei der Atomabsorptions-Spektrometrie, Anal. Chim. Acta 85, 4o7 (1976).

(41) H. Massmann, Z. El Gohary and S. Güçer, Analysenstörungen durch strukturierten Untergrund in der Atomabsorptionsspektrometrie, Spectrochim. Acta 31B, 399 (1976).

(42) J. B. Dawson and M. J. Keir, Interactions in Flameless Atomic Absorption, Analysis 4, 273 (1976).

(43) M. Stoeppler et al., Zur Blei- und Cadmiumanalytik in biologischen Matrices, Proc. Internat. Symposium, Recent Advances in the Assessment of the Health Effects of Environmental Pollution, Paris, June 24-28, 1974, Luxembourg (1975).

(44) H. W. Nürnberg, M. Stoeppler, and P. Valenta, On the Accuracy and Reliability of Trace Metal Determinations in Environmental Matrix Types by Advanced Polarographic and Spectroscopic Techniques, Thalassia Yugosl. 11, 85 (1975).

(45) H. Jönsson, Determination of Lead and Cadmium in Milk with Modern Analytical Methods, Z. Lebensm. Unters.-Forsch. 16o, 1 (1976).

(46) P. Valenta, H. Rützel, H. W. Nürnberg, and M. Stoeppler, Trace Chemistry of Toxic Metals in Biomatrices. II. Voltammetric Determination of the Trace Content of Cadmium and Other Toxic Metals in Human Whole Blood, Fresenius Z. Anal. Chem. 285, 25 (1977).

(47) J. Golimowski, P. Valenta, M. Stoeppler, and H. W. Nürnberg, Ein Verfahren zur raschen voltammetrischen Simultanbestimmung der toxischen Spurenmetalle Cd und Pb in Harn, Fresenius Z. Anal. Chem. 29o, 1o7 (1978).

(48) H. Berndt and E. Jackwerth, Automatische Mikromethode der Flammen-Atomabsorptionsspektrometrie für die Bestimmung von 8 Elementen in Humanseren, Fresenius Z. Anal. Chem. 29o, 1o5 (1978).

(49) C. J. Pickford and G. Rossi, Analysis of High Purity Water by Flameless Atomic Absorption Spectroscopy, Part I. The Use of an Automated Sampling System, Analyst 97, 647 (1972).

(50) K. G. Brodie, N. K. Pradhan, B. J. Steens, The Measurement of Whole Blood Lead and Cadmium Using an Automatic Sample Dispenser and Furnace Atomic Absorption, in: Clinical Chemistry and Chemical Toxicology of Metals, S.S. Brown (ed.), Elsevier/North Holland Biochemical Press, 1977, p. 357.

(51) S. Sabet, J. M. Ottaway, and G. S. Fell, Comparison of the Delves Cup and Carbon Furnace Atomisation Used in Atomic-absorption Spectroscopy for the Determination of Lead in Blood, Proc. Analyst. Div. Chem. Soc., 1977, 3oo.

(52) M. Stoeppler, to be published.

(53) I. A. Kuzuvlev, J. N. Kuznecow, and O. A. Sverdlina, Verbesserung der Technik und Methode bei der Atomabsorption mit Graphit-Küvette, Zavodsk. Lab. 39, 428 (1978).

(54) H. M. Ortner and E. Kanutscher, Metallsalzimprägnierung von Graphitrohren zur verbesserten AAS-Siliziumbestimmung, Talanta 22, 581 (1975).

(55) T. Stiefel, K. Schulze, G. Tölg, and H. Zorn, A Combined Method for the Determination of Beryllium in Biological Materials by Means of Flameless Atomic Absorption Spectrometry, Anal. Chim. Acta 87, 67 (1976).

(56) V. J. Zatka, Tantalum Treated Graphite Atomizer Tubes for Atomic Absorption Spectrometry, Anal. Chem. 5o, 538 (1978).

(57) D. D. Siemer, R. Woodriff, and B. Watne, A Simple Technique for Coating Carbon Atomic Absorption Atomizer Components with Pyrolytic Carbon, Appl. Spectroscopy 28, 582 (1974).

(58) S. A. Clyburn, T. Kantor, and C. Veillon, Pyrolysis Treatment for Graphite Atomization Systems, Anal. Chem. 46, 2213 (1974).

(59) K. C. Thompson, R. G. Godden, and D. R. Thomerson, A Method for the Formation of Pyrolytic Graphite Coatings and Enhancement by Calcium Addition Techniques for Graphite Rod Flameless Atomic Absorption Spectrometry, Anal. Chim. Acta 74, 289 (1975).

(60) D. C. Manning and R. D. Ediger, Pyrolysis Graphite Surface Treatment for HGA-2100 Sample Tubes, At. Absorpt. Newslett. 15, 42 (1976).

(61) G. Tölg, On the Analysis of Trace Elements in Biological Material, Fresenius Z. Anal. Chem. 283, 257 (1977).

(62) E. I. Hamilton, Review of the Chemical Elements and Environmental Chemistry-Strategies and Tactics, The Science of the Total Environment 5, 1 (1976).

(63) M. Murozumi, T. J. Chow, and C. C. Patterson, Chemical Concentrations of Pollutant Lead Aerosols, Terrestrial Dusts, and Sea Salts in Greenland and Antartic Snow Strata, Geochim. Cosmochim. Acta 33, 1247 (1969).

(64) R. Elias, Y. Hirao, and C. C. Patterson, Impact of Present Levels of Aerosol Lead Concentration on Both Natural Ecosystems and Humans, in: International Conference on Heavy Metals in the Environment, Toronto, Canada, 1975, T. C. Hutchinson (ed.), Vol. II, 1 257 (1977).

(65) P. D. LaFleur, Biological Matrix Standard Reference Materials for Trace Element Determinations, J. Radioanal. Chem. 19, 227 (1974).

(66) M. Stoeppler, Choice of Species, Sampling and Sample Pretreatment for Subsequent Analysis and Banking of Marine Organisms Useful for Hg, Pb and Cd Monitoring, Int. Workshop on Monitoring Environmental Materials and Specimen Banking, Berlin, 23-28 Oct. 1978 (preprint).

(67) D. Adler and M. Stoeppler, Radiochemical and Methodological Studies on the Recovery and Analysis of Nickel in Urine, Journ. Anal. Toxicol. 1, 252 (1977).

(68) D. Horbach and H. Diehl, Comparison of Sample Preparation Methods for Quantitative Determination of Heavy Metals in Fruit Juices with Flameless Atomic Absorption Spectroscopy, Fresenius Z. Anal. Chem. 29o, 145 (1978).

(69) B. V. L`vov, Trace Characterization of Powders by Atomic Absorption Spectrometry, Talanta 23, lo9 (1976).

(70) B. Sansoni and W. Kracke, Aufschluss und Veraschung organischer Substanzen durch Radikale in wässriger Lösung, II. Nassveraschung von Nahrungsmitteln und biologischem Material mit $H_2O_2/Fe2^+$, Fresenius Z. Anal. Chem. 243, 209 (1969).

(71) W. Matthes, T. Kiss, and M. Stoeppler, Zur Jodbestimmung im Nanogrammbereich nach der Sandell-Kolthoff-Reaktion, Fresenius Z. Anal. Chem. 267, 89 (1973).

(72) F. Knapp, Mechanized System for the Performance of Wet Decomposition Methods for Organic Matrices, Fresenius Z. Anal. Chem. 274, 271 (1975).

(73) G. D. Martinie and A. A. Schilt, Investigation of the Wet Oxidation Efficiencies of Perchloric Acid Mixtures for Various Organic Substances and the Identities of Residual Matter, Anal. Chem. 48, 7o (1976).

(74) D. Kunkel, Advances in Wickbold Combustion Techniques, Microchim. Acta 9 (1976).

(75) B. Bernas, A New Method for Decomposition and Comprehensive Analysis of Silicates by Atomic Absorption Spectrometry, Anal. Chem. 4o, 1682 (1968).

(76) F. J. Langmyhr and P. E. Paus, A Bomb for the Hydrofluoric Decomposition of Inorganic Materials, Anal. Chim. Acta 49, 358 (197o).

(77) W. Holak, B. Krinitz, and J. C. Williams, Simple Rapid Digestion Technique for the Determination of Mercury in Fish by Flameless Atomic Absorption, IAOAC 55, 741 (1972).

(78) L. Kotz, G. Kaiser, P. Tschöpel, and G. Tölg, Aufschluss biologischer Matrices für die Bestimmung sehr niedriger Spurenelementgehalte bei begrenzter Einwaage mit Salpetersäure unter Druck in einem Teflongefäss, Fresenius Z. Anal. Chem. 26o, 2o7 (1972).

(79) A. M. Hartstein, R. W. Freedman, and D. W. Platter, Novel Wet-Digestion Procedure for Trace Metal Analysis of Coal by Atomic Absorption, Anal. Chem. 45, 611 (1973).

(80) M. Stoeppler and F. Backhaus, Pretreatment Studies with Biological and Environmental Materials, I. Systems for Pressurized Multisample Decomposition, Fresenius Z. Anal. Chem. 291, 116 (1978).

(81) S. Gomiscek, V. Hudnik, and V. Veber, Present Status and Some Problems of Pressurized Sample Decomposition in Trace Analysis of Biological Materials, in: Clinical Chem. and Chem. Toxicol. of Metals, S. S. Brown, (ed.), Elsevier, North-Holland, Biochemical Press, 1977, p. 319.

(82) C. A. Evans, Jr., and G. H. Morrison, Trace Element Survey Analysis of Biological Materials by Spark Source Mass Spectrometry, Anal. Chem. 6, 869 (1968).

(83) G. Kaiser, P. Tschöpel and G. Tölg, Aufschluss mit aktiviertem Sauerstoff bei Bestimmung extrem niedriger Spurenelementgehalte in organischem Material, Fresenius Z. Anal. Chem. 253, 177 (1971).

(84) J. Fabry and N. Nangniot, La Calcination a Basse Temperature, Analysis 1, 117 (1972).

(85) S. R. Koirtyohann and J. W. Wen, Critical Study of the APDC-MIBK Extraction System for Atomic Absorption, Anal.Chem. 45, 1986 (1973).

(86) J. D. Kinrade and J. C. van Loon, Solvent Extraction for Use with Flame Atomic Absorption Spectrometry, Anal. Chem. 46, 1894 (1974).

(87) M. Stoeppler and K. Brandt, Determinations of Lead and Cadmium in Whole Blood by Electrothermal Atomic Absorption Spectroscopy, in: Diagnosis and Therapy of Porphyrias and Lead Intoxication, M. Doos (ed.), Springer-Verlag Berlin Heidelberg 1978, p. 185.

(88) M. Stoeppler and H. W. Nürnberg, Toxische Spurenmetalle in der Umwelt, KFA, annual report 1974, updated reprint (1978).

(89) H. W. Nürnberg, Potentialities and Applications of Advanced Polarographic and Voltammetric Methods in Environmental Research and Surveillance of Toxic Metals, Electrochim. Acta 22, 935 (1977).

(90) H. W. Nürnberg, Polarography and Voltammetry in Studies of Toxic Metals in Man and his Environment, Sci. Tot. Env. (1978) (in press).

(91) M. Stoeppler and K. Brandt, On the Importance of Hair as an Indicator of Environmental and Occupational Exposure to Toxic Trace Elements, Report to the IAEA Advisory Group on Accelerator-based Techniques for the Analysis of Pollutants in Man (1978).

(92) T. J. Chow, C. C. Patterson, and D. Settle, Occurrence of Lead in Tuna, Nature 251, 159 (1974).

(93) C. C. Patterson and D. M. Settle, The Reduction of Order of Magnitude Errors in Lead Analysis of Biological Materials and Natural Waters by Evaluating and Controlling the Extent and Sources of Industrial Lead Contamination Introduced During Sample Collection and Analysis, in: Accuracy in Trace Analysis, Sampling, Sample Handling, Analysis, P. D. LaFleur, (ed.), NBS Spec. Publ. 422, 321 (1976).

(94) M. Stoeppler, K. Brandt (to be published).

(95) K. Hilpert, J. D. Schladot, E. Waidmann, H. W. Nürnberg (to be published).

(96) R. Breder, H. W. Nürnberg (to be published).

(97) J. Golimowski, P. Valenta, R. Breder, H. W. Nürnberg (to be published).

(98) M. Stoeppler, K. Brandt, C. Mohl, R. Flucht (to be published).

(99) M. Stoeppler, K. Brandt (to be published).

(100) P. T. S. Wong, Y. K. Chan and P. L. Luxon, Methylation of Lead in the Environment, Nature 253, 263 (1975).

(101) A. W. P. Jarvie, R. N. Markall, and H. R. Potter, Chemical Alkylation of Lead, Nature 255, 217 (1975).

(102) U. Schmidt and F. Huber, Methylation of Organolead and Lead(II) Compounds to $(CH_3)_4Pb$ by Microorganisms, Nature 259, 157 (1976).

(103) Y. K. Chau, P. T. S. Wong, and P. D. Goulden, Gaschromatography-Atomic Absorption Spectrometry for the Determination of Tetra-alkyllead Compounds, Anal. Chim. Acta 85, 421 (1976).

(104) G. R. Sirota and J. F. Uthe, Determination of Tetraalkyllead Compounds in Biological Materials, Anal. Chem. 49, 823 (1977).

Biotransformation and Toxicity of Lead in the Aquatic Environment

Y. K. Chau and P. T. S. Wong
Canada Centre for Inland Waters, Burlington, Ontario, Canada

ABSTRACT

Volatile Tetramethyllead was detected in culture flasks in which sediments from various locations in Ontario were incubated with and without the addition of extraneous organic and inorganic lead compounds. Under optimum laboratory conditions, the maximum rate for the conversion was ca. 6% of the lead at 20°C. The conversion rate was lower at lower temperatures. Several pure bacterial cultures were found to methylate Trimethyllead acetate but not inorganic lead to Me_4Pb. Chemical disproportionation reactions of Me_3PbOAc to Me_4Pb accounted for ca. 15-20% as compared with the biological reactions. Me_4Pb was much toxic to several algae than organic and inorganic lead compounds.

INTRODUCTION

Lead is one of the most abundant toxic pollutants in the environment and its fate is not thoroughly understood (Ref. 1). The discovery that microorganisms in some lake sediments were able to methylate mercury compounds to a highly neurotoxic methylmercury species stimulated a number of studies on the methylation of other elements in several laboratories. Furthermore, methylation also plays an important role in the transport of elements from sediment to water. Methylation mechanisms of certain elements have been studied (Ref. 2), and their ecological implications discussed (Ref. 3, 4). Methylation of lead has been reported by several laboratories (Ref. 5, 6, 7) and occurrence of tetraalkyllead in fish has also been reported (Ref. 8).

EXPERIMENTAL

Methods

Conical filter flasks (250 ml) with side-arms were used for incubation of 50 g wet sediment and 150 ml lake water from various lakes in Ontario, Canada. Nutrient broth (0.5%) and glucose (0.1%) were added to stimulate the microbial growth. The flasks were capped and sealed to generate anaerobic condition by the growth of the microorganisms. After incubation of about two weeks at 20°C, the headspace gases in the incubation flasks were withdrawn through the side-arm by means of a peristaltic pump to a U-tube containing 3% OV-1 on Chromosorb W at -70°C. The gases trapped in the U-tube were later on swept into a gas chromatograph - atomic absorption spectrophotometer system for the separation and analysis of the volatile lead compounds with 0.1 ng sensitivity (Ref. 9). The tetramethyllead (Me_4Pb) in the culture flask was identified by the retention time with reference to the authentic compound, and was further confirmed by mass spectrometry.

For studying the conversion of lead compounds to Me_4Pb, either lead nitrate or trimethyllead acetate (Me_3PbOAc) at 5 mg l^{-1} as Pb was added to the incubation flasks with lake water and sediment. In all these experiments, appropriate controls were prepared either by omitting the lead compounds or by sterilizing the medium.

Biologically generated Me_4Pb was used to test its toxicity on algal growth since the commercially-prepared compound is generally supplied in organic solvents and the influences of organic solvents and other impurities are difficult to assess. Also, Me_4Pb is highly volatile and insoluble in water, making the dosing of the chemical to the algae difficult. Bacterial inoculum (Aeromonas sp. 150 ml) was added to 1350 ml of the fresh nutrient medium with and without Me_3PbOAc (final conc. 5 mg l^{-1} as Pb) in a 4 l culture flask (Fig. 1). When sediment was used, 500 g sediment and one litre of lake water with and without Me_3PbOAc were incubated with glucose (0.1%). After about 10 days incubation, the headspace gases were analyzed for the presence of Me_4Pb before the culture flask was connected to a test flask (4 litre) containing 1.4 l of fresh CHU-10 medium and 100 ml of algal inoculum. The generated Me_4Pb was sucked through the algal culture flask by a peristaltic pump running at the rate of ca. 150 ml min^{-1}. An identical set-up except without the addition of Me_4Pb was used as a control. The effects of Me_4Pb on algal growth and primary productivity were determined (Ref. 10).

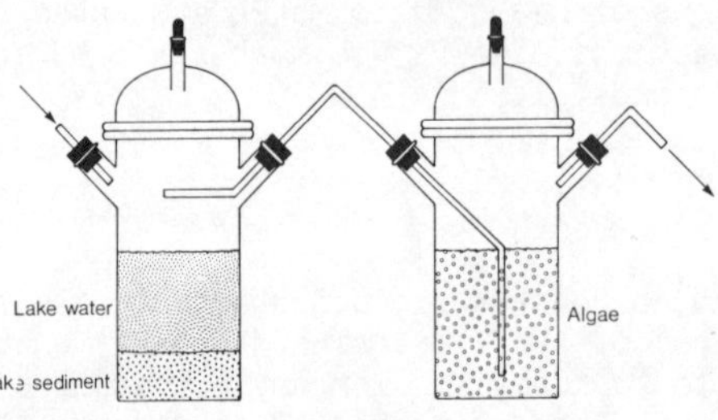

Fig. 1. Experimental design for testing the toxicity of biologically generated Me_4Pb to freshwater algae.

RESULTS AND DISCUSSION

It was observed that incubation of certain lake sediments produced Me_4Pb even without the addition of extraneous lead compounds (Table 1). There was no direct relationship between lead concentration in the sediment and the amount of Me_4Pb produced. Additions of inorganic lead, $Pb(NO_3)_2$ or organic lead, Me_3PbOAc at 5 mg l^{-1} (as Pb) greatly increased the productivity of Me_4Pb.

The conversion of Me_3PbOAc to Me_4Pb was observed in all experiments but that of lead nitrate was only sporadic. There was little or no transformation if the culture system was sterilized by autoclaving. Jarvis et al. (Ref. 6) reported similar findings and suggested that the mechanism of methylation was a chemical process due to the formation of trimethyllead sulfide which subsequently decomposed to yield Me_4Pb. However, these workers did not regard the microbial production of sulfide in the aquatic environment as a biological process.

TABLE 1 Methylation of Lead in Lake Sediments

Lake	Total Pb conc. in sediment (mg/kg dry wt.)	μg Me_4Pb generated from sediment supplemented with		
		No addition	$Pb(NO_3)_2$	Me_3PbOAc
Mitchel Bay	110	1.20	2.20	4.7
Erieau Harbor	60	0	0	2.9
Port Stanley	69	0	0	5.2
Long Lake	116	0	0.09	8.6
Kelly Lake	285	1.80	1.10	4.7
Lunch Lake	47	0.16	0.14	4.0
Robinson Lake	48	0	2.10	21.0
Dill Lake	47	0.71	0.55	7.6
Norway Lake	48	0	0	2.4
Babine Lake	43	0	0	1.4
Hamilton Harbor	273	0.01	0.13	6.4

50 gm sediment (wet wt.), 150 ml lake water, 0.5% nutrient broth and 0.1% glucose with and without the addition of 1 mg Pb as $Pb(NO_3)_2$ or Me_3PbOAc. Final Pb concentration 5 ppm.

Chemical disproportionation reactions of Me_3Pb^+ salts are known to generate Me_4Pb. Such reactions were accounted for and corrected by the control with a pure chemical system and were responsible only for about 15-20% of the total Me_4Pb production (Fig. 2). Ultraviolet light irradiation did not cause further chemical conversion of Me_3PbOAc to Me_4Pb in the absence of microorganisms. Furthermore, no specific requirements for nutrient (glucose), vitamin B-12, and anaerobic conditions were found (Table 2).

Weekly sampling of the atmosphere in the incubation flasks was carried out to determine the rate of conversion of Me_3PbOAc to Me_4Pb. With 10 mg of lead (as Me_3PbOAc) in the sample, 125 μg of lead (as Me_4Pb) was detected after incubation for one week. The conversion rate increased to 642 μg in the second week and declined to 550 μg and 256 μg of lead in the third and fourth weeks, respectively. The decrease in lead conversion after two weeks could be due to the exhaustion of nutrients, accumulation of toxic metabolic products, or changes in pH in the medium. Assuming that 642 μg of lead is the maximum rate for the conversion, the microorganisms were able to convert about 6% of the lead under optimum laboratory conditions (Fig. 3). The conversion rate was temperature dependent. At 10°C and 4°C the conversion rate was reduced to 30% and 15%, respectively, as compared with that at 20°C.

The toxicity of Me_4Pb on several freshwater algae (*Scenedesmus quadricauda*, *Ankistrodesmus falcatus*, *Chlorella pyrenoidosa*) was tested by bubbling the biologically generated Me_4Pb into the culture flasks where the test alga was grown (Fig. 1). Because of the high volatility and insolubility of the lead compounds, the exposure of the alga to the lead compound was only momentary. It was estimated that less than 0.5 mg od Pb (Me_4Pb) had passed through the culture medium. The toxic effect of Me_4Pb on *Scenedesmus quadricauda* was indicated by an 85% decrease in the algal primary productivity and 32% drop in cell growth (dry weight) as

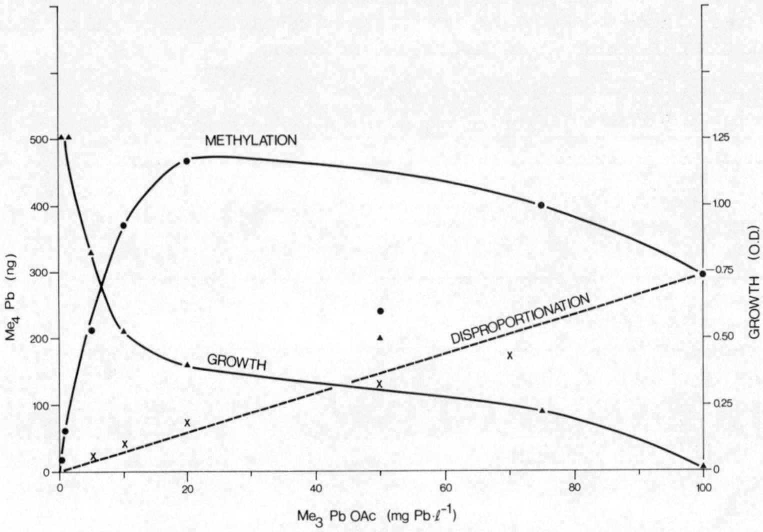

Fig. 2. Production of Me_4Pb by chemical disproportionation and biological methylation, and their relationships to growth of Aeromonas species. Nutrient broth was used as medium with addition of 0-100 mg Pb/l of Me_3PbOAc. An identical set of samples was sterilized for determining the chemical disproportionation reactions. After 3 weeks incubation, the bacterial growth was measured and the amounts of Me_4Pb in the chemical and biological systems were quantified.

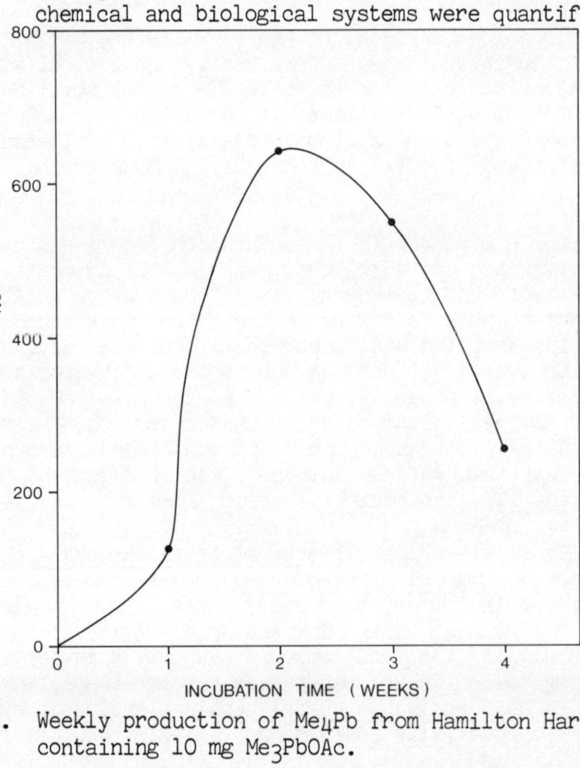

Fig. 3. Weekly production of Me_4Pb from Hamilton Harbor sediment containing 10 mg Me_3PbOAc.

TABLE 2 Factors Required for $(CH_3)_4Pb$ Formation

Incubation System			$(CH_3)_4Pb$ formation
Complete System*			+
"	"	(No glucose)	+
"	"	(No B_{12})	+
"	"	(No Nitrogen)	+
"	"	(No light)	+
"	"	(No sediment)	-

* Complete system contains 50 gm sediment (wet weight), 150 ml lake water, 0.2 gm glucose, 0.15 mg vitamin B_{12}, and 1 mg Pb as $(CH_3)_3PbOAc$ incubated in the presence of nitrogen and light.

compared to controls without exposure to laad (Table 3). To obtain similar inhibitions, twice as much Me_3PbOAc, and twenty times as much $Pb(NO_3)_2$ was required. Cells, after exposure, tended to clump together with striking alterations in cell fine structure (Ref. 10). An electron microscopic analysis by an energy dispersive spectrometer revealed the Pb ions had penetrated the cell and were deposited within concretion bodies. Similar observations were recorded with other test algae (Ref. 10).

TABLE 3 Toxicity of Tetramethyl Lead upon Three Green Algal Species

Species	Per cent decrease as compared to control: Growth[a]	Photosynthesis[b]
Chlorella pyrenoidosa [c]	74	83
Scenedesmus quadricauda [d]	32	85
Ankistrodesmus falcatus [c]	32	49

[a] Growth was determined by dry weight as well as by cell counts. 100% gave approximately 50 mg l^{-1} and 6.4×10^6 cells ml^{-1}.

[b] Photosynthesis was determined by ^{14}C-carbonate method.

[c] Me_4Pb was generated by *Aeromonas* sp. in a chemically defined medium.

[d] Me_4Pb was generated by microorganisms in a water-sediment system.

Direct chemical synthesis of methyllead compounds through alkylation of inorganic lead is very difficult because of the extreme instability of the postulated first intermediate monomethyllead salts ($MePb^{3+}$). The difficulties have also been explained in terms of the oxidation-reduction reaction involved in biomethylation (Ref. 2). From a chemical point of view, it is possible to envisage a special ligand which would form a stable monomethyllead complex before the successive methylation steps occur. Such may have been the case in observations that lead nitrate and lead chloride were converted to Me_4Pb in sediment. No Me_4Pb transformation has yet been detected from lead hydroxide, lead cyanide, lead oxide, lead bromide or lead palmitate. In none of these experiments, however, was the stable intermediate compounds, Me_3Pb^+ detected in the culture medium. Biological methylation of inorganic lead has been reported in water samples seeded with microorganisms from an aerated aquarium (Ref. 7). These investigators suggested that microorganisms could transform certain organolead and Pb(II) compounds to Me_4Pb.

The transformation of Me_3Pb^+ salts to Me_4Pb proceeded readily with pure species of bacterial isolates from Lake Ontario even without the presence of sediment. Species of *Pseudomonas*, *Alcaligenes*, *Acinetobacter*, *Flavobacterium* and *Aeromonas* growing in the presence of nutrient broth (0.5%) and yeast extract (0.1%) transformed Me_3PbOAc to Me_4Pb. Methylation of Me_3PbOAc to Me_4Pb by bacterial culture is also pH dependent. Maximum Me_4Pb production was observed at lower pH (pH 5) and decreased with increasing pH (pH 9), concurrent with the bacterial growth. However, none of the isolates was able to transform inorganic lead in the chemically-defined media.

Lead methylation is anologous to mercury in some aspects. It is dependent on temperature, pH, and microbial activities of the medium but independent of the concentration of lead in the sediment. So far there are only three reports of laboratory experiments on lead methylation. Evidence of its occurrence in the environment is still lacking. However, existence of significantly high ratios of tetraalkyllead to total lead in certain fishery products (Ref. 8) indicated the possibility of methylation in sediment and/or in fish tissue. The factors and the occurrence of in situ lead methylation in the natural environment are being investigated further.

REFERENCES

(1) Lead in the Canadian Environment. Publication No. BY73-7(ES) Ch. 2, *National Research Council*, Canada (1973).

(2) W. P. Ridley, L. J. Dizikes and J. M. Wood, Biomethylation of toxic elements in the environment, *Science* 197, 329-332 (1977).

(3) A. Jernelov and A. L. Martin, Ecological implications of metal metabolism by microorganisms, *Ann. Rev. Microbiol*. 29, 61-77 (1975).

(4) J. Saxena and P. H. Howard, Environmental transformation of alkylated or inorganic forms of certain metals, *Adv. in Appl. Microbiol*. 21, 185-226 (1977).

(5) P. T. S. Wong, Y. K. Chau and P. L. Luxon, Methylation of lead in the environment, *Nature* 253, 263-264 (1975).

(6) A. W. P. Jarvis, R. N. Markall and H. R. Potter, Chemical alkylation of lead, *Nature* 255, 217-218 (1975).

(7) U. Schmidt and F. Huber, Methylation of organolead and lead (II) compounds to $(CH_3)_4Pb$ by microorganisms, *Nature* 259, 157-158 (1975).

(8) G. R. Sirota and J. F. Uthe, Determination of tetraalkyllead compounds in biological materials, *Anal. Chem.* 49, 823-825 (1977).

(9) Y. K. Chau, P. T. S. Wong and P. D. Goulden, Gas chromatography-atomic absorption spectrometry for the determination of tetraalkyllead compounds, _Anal. Chim. Acta_ 85, 421-424 (1976).

(10) B. A. Silverberg, P. T. S. Wong and Y. K. Chau, Effects of tetramethyllead on freshwater green algae, _Arch. Environ. Contam. Toxicol_. 5, 305-313 (1977).

The Acute Toxicity and Bioaccumulation of Some Lead Alkyl Compounds in Marine Animals

B. G. Maddock and D. Taylor
Imperial Chemical Industries Ltd., Brixham Laboratory, Devonshire, U.K.

ABSTRACT

The acute toxicity of a range of alkyl lead compounds including tetra methyl and tetra ethyl lead and their tri alkyl and di alkyl degradation products has been determined with four marine phyla. An alga (*Phaeodactylum tricornutum*), a mollusc (*Mytilus edulis*), a crustacean (*Crangon crangon*) and a fish (*Pleuronectes platessa*).

The accumulation of the relatively stable trialkyl lead compounds has also been determined in a mollusc (*Mytilus edulis*) and a fish (*Limanda limanda*).

The evidence suggests that the major environmental impact of a release of tetra methyl or tetra ethyl lead into the marine environment is likely to be due to acutely toxic effects rather than to those of bioaccumulation.

INTRODUCTION

Alkyl lead compounds are used as additives to petrol as a convenient and economic method of increasing the octane rating of fuels for use in high compression internal combustion engines.

The United Kingdom is the largest European producer of these materials, some 11% (40,000 tonnes) of the total amount of lead used in the United Kingdom is converted into these additives, and approximately 80% of the UK production is exported by sea (Ref. 1).

In the last few years, increasing concern has been expressed about the possible harm to aquatic life that might be caused by the accidental release of dangerous materials carried in cargo ships due to shipwreck or collision. An IMCO Working Party is preparing regulations that should minimise the risk of environmental damage caused by such accidents, but in order to do this effectively, adequate data is required about the effects of the various compounds on aquatic life.

In addition to the possibility of accidental releases of man-made lead alkyls, it has recently been shown (Ref. 2, 3, 4) that the possibility exists, in the aquatic environment, for the transformation of inorganic lead compounds into alkylated forms by chemical or bacterial mechanisms. Furthermore, Sirota and Uthe (Ref. 5) have demonstrated that tetra alkyl lead compounds can be detected in some Canadian fish products.

In terms of its acute toxicity to aquatic life, lead is one of the least studied of the heavy metals, and most of the data that has been reported concerns the effects of inorganic lead compounds. Prior to 1975, when the present study began, the only work reported in the literature, concerning the toxicity of alkyl lead compounds to fish was the paper by Turnbull et al. (Ref. 6) who reported that the 24 hr LC_{50} of Bluegill Sunfish (<u>Lepomis macrochirus</u>) exposed to tetra ethyl lead was 2.0 mg/l. He also quoted the figure of 0.2 mg/l as a "safe level", but without producing evidence to support that conclusion.

This data has been quoted extensively and used as the basis for assessment of the risk caused by spillages of tetra ethyl lead (Ref. 7). In addition to this study, there had been two other investigations of the effects of tetra ethyl lead on lower forms of aquatic life. Kozyura et al. (Ref. 8) had studied its effects on two species of blue-green algae; <u>Cladophera sp</u>. and <u>Scenedesmus sp</u>. and Siegal et al. (Ref. 9) had compared its toxicity with a range of organo-mercurials to a flat worm. Since 1975, two further papers have been published. Krupinska (Ref. 10) has studied the effects of tetra ethyl lead on two species of Bryophytes whilst Silverberg et al. (Ref. 11) have investigated the effects of bacterially produced tetra methyl lead on various species of green algae.

The results of all these studies, with the exception of Silverberg et al. must be interpreted with caution since the high volatility and degradability of these materials may have been underestimated. The object of the present study was to measure the acute toxicity of a range of alkyl lead compounds and to gain information on the potential bioaccumulation of these materials in marine animals.

It should be noted that throughout this paper, concentrations of alkyl lead compounds in water are quoted as mg/l total lead and all tissue concentrations refer to mg/kg (dry weight) unless stated otherwise.

MATERIALS AND METHODS

Alkyl Lead Compounds

The most important organic lead compounds in terms of scale of manufacture and use, are the lead alkyl derivatives. A wide range of these materials are manufactured, but the most significant are the tetra alkyl lead compounds: Tetra methyl lead (TML) and tetra ethyl lead (TEL). These two materials are unstable in sea-water and undergo progressive dealkylation eventually forming inorganic lead (Ref. 12). The initial step in this sequence proceeds very rapidly to produce the trialkyl lead chloride, although subsequent steps in the sequence are considerably slower. Thus the acutely toxic effects of a release of a tetra alkyl lead compound into the marine environment will initially be due to the parent compound, but within a few hours the toxic effects of the trialkyl lead compounds will become of considerable significance and in the longer term the effects of the di-substituted materials may be important. Consequently we have studied both the tetra alkyl lead compounds and the primary and secondary dealkylation products. In addition, since the mammalian toxicity of TML and TEL are significantly different (Ref. 13), we have investigated both the methyl and ethyl derivatives of these materials.

In the context of accumulation, the tetra alkyl lead compounds are of only minor interest since they have only a transitory existence in sea-water. Thus, although some information has been obtained on the accumulation of these materials during short term exposures to high concentrations, the majority of the work has been concerned with the accumulation of the trialkyl lead compounds during long term exposures.

The test materials were supplied by the Ellesmere Port Research Laboratories of the Associated Octel Company Limited and with the exception of the tetra methyl lead, all materials had a purity of 98%. Tetra methyl lead was tested as an anti knock formulation (TML-CB) containing dibromo and dichloroethane in addition to the lead alkyl. Throughout the investigations, the safe handling techniques recommended for these materials (Ref. 14) were followed.

Analytical Techniques

Water samples were analysed for lead alkyl compounds using the technique of Noden (Ref. 15). The total amount of the tetra, tri and di-alkyl lead components is obtained by reducing the tetra and tri alkyl lead to dialkyl lead by treatment with iodine monochloride, and then measuring the total dialkyl lead content spectrophotometrically as a complex with 4(2-Pyridylazo) resorcinol (PAR) in the presence of 1.2 Diamino-cyclohexane-NNN'N'-tetra acetic acid (CDTA) which masks inorganic lead, calcium, magnesium and other metal salts. The amount of the tri and dialkyl lead compounds is then measured in a similar manner on another portion of the solution which has been extracted with hexane to remove the tetra alkyl lead. Finally the dialkyl lead component is measured on a further portion of the extracted solution using the PAR complex. Trialkyl lead does not form a complex with PAR. The precision of this technique is ca. ± 5% at a detection limit of 0.05 mg/l. Periodic measurements of total lead were also carried out using atomic absorption techniques. During studies with the tetra alkyl lead compounds a lower detection limit (0.005 mg/l) was required and this was achieved by extracting the tetra alkyl lead from solution with hexane, converting to inorganic lead with bromine in carbon tetrachloride, followed by determination of the lead using atomic absorption spectrophotometry with deuterium arc background correction.

During the investigation of the formulated lead alkyl (TML-CB), measurements were made of the dibromoethane and dichloroethane content of the test solutions. The method adopted was a GLC technique involving direct injection of aqueous samples.

Dissected tissue samples (0.1g - 10g) were ashed at 100°C in Aristar grade concentrated nitric acid for several hours prior to measurement of lead by atomic absorption spectrophotometry using background correction. The precision of the technique was ± 10% at a detection limit of 0.5 mg/kg. The method has been used to analyse two standard biological reference materials, Bowen Standard Kale (Ref. 16) and Monoco Oyster (Ref. 17). In each case the method produced a value within 25% of the accredited value.

Experimental Animals

Five species, representing four phyla were selected for investigation (Table 1). With the exception of the algae, the species are of commercial importance and could provide a pathway for the transfer of lead alkyls from sea-water to man. In addition, all the species are commonly found in estuarine and coastal areas where there is a greater risk of a serious spillage of lead alkyl compounds than in deeper oceanic waters. All the test animals were obtained from areas free from the effects of industrial and domestic effluent discharges, and, at the commencement of the exposure tests, contained lead concentrations equivalent to those quoted for uncontaminated animals (Ref. 19). All animals were maintained in flowing sea-water at the laboratory for a minimum of seven days prior to their exposure to test materials.

TABLE 1 Experimental Animals

Phylum	Species	Mean Length (cm)	Mean Weight (g)	Source	Mode
Algae	Phaeodactylum tricornutum	-	-	Plymouth Collection	Sub-culture on modified Erd-Schrieber (Ref. 18)
Mollusc	Mytilus edulis (Common Mussel)	6.4	28.5	Teign Estuary, S. Devon	Hand Collection
Crustacean	Crangon crangon (Brown Shrimp)	4.8	1.1	Tor Bay, S. Devon	Push Net
Fish	Pleuronectes platessa (Plaice)	5.2	2.8	Tor Bay, S. Devon	Otter Trawl
	Limanda limanda (Dab)	18.9	191	Tor Bay, S. Devon	Otter Trawl

BIOASSAY TECHNIQUES

Acute Toxicity Experiments

The techniques used are conveniently subdivided into three sections dealing with the tetra alkyl compounds, the degradation products and the algal assay procedure.

(a) Tetra Alkyl Lead Compounds. The tetra alkyl lead compounds are volatile and relatively insoluble in seawater, hence they are readily lost from solution to the atmosphere. Secondly, they rapidly undergo dealkylation in seawater to produce the trialkyl lead derivatives (Ref. 12). Consequently it was necessary to design a test system that would permit the exposure of aquatic animals to constant levels of the materials for periods of up to 100 hours. The system used is shown in Fig. 1 and 2. In view of the high mammalian toxicity and high volatility of the tetra alkyl lead compounds, the test equipment was constructed in a self contained and vented enclosure and the waste discharged into a sealed drainage system. Seawater (salinity 34.9°/oo) was pumped directly from the sea at Brixham and supplied at a constant temperature (15°C) to the system. At the design flow rate of 100 ml/min, the 90% replacement time, was 7.5 hours and, with the exception of small pieces of silicone rubber in the peristaltic pumps, those parts of the system which were in contact with the toxicant were made of either glass or PTFE. This system ensured minimum contact between the test solution and the atmosphere and hence minimal losses due to volatilisation.

Stock solutions were prepared by gently agitating 22 litres volumes of water with 10g of the lead alkyl for four hours, allowing the excess to settle for two hours and then pumping the top 20 litres into a sealed stock vessel. This procedure enabled stock solutions of seawater saturated with lead alkyls to be produced without the use of solvents. The stock solutions were used to supply five test vessels simultaneously, and the total rate of utilisation of the stock solution was maintained, with the aid of a bleed line to waste, at a constant rate of 1.2 l/hr. This ensured that the stock solutions were always replaced before decomposition of

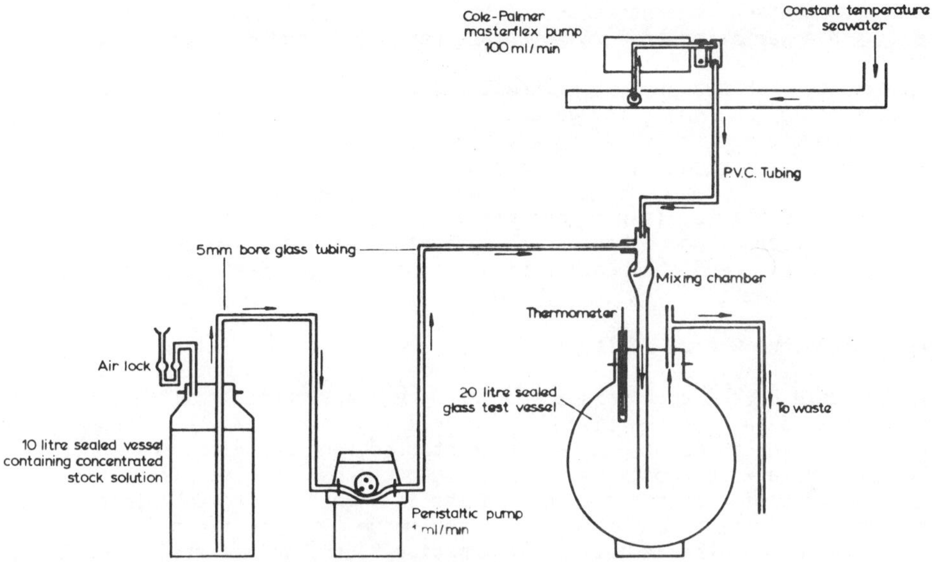

Fig. 1. Dosing System.

Fig. 2. Bioassay System.

the lead alkyl became significant. Since the dealkylation of the tetra alkyl lead compounds is promoted by light, the stock vessels were painted black and the complete test system isolated from direct sunlight. Using these procedures it was possible to maintain concentrations of the tetra alkyl lead compounds in the test vessels with a variation of ± 10% over the range 0.01 mg/l to 0.7 mg/l.

The halogenated ethanes present in the TML formulation are very soluble in seawater by comparison to TML, and although the ratio of TML to halogenated ethane in the original formulation was 1:0.4, in a saturated solution this ratio can become 1:1000. In our experiments, it was not possible to maintain the 1:0.4 ratio, but by using peristaltic pump tubing which was more permeable to the halogenated ethanes than to the lead alkyls, it was possible to achieve ratios of 1:10 for dichloroethane and 1:4 for dibromoethane. The toxicological significance of the presence of these materials, and the dealkylation products in the test vessels is discussed in a subsequent section.

The test procedure was as follows:

Ten individuals of the appropriate species were transferred from the laboratory holding tanks to each of the five test vessels and maintained in flowing seawater for 24 hours. The injector pump was then started, and preselected amounts of the stock solution were fed continuously into four of the vessels. The fifth vessel acting as a control. This procedure results in a gradual increase in the concentration of the toxicant from zero to its set value over the first 3-4 hours of exposure, but allows the animals to acclimatise to the actual test conditions.

The test vessels were regularly monitored for the deaths of the individual animals, and the time of death recorded. Water samples were taken from the stock solutions and the test vessel overflow lines twice daily and analysed for the tetra alkyl component and its degradation products and where appropriate for the halogenated ethanes. In addition the temperature, pH and dissolved oxygen content of the water in each vessel was measured daily.

The test exposure terminated after 96 hours. The test animals were fed prior to but not during the exposure.

(b) Tri and Di Alkyl Lead Compounds. The methyl and ethyl derivatives of the tri and di alkyl lead chlorides are all soluble in water at >2g/100 ml.(Ref. 20). They are not readily lost from solution by volatilisation, adsorption or dealkylation, the latter process proceeding only slowly in seawater . Consequently, a much simpler bioassay procedure was used for these materials.

A series of 15 litre volumes of sea-water containing a range of concentrations of the appropriate material were prepared in 30 litre glass tanks. Each tank was fitted with an aeration system supplying a flow of 300 ml/min of air, covered with a loose fitting lid and maintained at 15°C in a constant temperature room. Experimental animals, in groups of ten, were transferred to the test vessels from the laboratory holding tanks and observed at regular intervals over a period of 96 hours. Dead animals were removed, deep frozen for subsequent analysis, and the time of death recorded. The levels of the toxicant, its dealkylation product and the pH, temperature and dissolved oxygen levels were measured daily and the test solutions changed at 24 hour intervals.

This procedure permitted concentrations of tri and di alkyl lead chlorides to be maintained to an accuracy of ± 5% over the range 0.1 - 420 mg/l. The level of the dealkylation products i.e. dialkyl lead dichloride was always less than the detection limit of the method used i.e. 0.05 mg/l.

(c) Algal Assay. The procedure used to study the effects of these materials on algal productivity was based on the technique developed by Johnstone (Ref. 21) from the organic productivity assessment technique of Steeman-Nielsen (Ref. 22). A standardised quantity of algae, in this case Phaeodactylum tricornutum was incubated for 6 hours, under controlled conditions, in the presence of a range of concentrations of the toxicant and a standard amount of C^{14} labelled carbonate. The accumulation of C^{14} by the control cultures, after correction for adsorption indicates the "normal" activity of the algae which can be compared to the activity of the algae exposed to the toxicant.

Accumulation Experiments

Two accumulation experiments have been undertaken involving the exposure of mussels (Mytilus edulis) and Dabs (Limanda limanda) to seawater (salinity 34.9°/oo) containing either trimethyl or triethyl lead chloride. Since these materials are soluble and relatively stable in sea-water, the relatively simple experimental system, shown diagrammatically in Fig. 3 was used.

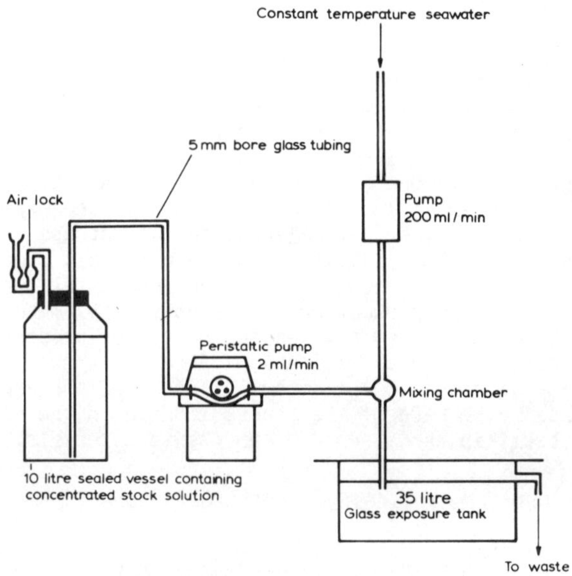

Fig. 3. Test apparatus for mussel accumulation.

In the experiment with mussels, constant temperature seawater (14°C) was supplied at a rate of 200 ml/min to exposure tanks consisting of 35 litre all glass aquaria, fitted with glass lids. This produced a 95% replacement time of 6 hours. The dabs were exposed in substantially larger (i.e. 300 litre) glass fibre tanks and these received a flow of 900 ml/min of constant temperature sea-water (12°C) producing a 95% replacement time of 13 hours. In order to maintain a homogenous toxicant distribution, it was necessary to equip each of these large tanks with a low volume aeration system. All the apparatus which was in contact with the toxicant was made of glass, fibre glass or PTFE with the exception of peristaltic pump tubing.

A standard procedure was used. An appropriate number of animals (60 mussels, 50 dabs) were transferred from the laboratory holding tanks to each of four test tanks (two for triethyl and two for trimethyl exposure) and a control tank. These animals were than maintained in flowing seawater for at least 48 hours prior to the commencement of the experiment. At the start of the exposure period, the peristaltic pumps were started and then a suitable volume of stock solution was added to each tank, with gentle agitation, to produce the appropriate test concentration. During the first few days of the exposure period, the trialkyl and dialkyl lead content of the test solutions was monitored at three hourly intervals, subsequently the frequency was reduced to daily measurements of trialkyl lead with occasional checks on the dialkyl lead content. The level of this latter material in the test vessels was always <0.05 mg/l. The temperature of the test tank water was recorded continuously and the pH, dissolved oxygen content and the flow rates of sea-water and stock solution were measured daily. At the end of the exposure period, the peristaltic pumps supplying the toxicant were stopped and 95% of the test tank water replaced immediately with clean seawater at the same temperature.

The mussels were able to obtain sufficient food from the seawater supply used. However, the fish were fed three times per week on a diet of ox liver. At each feed, the total amount of food given was equivalent to 1% of the total weight of the experimental animals.

Animals were removed from the tanks at regular intervals. On each occasion three individuals were selected at random, dissected and the tissues analysed immediately for total lead content. Four tissues from the mussel (gill, digestive gland, gonad and foot) and two tissues from dabs (liver and flesh) were examined.

In any long term experimental work with animals, aimed at detecting chronic or sublethal effects of pollutants, it is necessary to know if the experimental animals are under excessive stress. Bayne (Ref. 23) has defined stress in this context as "A measurable alteration of a physiological (or behavioural, biochemical or psychological) steady state which is induced by an environmental change and which renders the individual (or the population or the community) more vulnerable to further environmental change". Several indices have been proposed for describing this physiological condition and an estimation of one of these indices gives an easily determined general guide to the metabolic state of the experimental animals and allows an estimate to be made of the stress induced in the control population by the test system itself in the absence of any toxic materials.

In this work we have used the following condition factors:

$$\text{Condition Factor}_{\text{Mussel}} = \frac{\text{Flesh Weight}}{\text{Total Body Weight}} \times 100$$

$$\text{Condition Factor}_{\text{Dab}} = \frac{\text{Liver Weight}}{\text{Total Body Weight}} \times 100$$

The factor for mussels has been used previously by several workers, e.g. Bayne and Thompson (Ref. 24); Roberts (Ref. 25). The factor for dabs was suggested by Thompson (Ref. 26).

RESULTS AND DISCUSSION

Acute Toxicity

The acute toxicity of four alkyl lead compounds (tetra methyl lead, tetra ethyl lead, trimethyl lead chloride and triethyl lead chloride) have been measured to three marine species (mussel, shrimp and plaice) and the effect of these materials on the photosynthetic activity of *Phaeodactylum tricornutum* has been established. In addition to toxicity of dimethyl and diethyl lead dichlorides have been measured to plaice.

Results

The results, as 96 hour LC_{50} values are presented in Table 2 and diagrammatically in Fig. 4. All the 96 hour LC_{50} values have been calculated after the method proposed by Doudoroff et al. (Ref 29) and in addition, where appropriate data was available, by the method of Litchfield and Wilcoxon (Ref. 30). Since the values calculated using the Doudoroff technique fell within the 95% confidence limits of the other method, only the Doudoroff results are presented.

TABLE 2 Summary of 96 Hr. LC_{50} Values (mg/l)

Species	Shrimp	Mussel	Plaice	Algae*
Tetra Methyl Lead (TML-CB)	0.11	0.27	0.05	1.3
Tetra Ethyl Lead (TEL)	0.02	0.10	0.23	0.1
Tri Methyl Lead Chloride	8.8	0.5	24.6	0.8
Tri Ethyl Lead Chloride	5.8	1.1	1.7	0.1
Di Methyl Lead Dichloride			300	
Di Ethyl Lead Dichloride			75	
Inorganic Lead	(375)[1]	(>500)[1]	(180)[2]	>5.0

* Parameter measured is a reduction in photosynthetic activity (6 hr EC_{50}).

Figures in brackets refer to related species: 1. Portman and Wilson (Ref. 27).
2. Jackim et al. (Ref. 28)

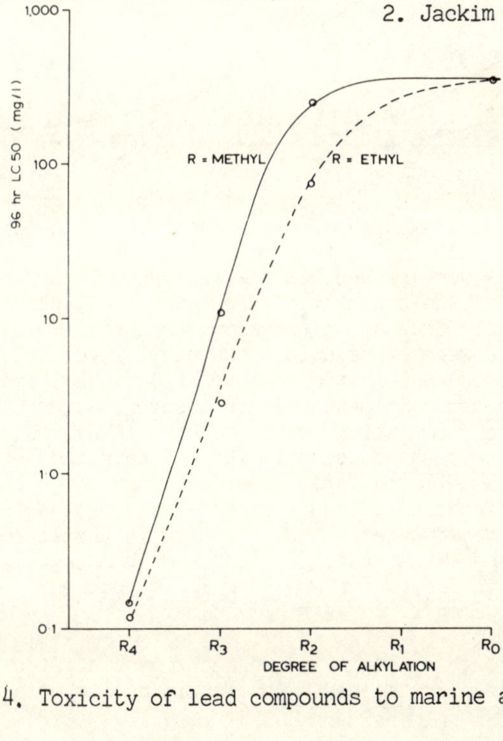

Fig. 4. Toxicity of lead compounds to marine animals.

It can be clearly seen from the results that in general the ethyl derivatives are more toxic than the corresponding methyl derivatives and that the toxicity decreases with decreasing degree of alkylation; the dialkyl lead dichlorides having a similar order of toxicity to inorganic lead. There are two exceptions to this order, mussels are more sensitive to trimethyl than triethyl lead chloride, although the difference in toxicity is so slight that it is probably justifiable to consider them of similar order of toxicity. The other exception is that plaice seem to be rather resistant to the effects of tetra ethyl lead.

The results of the algal exposures are shown in greater detail in Fig. 5, from which it can be seen that the nature of the alkyl group has much more effect than the degree of alkylation.

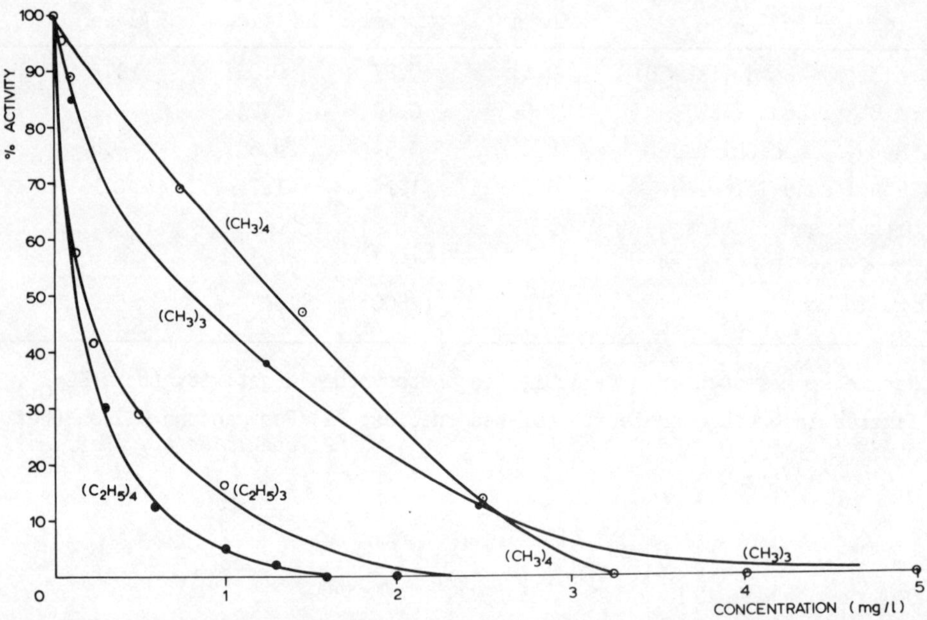

Fig. 5. Reduction of algal photosynthetic activity.

Several other materials were present in the test vessels during part of the study. In the investigations of the toxicity of the tetra alkyl compounds, trialkyl lead chlorides were inevitably present as breakdown products. However, in all but one instance (where mussels were exposed to tetra ethyl lead), the concentration of trialkyl lead chloride present in the test solution was always well below the level that had been shown to cause no deaths in that species over 96 hours. In the study of the tetramethyl lead formulation, dibromoethane (EDB) and dichloroethane (EDC) were also present in the test solutions. The concentration of EDC in the test solution was always <5 mg/l and Taylor (Ref. 31) has shown that levels of 60 mg/l had no apparent effect on dabs (Limanda limanda) and prawns (Palaemon serratus) respectively in 96 hour exposures. The only data in the literature on EDB is the report by Davis and Hardcastle (Ref. 32) that the 48 hour LC_{50} to Bluegill sunfish (Lepomis macrochirus) in freshwater was 18 mg/l. Compared to our test concentrations in seawater of <2 mg/l.

We conclude that, although synergistic and antagonistic effects cannot be ruled out, the contaminants measured in the test vessels did not materially affect the recorded toxicity of the lead alkyl compounds with the possible exception of mussels exposed to tetra ethyl lead, where the presence of triethyl lead chloride may have had a minor effect.

Discussion

The toxic effects of inorganic lead compounds on aquatic animals are largely governed by the chemical form in which the lead is present. In freshwater the dominant controlling factor has been shown to be the hardness of the water (Ref. 33). Lead has a low solubility of 0.5 mg/l in soft water and an even lower solubility (0.003 mg/l) in hard water (Ref. 34), although considerably higher concentrations of suspended and colloidal lead may remain in the water. In hard water, the lead will be mainly present as carbonate salts and the amount of "free ionic" lead i.e. uncomplexed, which is thought to be the most toxic species, will be relatively small compared to the situation is soft water. This has been confirmed by Davis et al. (Ref. 35), who were able to monitor electrochemically the levels of "free ionic" lead in their test solutions in addition to the total lead concentration. They reported that on the basis of the "free ionic" lead concentration there was no difference in the toxicity in hard and soft waters, whereas in terms of total lead concentration the toxicity differed by a factor of 100.

As in the case with many materials, juveniles are more sensitive than adult animals. McKim et al. (Ref 36) state that the 96 hour LC_{50} for juvenile rainbow trout (*Salmo gairdneri*) is 0.14 mg/l, i.e. 10% of the value recorded for adults by Brown (Ref. 37).

There have been few studies of the acute toxicity of lead to marine animals, primarily because of the difficulty of maintaining solutions of lead in seawater at concentrations large enough to cause toxic effects. The most reliable data available at the present time is that of Portmann and Wilson (Ref. 27) who report 48 hour LC_{50} values for the Pink Shrimp (*Pandalus montagui*) and the cockle (*Cardium edule*) of 375 and 500 mg/l respectively.

The acute toxicity of inorganic lead to aquatic animals can be summarised as follows: In soft freshwater, the acute toxicity, as indicated by the 96 hour LC_{50} value fall between 1-10 mg/l. In hard freshwater or in saline waters the acute toxicity falls between 300-500 mg/l.

No data has been published relating to the aquatic toxicity of the trialkyl or dialkyl lead compounds and in general, the data published on the toxicity of the tetra alkyl lead compounds is not directly comparable with the present study. The 24 hour LC_{50} value of 2.0 mg/l quoted by Turnbull et al. (Ref. 6) for the Bluegill sunfish (*Lepomis macrochirus*) in freshwater may be an underestimate when compared to data reported in the present study for marine fish (i.e. 0.2 mg/l) which would be expected to be less sensitive. This may be due to losses of the toxin from the test environment, by volatilisation and degradation, in previous studies.

The reduction in algal photosynthesis was studied by Silverberg et al. (Ref. 11) with the method similar to that used in the present study. They report a 4 hour EC_{50} value of < 0.3 mg/l for the freshwater alga *Ankistrodesmus folcatus* compared to the 6 hour EC_{50} of 1.3 mg/l for the marine alga *Phaeodactylum tricornutum* determined in the present study.

The toxicity of inorganic lead compounds to aquatic animals was originally attributed to coagulation film anoxia (Ref. 38). This phenomenon, which was first reported for lead by Carpenter (Ref. 39), involves the formation, by the action of a toxicant, of a veil like film of coagulated mucus on the body surface of the fish. If this film affects the gill tissue, the fish suffers acute respiratory distress and dies from lack of oxygen. This type of mucus development was not

detected in the present study. More recently, other workers (Ref. 28) have indicated that direct action of the metal on enzyme systems may be responsible for their acute effects, and such responses may be involved in the case of the alkyl lead compounds.
In mammals, the toxicity of tetraethyl lead is several times greater than the toxicity of tetramethyl lead. This has been demonstrated in studies with rats (Ref. 13, 40) and in man (Ref. 41). However, the toxicity is not simply dependant on the size of the alkyl group since Cremer and Callaway (Ref. 42) have shown that tetrapropyl lead is less toxic than tetraethyl lead. In the present study a similar order has been established for marine animals i.e. tetra ethyl lead is more toxic than tetramethyl lead. However, in mammals, the trialkyl lead compounds are more toxic than tetra alkyl lead compounds and the presently accepted view (Ref. 43) is that the tetra alkyl lead compounds are rapidly broken down in the liver of mammals to produce trialkyl lead derivatives which are then responsible for the toxic effects, primarily to the central nervous system. This study, however, appears to show the reverse situation in marine animals since the trialkyl compounds are shown to be considerably less toxic than the tetra alkyl compounds.

This is probably due to the fact that the hydrophobic tetra alkyl lead compounds can be readily absorbed into the aquatic animal where they may exert their toxic action after dealkylation as suggested by Cremer (Ref. 43), whereas the more hydrophilic and ionic trialkyl lead chlorides have more difficulty in penetrating the gill membranes.

It is desirable, for monitoring and control purposes, to obtain an estimate of the "safe" concentration of any contaminant. "Safe" is defined here to mean that concentration of the contaminant which is not expected to cause any acutely toxic response in the animal. The "safe" concentration can be obtained from the 96 hour LC_{50} value using an appropriate "application factor". This factor, which may range from 0.1 - 0.001, is intended to allow for the existence of more sensitive species than those studied and to compensate for the difference between the 96 hour LC_{50} value and the incipient LC_{50} or lethal threshold value which is the concentration just lethal to 50% of the animals after indefinite exposure.

A more satisfactory procedure is to obtain the incipient LC_{50} value from a plot of the concentration against the median time of death of the animal (Median Survival Period - MSP curve) and apply an appropriate safety factor to this value. This procedure, which is applied below, eliminates errors caused by the variable slope of the MSP curve. Insufficient data is available from this study to allow the complete median survival period curve to be plotted, since in some of the experiments, the maximum exposure level that could satisfactorily be obtained, did not cause sufficient mortality within 96 hours. However, sufficient data is available to produce points on the curve representing 24, 48 and 96 hour LC_{50} values (Fig. 6). This data shows that in most cases, a threshold value had not become evident after 96 hours. This is in agreemnnt with the work of Cairns (Ref. 44) who reported similar findings with tetra ethyl lead. However, the data can be used to predict threshold values for plaice of 0.04 mg/l for TML-CB and 1.0 mg/l for triethyl lead chloride. This information, coupled with a reasonable application factor of 0.1(Ref. 45) produces estimated safe levels of 0.004 mg/l for TML-CB and 0.1 mg/l for triethyl lead chloride. This latter figure will also apply to trimethyl lead chloride, since although no threshold value can be predicted from the data, its toxicity is far less than the triethyl derivative. Similarly since the toxicity of TEL is similar to that of TML-CB and ca. 70-80 times that of the trialkyl compounds a "safe" level in the region of 0.001 mg/l for both the tetra alkyl compounds seems reasonable. Support for these levels is provided by the data obtained during the accumulation experiments, which are discussed in the next section. In these studies, marine animals were maintained for periods in excess of 30 days without apparent ill effect, in concentrations of trialkyl lead chlorides ranging from 0.1-2.0 mg/l.

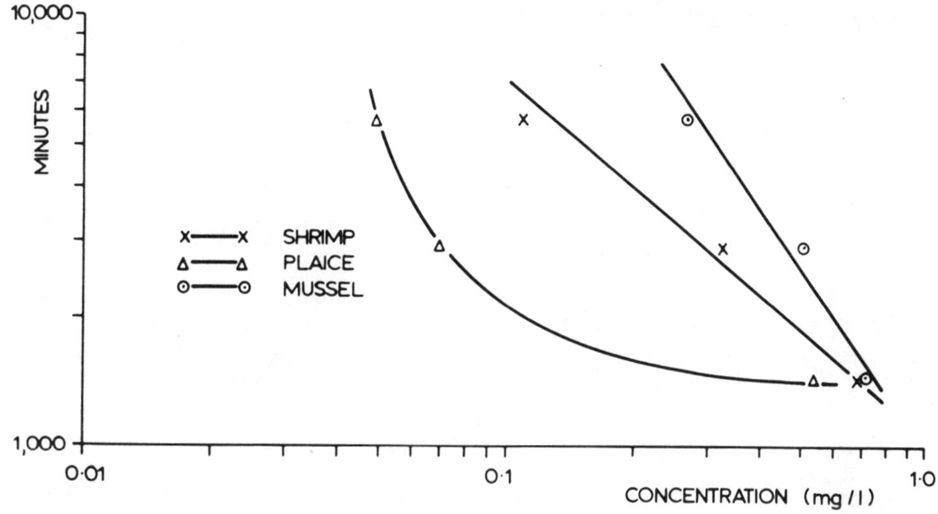

Fig. 6a. Median survival period plot - TML-CB.

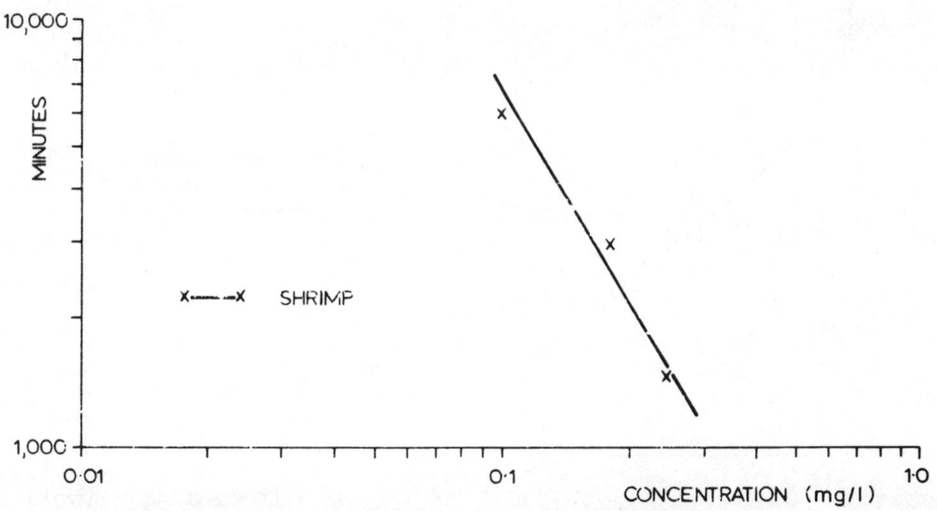

Fig. 6b. Median survival period plot - TEL.

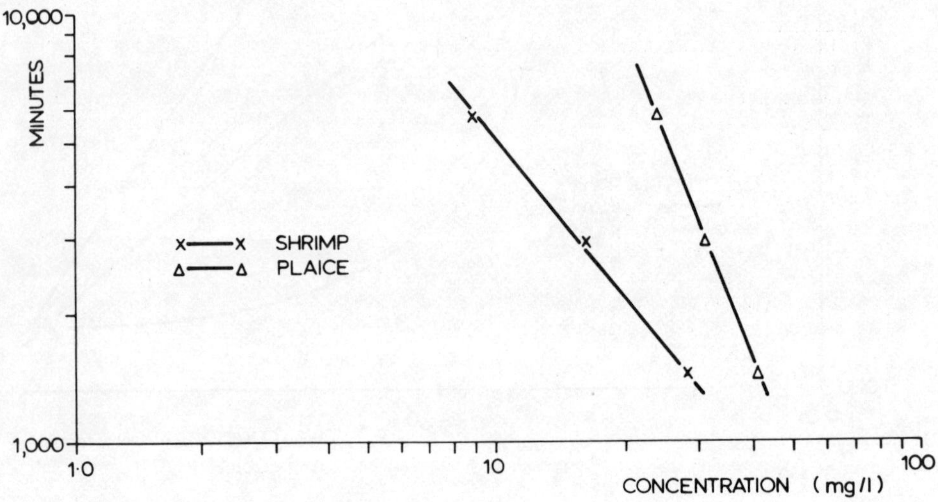

Fig. 6c. Median survival period plot - trimethyl lead chloride.

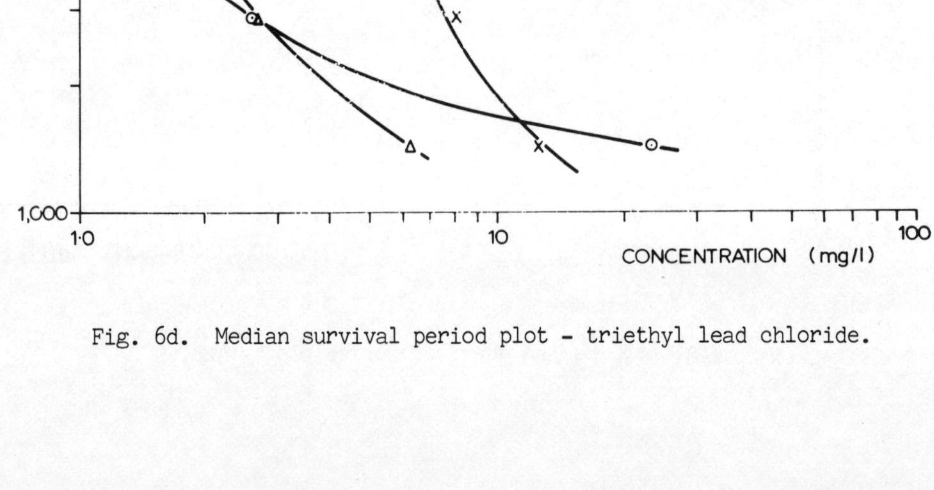

Fig. 6d. Median survival period plot - triethyl lead chloride.

Thus in the light of existing evidence we can propose tentative "safe" levels for the avoidance of acutely toxic effects in the marine environment of 0.001 mg/l for tetra alkyl lead compounds and 0.1 mg/l for trialkyl lead compounds. Since the tetra alkyl lead compounds are rapidly degraded to the trialkyl lead derivatives it is the latter figure which is of most importance.

Bioaccumulation

There are two principal routes by which aquatic animals may accumulate a contaminant, via food or via water. Exposure of animals to contaminants in food is usually used to assess the significance of food chain transfer and is usually considered to be secondary to accumulation direct from the water. At this stage in the investigation, the water route was considered to be of primary importance, since if no significant accumulation occurred by this route, the possibility of food chain transfer and magnification would be relatively small.

The tetra alkyl lead compounds have only a short lifetime in seawater and therefore they are unlikely to be accumulated over long periods by marine life. Thus, although some information has been obtained on the accumulation of lead from these materials during short-term exposures to high concentrations, most of the experimental work discussed is concerned with the accumulation of lead from trialkyl lead compounds during long term exposures.

Results

(a) Short Term Exposure. The animals, which had been exposed for 96 hours to a wide range of concentrations of four alkyl lead compounds during the acute toxicity investigations, were subsequently analysed for their lead content. The data obtained from the experiment with shrimps (*Crangon crangon*) is shown in Fig. 7.

The results obtained must however, be treated with caution since at exposure concentrations in excess of the 96 hour LC_{50} value a significant proportion of the animals would have died before the end of the test period. The behaviour of these animals under high stress, such as immediately prior to death, may have caused anomalies in the rate of accumulation of the material being investigated, and, in addition, since it was not practicable to remove dead animals from the test system the analytical data obtained for these species may be unrepresentative due to decomposition. The data obtained is of little use for predicting the results of long term exposures, but can be used to indicate the likely concentrations of lead in the tissues of animals exposed to high levels of lead alkyls in the immediate vicinity of a spillage.

The results of these short term exposures are summarised in Table 3. The revelant tissue levels, as far as hazards to fish consumers are concerned is that produced by exposure to the 96 hour LC_{50} value since higher exposure levels will lead to substantial mortalities in the exposed population. As can be seen the concentration factors (see Table 3) measured were all relatively small with the tetra alkyl compounds showing a greater tendency to accumulate.

(b) Long Term Exposures. Two species, the common mussel (*Mytilus edulis*) and the dab (*Limanda limanda*) both of which are of commercial importance and provide a pathway for the transfer of lead alkyls to man were chosen for investigation. The test conditions were selected to represent typical seawater and the exposure levels set at approximately 10% of the 96 hour LC_{50} values.

1. Mussel Experiment

The results of this experiment are summarised in Table 4. The shape of the accumulation curves was similar for both compounds and both concentrations tested and a typical example is shown in Fig. 8.

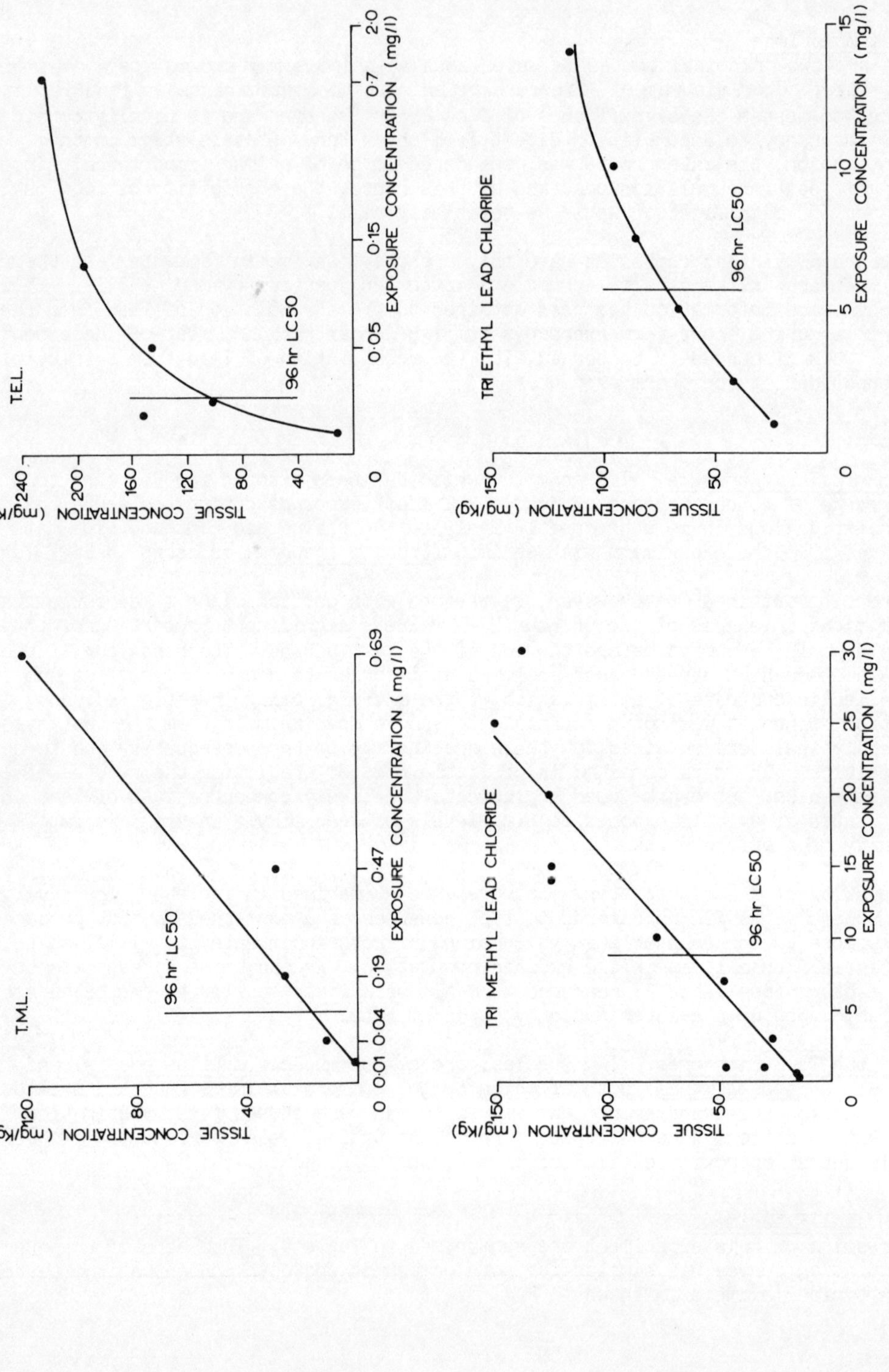

Fig. 7 Accumulation of lead alkyls by shrimps (<u>Crangon crangon</u>) during 96 hr exposures

TABLE 3 Summary of Accumulation Data from 96 Hr. Exposures

Compound	Species	Exposure Range (mg/l)	96 Hr LC_{50} (mg/l)	Concentration Factor[1]
1. Tetra Methyl Lead (TML-CB)	Shrimp	0.01 0.7	0.11	20x
	Mussel[2]	0.05 - 0.7	0.27	170x
	Plaice	0.02 - 0.7	0.05	60x
2. Tetra Ethyl Lead	Shrimp	0.01 - 0.2	0.02	650x
	Mussel[2]	0.01 - 0.2	0.10	120x
	Plaice	0.02 - 0.2	0.23	130x
3. Tri Methyl Lead Chloride	Shrimp	0.2 - 30	8.8	1x
	Mussel[2]	0.1 - 3.0	0.5	24x
	Plaice	10 - 50	24.6	1x
4. Tri Ethyl Lead Chloride	Shrimp	1.0 - 15.0	5.8	2x
	Mussel[2]	0.1 - 25.0	1.1	10x
	Plaice	1.0 - 10.0	1.7	2x

1. $\text{Concentration Factor} = \dfrac{\text{Concentration in Animal Tissue (mg/Kg. Wet Wt.)}}{\text{Concentration in water (mg/l)}}$

After exposure at 96 hr. LC_{50} Concn.

2. Mussel figures are mean of 4 tissue (Digestive gland, Foot, Gill, Gonad).

TABLE 4 Mussel Experiment - Summary of Accumulation Data

		Control	Trimethyl Lead Chloride			Triethyl Lead Chloride		
Exposure								
Exposure Concn. (mg/l)		0	0.01	0.05	0.10	0.01	0.05	0.10
Exposure Time (Days)		35	9	35	21	9	35	21
Time to Equilibrium (Days)		-	-	9	7	-	6	5
Tissue Concn. (mg/Kg Dry Wt.) at Equilibrium	Gill	6	10	39	99	20	52	60
	D. Gland	6	10	32	69	11	33	56
	Gonad	9	15	21	67	10	15	42
	Foot	3	5	19	39	6	18	35
	Mean	6	10	27	68	18	29	48
Concentration Factors								
(a) Tissue (Wet Wt.)/Water		-	60	60	90	90	66	60
(b) Exposed/Control		-	1.6	4.5	11	2.0	4.8	8.0
Depuration								
Half Life (Days)	Gill			2	1		2	-
	D. Gland			2	4		12*	5
	Gonad			4	2		4	2
	Foot			4	6		5	3
	Mean			3	3		4	3
Total Experiment Duration (Days)		62	9	62	48	9	62	48

* Omitted from Mean Value

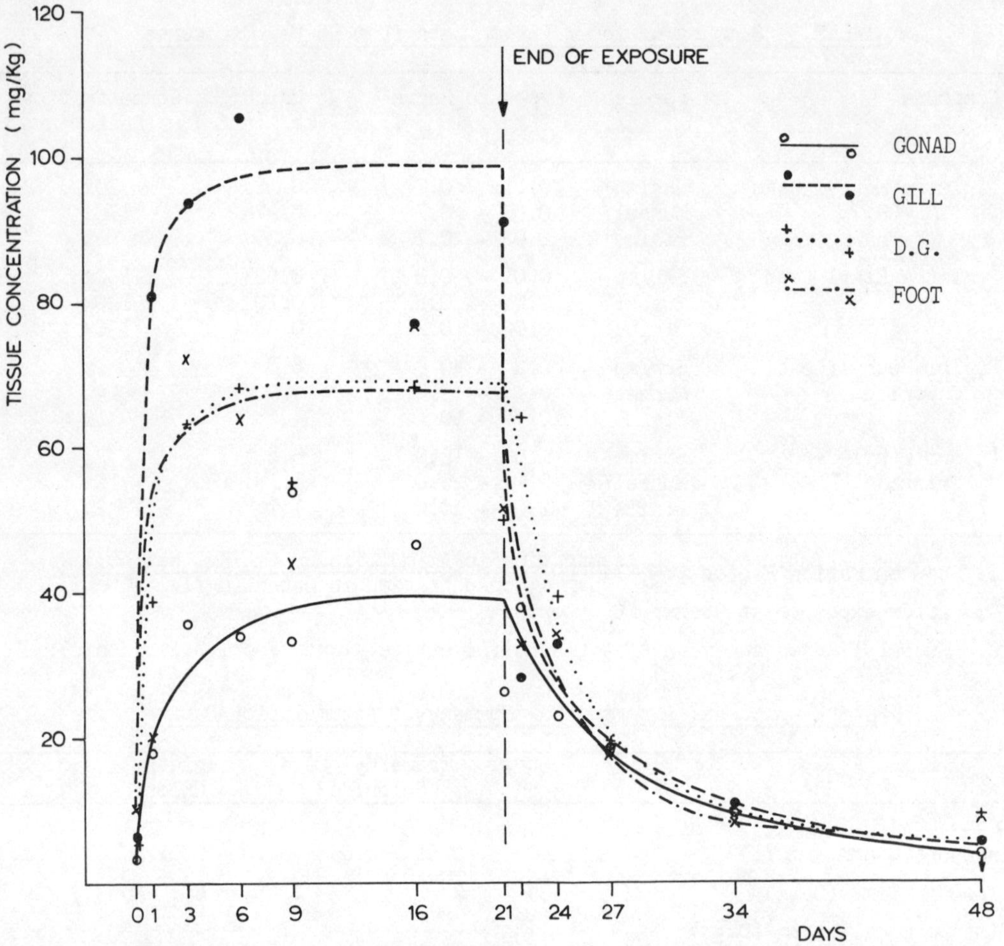

Fig. 8. Accumulation of lead by mussels exposed to 0.1 mg/l tri methyl lead chloride.

After an initial rapid accumulation of lead, during which intake was greater than loss, the mussels reached an equilibrium. The time taken to reach this plateau varied from 5-9 days but was apparently independent of the concentration of alkyl lead. No further accumulation took place although the exposures were continued for up to 35 days. After termination of the exposure the accumulated lead was rapidly lost with a half life of ca. 3 days.

At equilibrium, the average tissue level of lead was directly proportional to the exposure concentration with a concentration factor of ca. 70x. The maximum lead concentration, 68 mg/Kg, was measured in the mussels that had been exposed to 0.1 mg/l of trimethyl lead chloride (20% of the 96 hour LC_{50} value) and represented an order of magnitude increase in lead concentration compared to the control population.

The distribution of lead within the mussels could be represented as follows:

GILL > DIGESTIVE GLAND > FOOT > GONAD

The condition factor of the control population indicated no significant signs of stress and, this is confirmed by the low cumulative mortality of $< 5\%$, observed over the 60 day experimental period. There was some evidence of a decline in condition in the exposed animals for the first ten days of the experiment but this was within the natural variation.

2. Dab Experiment

A summary of the data is shown in Table 5 and the accumulation curves are shown in Fig. 9 and 10. In this animal, two tissues were analysed, muscle since it constitutes the major part of the animal and is eaten by human beings and the liver since this is both a storage organ and a major part of the animals detoxification system.

TABLE 5 Dab Experiment - Summary of Accumulation Data

	Tissue	Control	Trimethyl Lead Chloride		Triethyl Lead Chloride	
Exposure						
Exposure Concn. (mg/l)			1.0	2.0	0.1	0.2
Exposure Time (Days)		41	41	41	41	41
Time to Equilibrium (Days)	Liver		20	20	N.D.	N.D.
	Muscle		N.D.	N.D.	N.D.	N.D.
Tissue Concn. (mg/kg Dry Wt.) after 41 days	Liver	1.9	17	32	9	21
	Muscle	1.4	22	33	10	18
	Mean	1.6	20	32	10	20
Concentration Factors 41 days						
(a) Tissue (Wet Wt.)/Water			2.5	2.0	12	13
(b) Exposed/Control			12	20	6	12
Depuration						
Half Life (Days)	Liver		70	62	N.D.	N.D.
	Muscle		N.D.	41	56	N.D.
Total Experiment Duration (Days)		97	90	90	97	97

N.D. = Could not be determined

In contrast to the study with mussels, the increase in tissue concentration of lead, with one exception, was linear with respect to time, and there was no evidence for an equilibrium being reached even though the exposure was continued for 41 days. The exception is illustrated in Fig. 9 where, in the liver of fish exposed to trimethyl lead chloride, an equilibrium was reached after ca. 20 days. The reason for this is not known.

The rate of accumulation of lead into the fish tissues was very low. After 41 days, the concentration factors were 2x for trimethyl and 12x for triethyl lead chloride. The liver and muscle tissues accumulated similar levels of lead with the average concentration reaching a maximum 30 mg/Kg.

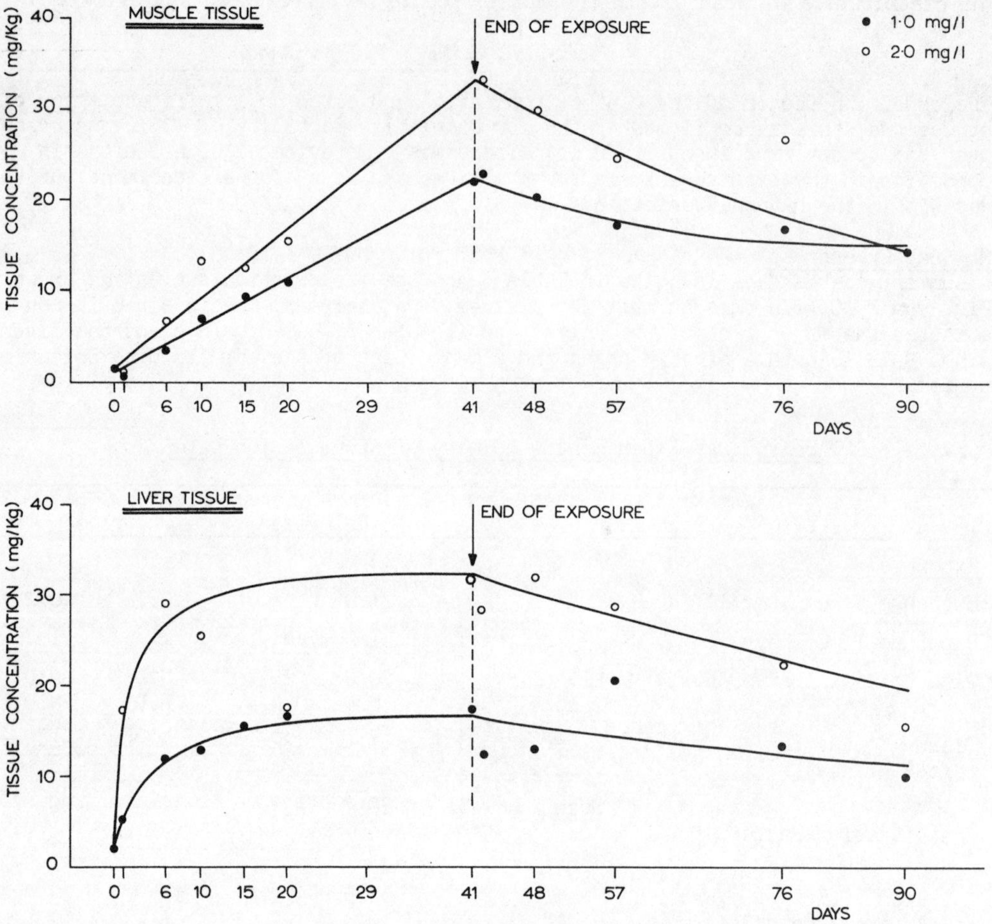

Fig. 9. Accumulation of lead by dabs exposed to tri methyl lead chloride.

In contrast to the mussel experiment, the rate of loss of lead from the fish, after termination of the exposure, was very slow with half lives in excess of 41 days. This low rate of loss was expected, since the linear accumulation curves had implied that the rate of loss was lower than the rate of accumulation. Since the accumulated lead levels in liver and muscle were very similar it seems unlikely that the liver is involved in any major detoxification mechanism in the animal, although the observation that the liver concentrations tended to increase after the exposure had terminated (Fig. 10) implies that some translocation of lead to the liver may take place.

During the course of the experiment, there was a gradual but minor decline in condition of all the experimental animals. It is interesting to note that the mean deterioration in condition of the fish exposed to triethyl lead chloride was identical to that of the control population (0.004%/day) whereas the corresponding figure for fish exposed to trimethyl lead chloride was 0.008%/day.

Several workers have studied the accumulation of inorganic lead compounds by members of the Mytilus family (Ref. 46-49). Majori and Petronio (Ref. 47), using a

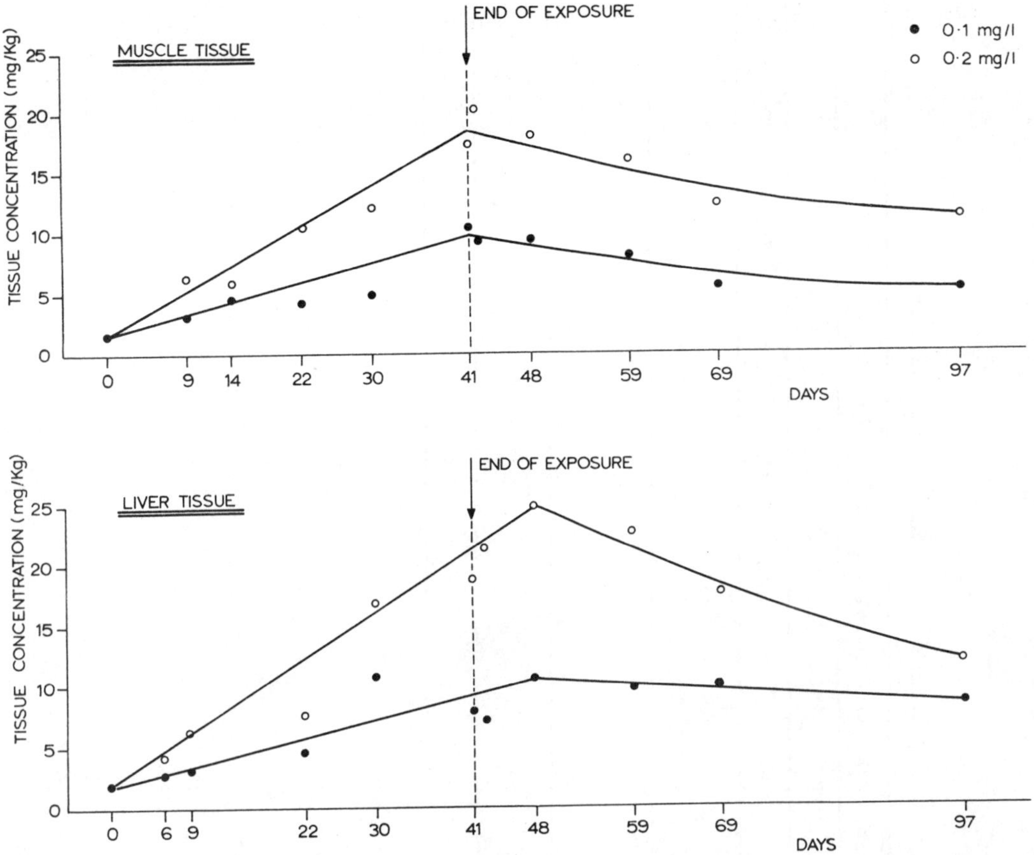

Fig. 10. Accumulation of lead by dabs exposed to tri ethyl lead chloride.

Mediterranean species Mytilus galloprovincialis reported concentration factors of 200-300x at exposure levels of 0.05 - 0.10 mg/l. However, most other workers have found the accumulation of inorganic lead to be substantially greater in mussels than these authors have reported.

The most extensive study has been made by Schulz-Baldes (Ref. 48, 49) with Mytilus edulis and a direct comparison of his results with the present study is presented in Table 6. The major difference between inorganic lead, as reported by Schultz-Baldes, and trialkyl lead, as reported here, is that for inorganic lead there is no evidence of a steady state being achieved. Thus, although the tissue levels recorded by both groups after 7 days exposure are similar, after 35 days exposure the mussels exposed to trialkyl lead chloride have not accumulated any more lead (Concentration Factor (Dry wt) = 400) whereas the mussels exposed to inorganic lead have maintained their initial high rate of accumulation to reach a substantially higher level (Concentration Factor (Dry wt) = 6000). These results are shown diagrammatically in Fig. 11.

These high concentration factors for inorganic lead have been confirmed by Coombs (Ref. 46) who reported values of 3,000 in tissues of Mytilus edulis exposed to 0.1 mg/l inorganic lead in seawater.

TABLE 6 Comparison of the Accumulation of Inorganic and Trialkyl Lead by Mussels

	Exposure Concentration (mg/l)								
	0.01			0.05			0.10		
	Inorganic*	Trimethyl	Triethyl	Inorganic*	Trimethyl	Triethyl	Inorganic*	Trimethyl	Triethyl
Tissue Concentration After Exposure for 7 days (mg/Kg Dry Wt.)	10	10	9	50	25	29	110	68	48
Tissue Concentration After Exposure for 35 days (mg/Kg Dry Wt.)	60	10	9	300	27	29	600	68	48
Concentration Factor 35 days (Tissue/Water)	6000	400	300	6000	420	460	6000	620	420
Half Life (days)	>35	-	-	>35	3	3	>35	4	3

* Inorganic Data from Schulz-Bades (Ref. 49).

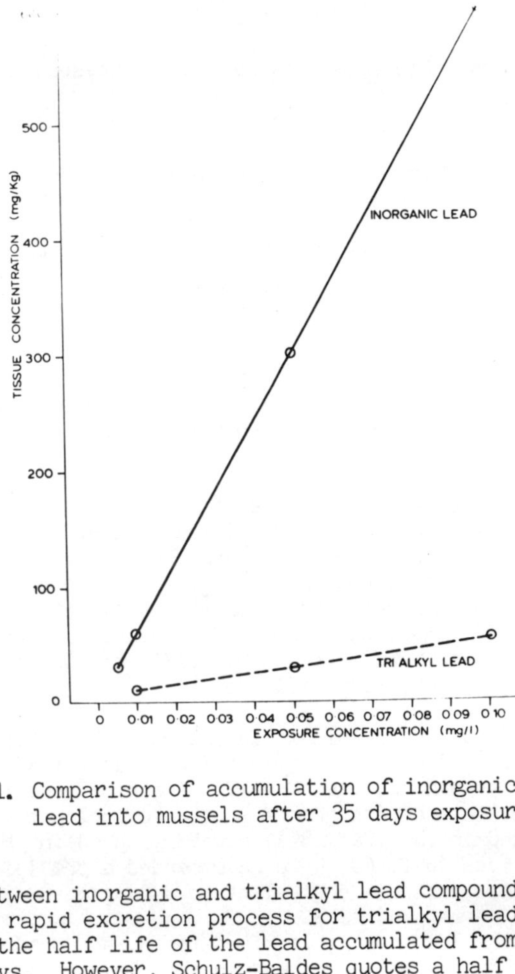

Fig. 11. Comparison of accumulation of inorganic and tri alkyl lead into mussels after 35 days exposure.

This difference between inorganic and trialkyl lead compounds appears to be due to the existence of a rapid excretion process for trialkyl lead in this animal. In the present study the half life of the lead accumulated from trialkyl lead solutions was ca. 3 days. However, Schulz-Baldes quotes a half life for inorganic lead of >35 days and it has been reported (Ref. 50) that the half life of lead in mussels could be as long as 300 days. At the present time the form in which lead is present in those animals that have been exposed to trialkyl lead chlorides is not known, but it would appear from these results that it is not present simply as inorganic lead.

Some similarities do exist between the accumulation of lead from inorganic and trialkyl lead compounds. In both cases the tissue residue levels at a fixed time are directly proportional to the exposure concentration (Fig. 11). Also, the tissue distribution of lead within the mussel appears to be unaffected by the type of lead compound to which the animal is exposed. Coombs and Schulz-Baldes both found the same ranking order for the tissue distribution as was found in this study.

No data has been found on the accumulation of inorganic lead in marine fish that can be contrasted with the present study. Some work has been reported with freshwater fish (Ref. 51-53) but comparisons must be made with care. Merlini and Pozzi (Ref. 52, 53) showed that the accumulation of inorganic lead by the sunfish (*Lepomis gibbosus*) was directly related to the species of lead present in the water and they quote concentration factors of 400-900x after 27 day exposures (compared to 2-10x in the present study). They also report in contrast to this study that the liver tissue contained from 20-30x more lead than the muscle tissue.

It is interesting to compare the data obtained in the present study with that produced in the case of mercury accumulation. Several workers have studied the accumulation of inorganic mercury into fish (Ref. 54, 55) and shellfish (Ref. 56, 57) and the generally agreed concentration factor is ca. 20-60x. However, experiments by Burkitt (Ref. 58) and Kopfler (Ref. 56) indicate that the concentration factor for alkyl mercury compounds in fish and shellfish is approximately an order of magnitude greater at ca. 500-800x. Thus alkyl mercury compounds have higher concentration factors in fish and shellfish than inorganic mercury compounds whereas the reverse situation applies in the case of lead.

CONCLUSIONS

The aim of the present study was to provide data which could be used to assess the risks caused by an accidental release of lead alkyl compounds in the marine environment.

It has been demonstrated that the toxicity of the alkyl lead compounds is directly related to the degree of alkylation and to a lesser extent to the nature of the alkyl group. The tetra alkyl lead compounds are the most toxic with 96 hour LC_{50} values of approximately 0.1 mg/l, but these rapidly degrade in seawater to the trialkyl lead chlorides which have 96 hour LC_{50} values in the range of 1-10 mg/l. Tentative "safe" levels for these two groups of compounds in the marine environment have been calculated as 0.001 mg/l and 0.1 mg/l, and these levels would not be expected to cause any acutely toxic effects.

The accumulation of lead by marine animals is important both because of the effect on the animal itself and due to the possible effect that the accumulated levels might have on predatory organisms, one of which is man. There is, at the present time, some considerable dispute with regard to the toxicological significance of lead residues in food and in man (Ref. 59), however, the Joint FAO/WHO Expert Committee on Food Additives (Ref. 60) have recommended a provisional tolerable weekly intake of 3 mg lead which is equivalent to 430 μg/day.

In the United Kingdom, the lead content of food for human consumption is regulated by either the Food and Drugs Act or the Lead in Food Regulations. These regulations have recently been reviewed by the Food Additives and Contaminants Committee of the Ministry of Agriculture Fisheries and Food (Ref. 61) and they have recommended in the light of all the available evidence that in future there should be a statutory lead limit of 2.0 mg/kg for fish and 5.0 mg/kg for shellfish (equivalent to 14.0 and 35.0 mg/kg on a dry weight basis).

It has long been known that lead is present in marine animals (Ref. 62), although due to the inaccuracies of early analytical methods, data prior to 1970 must be treated with caution. The distribution of lead in molluscs and particularly mussels obtained from the environment has recently been reviewed by a number of authors (Ref. 63-68) and their results are shown in Table 7(a) alongside data from the present study. The corresponding data for fish is presented in Table 7(b), although in this case the data is not as extensive and does not always refer to the dab, but to other similar species such as the plaice (*Pleuronectes platessa*) and flounder (*Pleuronectes flesus*).

The results of the present study indicate that the long term exposure of mussels and dabs to concentrations of trialkyl lead chlorides equivalent to 10% of the 96

TABLE 7 Distribution of Lead in Marine Animals

(A) Mussels

Author	Concn. Range (mg/kg Dry Wt.)	Source
Alexander and Young, 1976 (Ref. 63)	2.4 - 38	California
Chow et al., 1976 (Ref. 64)	0.2 - 42	California
MAFF 1972, 1975 (Ref. 65, 66)	<3.5 - 35*	England and Wales
Nickless et al.1972 (Ref. 67)	1 - 30	Bristol Channel
Topping 1973 (Ref. 68)	<1.4 - 38*	Scotland
This Study	70 - 250	Short Term Exposure
	9 - 68	Long Term Exposure
MAFF Recommended Limit (Ref. 61)	35	

* Converted from Wet Wt. Data

(B) Flatfish

Author	Species	Concn. Range (mg/kg Dry Wt.)	Source
Hardisty et al. 1974 (Ref. 69)	Flounder	14 - 28	Bristol Channel
MAFF 1971, 1975 (Ref. 65, 66)	All Flat Fish	<1.2-29*	England and Wales
	All Flat Fish	2.4-7.2*	Greenland
Topping, 1973 (Ref. 70)	Plaice	<1.4-5.6*	Scotland
This Study	Dabs	25-240	Short Term Exposure
	Dabs	10-32	Long Term Exposure
MAFF Recommended Limit (Ref. 61)		14	

* Converted from Wet Wt. Data

hour LC_{50} value produce lead residues equivalent to twice that permitted for human consumption, although the form of the accumulated lead is not known. The production of significantly higher residue levels would require exposure to higher concentrations of the lead alkyl which would cause significant mortality. For example, Table 7 indicates that residue levels of 7x the permitted maximum could be achieved by exposure of mussels to tetra methyl lead for 96 hours but under these conditions 50% of the exposed population would die.

Thus, the major environmental impact of a release of tetra alkyl lead compounds into the marine environment is more likely to be due to acutely toxic effects than those of bioaccumulation.

ACKNOWLEDGEMENT

The authors would like to express their thanks to the Associated Octel Company for the provision of materials and considerable technical assistance, Mr. J.E. Caunter and Mr. E. Gillings of the Brixham Laboratory for assistance with the experimental work and to Dr. L. Carter of the same establishment for his helpful and constructive criticism.

REFERENCES

(1) Anon, Lead in the environment and its significance to man - Pollution Paper No. 3, HMSO (1974).

(2) A. W. P. Jarvie, R. N. Markall and H. R. Potter, Chemical Alkylation of Lead, Nature (London) 255, 217 (1975)

(3) U. Schmidt and F. Huber, Methylation of organo lead and lead (II) compounds to $(CH_3)_4Pb$ by micro-organisms, Nature (London) 259, 157-158 (1976).

(4) P. T. S. Wong, Y. K. Chau and P. L. Lukon, Methylation of lead in the environment, Nature (London) 253, 263 (1975).

(5) G. R. Sirota and J. F. Uthe, Determination of tetra alkyl lead compounds in biological materials, Anal. Chem. 49, 823 (1977).

(6) H. Turnbull, J. G. De Mann and R. F. Weston, Toxicity of various refinery materials to freshwater fish, Ind. Eng. Chem. 46, 324 (1954).

(7) Anon, Oil and hazardous materials - Emergency procedures in the water environment, U.S. Dept. of Interior (1968).

(8) A. S. Kozyura, A. N. Smirnova and A. P. Mirnaya, Study of the possible decomposition of tetra ethyl lead under natural conditions, Trudy Ob` Edineyii 3, 55-59 (1961).

(9) S. M. Siegel, A. Eshleman, I. Uneno, N. Puerner and C. W. Smith, The general and comparative biology of toxic metals and their derivatives: Mercury and Lead, in Mercury in the Western Environment, ed. D. R. Buhler (1971).

(10) I. Krupinska, Influence of lead tetra ethyl on the growth of Funaria hygrometrica L. and Marchantial polymorpha L., Acta Soc. Botan. Pol. 45, 421 (1976).

(11) B. A. Silverberg, P. T. S. Wong and Y. K. Chau, Effects of tetramethyl lead on freshwater green algae, Arch. Env. Contam. Toxicol. 5, 305-313 (1977).

(12) J. R. Grove, Investigation into the formation and behaviour of aqueous solutions of lead alkyls, in Proceedings of International Experts Discussion on "Lead - Occurrence, Fate and Pollution in the Marine Environment, Rovinj, October 1977, M. Branica and Z. Konrad (eds.), Pergamon Press, 1979, 45-52.

(13) M. Magistretti, N. Zurio, F. Scolla and D. Pacillo, A comparative study of the toxicity of tetraethyl and tetramethyl lead, Med. Lav. 54, 486 (1963).

(14) Anon, Notes on the handling of lead alkyls in the laboratory, Associated Octel Company Ltd. (1968).

(15) F. G. Noden, The determination of tetraalkyl lead compounds and their degradation products in natural waters, in Proceedings of the International

Experts Discussion on "Lead - Occurrence, Fate and Pollution in the Marine Environment", Rovinj, October 1977, M. Branica and Z. Konrad (eds.), Pergamon Press, 1979, 83-91.

(16) H. J. M. Bowen, Comparative analysis of a standard plant material, Analyst 92, 124-131 (1967).

(17) R. Fukai, Personal Communication(1976).

(18) B. Foyn, Lebenszyklus, Cytologie and Sexualitat der Chlorophycae, Arch. Rotistenk 83, 1-56 (1934).

(19) H. J. M. Bowen, Trace Elements in Biochemistry, Academic Press, London (1966).

(20) G. Calingaert, H. Shapiro, F. J. Dykstra and L. Hess, The decomposition of alkyl lead compounds, J. Am. Chem. Soc. 70, 3902-3906 (1948).

(21) R. Johnston, Personal Communication (1974).

(22) E. Steeman-Nielsen, The use of radioactive carbon (^{14}C) for the measurement of organic productivity in the sea, J. Du. Conseil 18, 117 (1952).

(23) B. L. Bayne, Aspects of physiological condition of Mytilus edulis with special reference to the effects of oxygen tension and salinity, Proc. 9th European Marine Biological Symposium (Aberdeen) (1975).

(24) B. L. Bayne and R. J. Thompson, Some physiological consequence of keeping Mytilus edulis in the laboratory, Helgolander Wiss. Meeresunters.20, 526-552 (1970).

(25) D. Roberts, The assimilation and chronic effects of sub lethal concentrations of endosulfan on conditions and spawning in the common mussel (Mytilus Edulis), Mar. Biol. 16, 119-125 (1972).

(26) R. S. Thompson, Personal Communication, (1976).

(27) J. E. Portmann and K. W. Wilson, The toxicity of 140 substances to Brown Shrimp and other marine animals, MAFF, Shellfish Information Leaflet 22, HMSO (1971).

(28) E. Jackim, Influence of lead and other metals on fish - aminolevulinate dehydrase activity, J. Fish. Res. Bd. Can. 30, 560-562 (1973).

(29) P. Douderoff et al., Bioassay methods for the evaluation of acute toxicity of industrial wastes to fish, Sewage and Industrial Wastes 23, 1380-1397 (1951).

(30) J. T. Litchfield and F. Wilcoxon, A simplified method of evaluating dose effect experiments, J. Pharm. Exp. Ther. 96, 99-113 (1949).

(31) D. Taylor, Some investigations into the effect of 1.2. Dichloroethane (EDC) on marine life, Brixham Laboratory Report BL/B/1571 (1974).

(32) J. T. Davis and W. S. Hardcastle, Biological assay of herbicides for fish toxicity, Weeds 7, 397-404 (1959).

(33) Q. H. Pickering and C. Henderson, The acute toxicity of some heavy metals to different species of warm water fishes, J. Int. Air Water Poll. 10, 453-463 (1966).

(34) Anon, Water quality criteria. A report on the Committee of Water Quality Criteria for the United States Environment Protection Agency, National Academy of Science, New York (1972).

(35) P. H. Davies, J. P. Goettl, J. R. Sinley and N. F. Smith, Acute and chronic toxicity of lead to Rainbow Trout in hard and soft water, Water Research 10, 199-206 (1976).

(36) J. M. McKim, D. A. Benoit, K. E. Biesinger, W. A. Brungs and R. E. Siefert, Effects of pollution on freshwater fish, J. Wat. Poll. Cont. Fed. 47, 1711 (1975).

(37) V. M. Brown, The calculation of the acute toxicities of mixtures of poisons to Rainbow Trout, Water Research 2, 723-733 (1968).

(38) B. A. Westfall, Ecology 26, 283 (1945).

(39) K. E. Carpenter, The lethal action of soluble metallic salts on fish, J. Exp. Biol. 4, 378 (1927).

(40) G. W. H. Schepers, Tetraethyl Lead and Tetramethyl Lead, Arch. Env. Health 8, 277 (1964).

(41) R. T. P. De Treville, H. W. Wheeler and T. Sterling, Occupational exposure to organic lead compounds, Arch. Env. Health 5, 532 (1962).

(42) J. E. Cremer and S. Callaway, Further studies on the toxicity of some tetra and trialkyl lead compounds, Brit. J. Ind. Med.18, 277 (1961).

(43) J. E. Cremer, Biochemical studies on the toxicity of tetraethyl lead and other organic lead compounds, Brit. J. Ind. Med. 16, 191 (1959).

(44) J. Cairns, Environment and time in fish toxicity, Ind. Wastes 2, 1-5 (1957).

(45) J. B. Sprague, Measurement of Pollutant Toxicity to Fish Part III, Sub lethal effects and "safe" concentrations, Water Research 5, 245-266 (1971).

(46) T. L. Coombs, Metabolism and mechanism of uptake of heavy metals including lead by estuarine and freshwater invertebrates, EEC. Environmental Research Programme, Contract 086/087-74/7 Env. U. K., 1975.

(47) L. Majori and F. Petronio, Accumulation phenomena which takes place in a mussel (Mytilus gallo provincialis) grown in an artificially polluted environment, Ig. Mod. 66, 79-98 (1973).

(48) M. Schulz-Baldes, Toxizität und Anreicherung von Blei bei der Miesmuchel Mytilus edulis im Laborexperiment, Mar. Biol. 16 226-229 (1972).

(49) M. Schulz-Baldes, Lead uptake from seawater and food, and lead loss in the common mussel Mytilus edulis, Mar. Biol. 25, 177-193 (1974).

(50) P. Kauranen, T. Jarvenpaa, Biological Half Times of 210-Po and 210-Pb in some Marine Organisms, Annual Report, Dept. of Radiochemistry, University of Helsinki (1972).

(51) A. Berg, Lead in the aquatic environment, Euratom Report Eur. 5260 (1973).

(52) M. Merlini and G. Pozzi, Lead nad freshwater fish. 1. Lead accumulation and water pH, Env. Poll. 13 119-126 (1977).

(53) M. Merlini and G. Pozzi, Lead and freshwater fish. 2. Ionic lead accumulation. Env. Poll. 13, 119-126 (1977).

(54) J. C. Macleod and E. Pessah, Temperature effects of mercury accumulation toxicity and metabolic rate in Rainbow Trout, J. Fish. Res. Bd. Can. 30, 485-492 (1973).

(55) C. E. McKone, Rapid uptake of mercuric ion by goldfish, Env. Sci. Technol. 5, 1138 (1971).

(56) F. C. Kopfler, Accumulation of organic and inorganic mercury compounds by the Eastern Oyster Crassostrea virginica, Bull.Env.Cont.Toxicol. 11, 275 (1974).

(57) M. Y. Unlu, M. Heyraud and S Kečkeš, Mercury as a hydrospheric pollutant. 1. Accumulation and excretion of $HgCl_2$ in Tapes decussatus, in Marine Pollution and Sea Life, ed. Ruiso, FAO.

(58) R.D. Burkett, Influence of temperature on uptake of methyl mercury by Blunt Nose Minnows, Bull. Env. Cont. Toxicol. 12, 703 (1974).

(59) Anon, International Symposium on Environmental Health Aspects of Lead (Amsterdam), Commission of European Communities EUR 5004 (1972).

(60) FAO/WHO, Joint Expert Committee on Food Additives, WHO Tech. Report Series 505, (1972).

(61) Anon, Food Additives and Contaminants Committee. Review of the Lead in Food Regulations 1961, HMSO FAC/REP/21 (1975).

(62) A. P. Vinogradov, The elementary composition of marine organisms, Sears Foundation, Mar. Res. Men. 647 (1953).

(63) G. V. Alexander and D. R. Young, Trace Metals in South California Mussels, Mar. Poll. Bull 7, 7-9 (1976).

(64) T. J. Chow, H. G. Snyder and C. B. Snyder, Mussels as an indicator of lead pollution, Science of the Total Environment 6, 55-63 (1976).

(65) MAFF, Working Party on the Monitoring of Foodstuffs for Heavy Metals - Second Report, Survey of Lead in Food, HMSO (1972).

(66) MAFF, Working Party on the Monitoring of Foodstuffs for Heavy Metals - Fifth Report. Survey of Lead in Food - First Supplementary Report, HMSO (1975).

(67) G. Nickless, R. Stenner and N. Terille, Distribution of cadmium, lead and zinc in the Bristol Channel, Mar. Poll. Bull. 3, 188-190 (1972).

(68) G. Topping, Heavy metals in shellfish from scottish waters, Aquaculture 1, 379-384 (1973).

(69) M. W. Hardisty, S. Kartar and M. Sainsbury, Dietary habits and heavy metal concentrations in fish from the Severn Estuary and the Bristol Channel, Mar. Poll. Bull. 5, 61-63 (1974).

(70) G. Topping, Heavy metals in fish from scottish waters, Aquaculture 1, 373-377 (1973).

Some Morphological and Biochemical Hematological Parameters of Abnormal Lead Absorption in Fish

T. Beritić, J. Zibar-Šikić, D. Prpić-Majić and M. Tudor
Institute for Medical Research and Occupational Health, Zagreb
and Institute for Oceanography and Fishery, Split, Yugoslavia

ABSTRACT

Changes in basophilic stippled erythrocytes count and in δ-aminolevulinic dehydratase (ALAD) activity were studied in adult carp (*Cyprinus carpio*) and in adult European spotted dog-fish (*Scyliorhinus canicula*) both in *in vitro* and *in vivo* experiments, respectively. After sublethal concentrations, maximal average stippled cell count was obtained in carp on the 38th day. Inhibition of ALAD was found in response to the applied lead dose in the dog-fish, being an exponential decay function of added lead in vitro. Both tests proved valuable and simple biological tests for lead pollution in the hydroshpere.

INTRODUCTION

It was through this institution that a few years ago in Rovinj our interest in the effects of lead on sea fish was aroused. We were asked by a member of the Center for Marine Research who was preparing his Master's thesis on lead poisoning in *Blennius pavo* for help with the stippled cell count. For many years we had been concerned with the hematological aspects of the toxic action of lead in both clinical work and research. Not only did we readily fulfill the candidate's request, but we also hurried to tackle the problem, the more so because we felt that the hematology of fish could provide an extremely favorable basis for the transfer of experience gained in this field with regard to humans. As is well known, hemoglobin and erythrocytes of fish, owing both to the instability of the former and the preservation of substantial metabolism in the latter, are also, in fact, ideal models for the study of the biochemistry of changes in hemoglobin synthesis and of the morphological evidence of these changes.

On the other hand, as the problem of pollution of the hydrosphere grows, fish blood and their blood-forming organ appear to be a potential indicator of this pollution, as so well expressed by the fish hematologist, Klontz (Ref. 1): "Fish are so intimately associated with their aqueous environment that physical or chemical changes in this environment are often rapidly reflected as measurable physiological changes in the fish". The possible relationship between the changes in concentrations of lead in the hydrosphere, and the physiological response which could be detected by simple hematological methods was the main purpose of this preliminary work.

At this pint, a brief recapitulation of the relevant lead-induced blood changes in mammals should first be made. Lead has a great affinity for the blood-forming organ, which is bone marrow in mammals and the kidney in fish. There, it interferes with hemoglobin synthesis in the stem-cells of the red blood cells -- erythroblasts. At least two enzymes are inhibited by lead: cytoplasmic δ-aminolevulinic dehydratase and a mitochondrial enzyme ferrochelatase. Due to the disturbed biosynthesis of hemoglobin, defective erythrocytes may ensue, also in part morphologically evidenced in the form of small dots or ribosomes as (pathological) remnants of non-used ribonucleoprotein. Since in these dots or stipples no hemoglobin is synthesized, they remain basophilic, and are generally known as basophilic stippled cells (BSV). They have been considered and used as a classical sign of lead poisoning for many decades. When present in great numbers in the peripheral blood, or in the bone marrow of mammals, at a given lead exposure, they are a very reliable sign of lead poisoning. Some of their main characteristics have previously been studied and described (Ref. 2 and 3).

A much more sensitive index of lead exposure is the depressed activity of δ-aminolevulinic dehydratase (ALAD), the enzyme which is found in the peripheral erythrocytes of mammals as the (physiological) remnant of hemoglobin biosynthesis. The degree of activity, depression and the significance of this enzyme in various stages of lead action in humans has also recently been studied and reviewed (Ref.4).

Changes in these phenomena under the influence of lead have been studied in fish, and some preliminary observations are presented here.

ANIMALS AND METHODS

Adult carp, *Cyprinus carpio* and adult European spotted dog-fish, *Scyliorhinus canicula*, were used for this investigation. The carp averaged 200 grams in weight and the spotted dog-fish 250 grams. The carp were maintained throughout the experimental period in plastic tanks of 40 liters capacity with a permanent water and compressed air supply. The spotted dog-fish were maintained in similar tanks with sea water.

Lead was applied in the form of lead acetate [$Pb(C_2H_3O_2)$ 2.3 H_2O] dissolved in redistilled water. The mode of administration was intraperitoneal (in carp and dog-fish) or "oral" (in carp only). The dosage for carp was approximately 0.17 g of lead per kg, body weight for intraperitonal administration and 1 g per 100 g food, i.e., approximately 0.5 g lead per fish. Since our aim was to induce the formation of stippling in a certain period of time, we used sublethal concentrations, i.e., a dose large enough to induce changes without inducing the rapid death of the fish. The dog-fish received 0.1 ml of lead acetate intraperitonally in the *in vivo* experiments. The applied lead doses were 1-8 μg Pb/g body weight.

Blood samples of 1 ml were withdrawn by heart pucture in dog-fish into a syringe containing sodium heparinate (0.5 mg per ml blood). In carp, the blood samples (for smears) were taken from blood vessels in the caudal peduncle and a portion of the kidney was dissected away and used for making imprints. Blood smears were fixed and colored with methylene blue following standard hematological methods. BSC were counted per million red blood cells.

In vitro experiments with pooled dog-fish blood were carried out with 1 ml portions of the pooled blood. To each ml, 0.1 ml of lead acetate dissolved in water was added in various concentrations. The blood with lead acetate was incubated 30 minutes at 37 °C in order to make it possible for the lead to penetrate the membrane of the erythrocytes and to exert its action upon the enzyme, delta-aminolevulinic dehydratase.

ALAD activity in dog-fish blood, both in *in vivo* and *in vitro* experiments, was determined according to the European standardized method (Ref. 5).

RESULTS

During the entire experiment period, the fish which had received lead orally did not show any external signs of intoxication except a certain degree of restlessness. Maximal average stippled cell count was obtained in the peripheral blood on the 38th day. For the next 30 days the stippled cell count gradually decreased (Table 1).

TABLE 1 The Influence of Orally Administered Lead on the Incidence of Stippled Cell Count. Average Values were Obtained Making a Count During 75 Days.

Days	$\bar{X} \pm$ SE (N)
0	30.0 ± 21.3 (10)
7	390.0 ± 37.9 (10)
14	560.0 ± 63.6 (10)
21	800.0 ± 42.2 (10)
28	970.0 ± 79.0 (10)
33	1330.0 ± 84.4 (10)
38	1410.0 ± 69.0 (10)
43	1130.0 ± 59.7 (10)
51	900.0 ± 44.7 (10)
58	720.0 ± 57.3 (10)
65	530.0 ± 33.5 (10)
75	390.0 ± 34.8 (10)

In fish which had received lead interperitonally, signs of local inflammation appeared first and thereafter necrosis on the spot of the injection. The vast majority of fish (90%) died before the 30th day. Maximal average value of the stippled cells was calculated on the 9th day after lead intoxication (Table 2).

In dog-fish, inhibition of ALAD was found in response to the applied lead dose. Figure 1 represents ALAD inhibition by lead *in vitro* and Fig. 2 represents the results of the experiment *in vivo*. The decreasing activity is an exponential decay function of added lead *in vitro*. The experiment *in vivo* showed that the two-fold taking of blood samples increases ALAD in the organism by 4-10 %. After the dose of 1 μg/g body weight, ALAD activity has decreased with the increasing lead dose. The lowest lead concentration which could cause inhibition was 3 ppm.

DISCUSSION

The stippled cells induced by lead in the erythrocytes of the carp (*Cyprinus carpio*) correspond both morphologically and chromatologically to basophilic stippled cells in mammals. Stained with methylene blue they are seen as small blue dots, round or oval in form, scattered in the cytoplasm of the erythrocytes. They are general-

TABLE 2 The Influence of Interperitonal Administration of Lead on the Incidence of Stippled Cells. Average Values Obtained by Blood Analysis for 23 Days.

Days	$\bar{X} \pm$ SE (N)
0	$\pm$
3	110.0 $\pm$ 45.8 (10)
5	377.8 $\pm$ 83.0 (9)
7	1388.9 $\pm$ 182.9 (9)
9	1650.0 $\pm$ 187.1 (8)
12	1266.7 $\pm$ 111.6 (6)
16	950.0 $\pm$ 88.5 (6)
19	800.0 $\pm$ 70.7 (4)
23	633.3 $\pm$ 88.2 (3)

ly more scarce in fish than in mammals. In spite of the relatively high dosage of lead in the blood of the carp, the maximal number of stippled cells per million erythrocytes was 2,600 (up to 1,000 stippled cells per million in man is still considered a physiological or "normal" finding). However, since the total number of erythrocytes in fish blood is higher in relation to the blood of mammals (Ref. 6) and the values are expressed in percentages, the absolute stippled cell count in lead-poisoned fish is obviously greater. In severe lead poisoning in humans, bone marrow contains up to five times more basophilic stippled cells than in the peripheral blood (Ref. 7) but such a high number of stippled cells in the blood forming organs of the fish was not observed in any case. The experiments carried out in different seasons show that in summer, and at higher water temperatures, the same concentrations of lead induce a greater number of stippled cells in the peripheral blood (Table 3, Ref. 8). This is probably due to the well-known effect of temperature, but may also be the consequence of some cyclic changes in fish blood (Ref. 9, 10 and 11).

TABLE 3 Basophilic Stippled Cells (BSC) Count in Cyprinus Carpio after "Oral" Exposure to Lead.

Fish Group	Pb in food - g/100 g daily - 3 weeks	BSC/10^6E ($\bar{X} \pm$SE; N=10) 33 days after lead administration
Summer	0.545	1330.0 $\pm$ 84.4
	0.545	1070.0 $\pm$ 70.0
Winter	2.725	1150.0 $\pm$ 100.2
	5.450	1500.0 $\pm$ 88.1

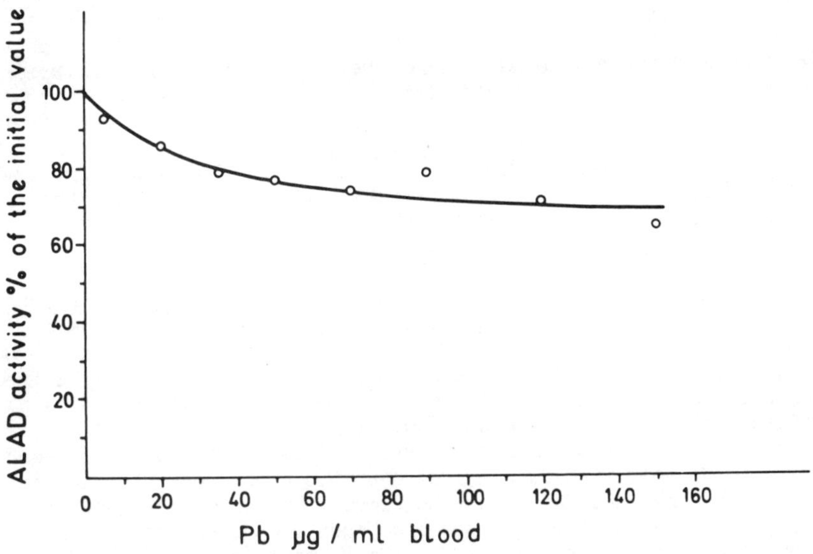

Fig. 1. _In vitro_ inhibition of ALAD activity by lead in blood of fish _Scyliorhinus canicula L._

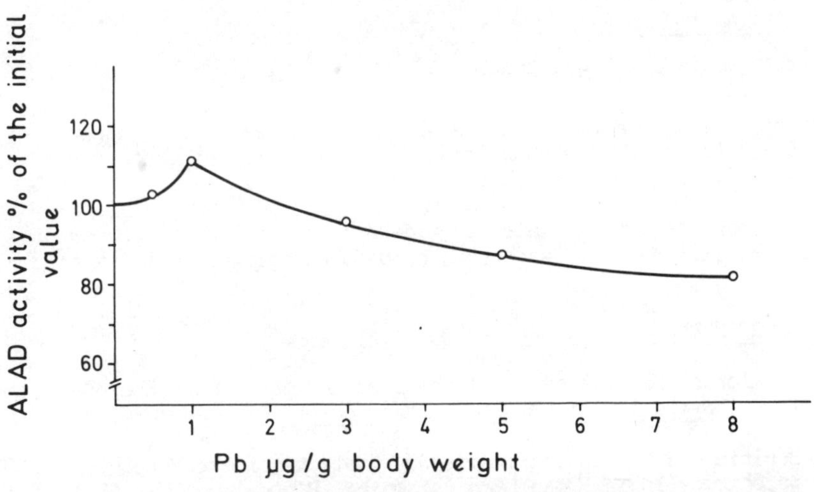

Fig. 2. _In vivo_ inhibition of ALAD activity by lead in blood of fish _Scyliorhinus canicula L._

A few years ago, Jackim (Ref. 12) studied the influence of lead and other metals on δ-aminolevulinate dehydratase activity by exposing mummichog (Fundulus heteroclitus) and winter flounder (Pseudopleuronestes americanus) to sublethal concentrations of lead nitrate. He found a considerable reduction of liver and kidney activity of the enzyme, but only after 4 weeks of exposure. He came to the conclusion that ALAD depression is one of the characteristic, though not specific, responses to lead exposure or toxicity in fish "although it is by no means a specific response to such stress". However, in regard to the very high degree of specificity of lead-induced ALAD inhibition in humans, this statement of Jackim`s should be clarified even more so because he confirmed that ALAD values in the control fish are rather consistent and not influenced markedly by the common environmental variability.

More recently, Hodson (Ref. 13) exposed rainbow trout (Salmo gairdneri) to nominal lead concentrations of 0, 10, 18, 32, 56 and 100 μg/liter in the form of lead nitrate. His results are at variance with those of Jackim`s since he found reduced enzyme activity in the blood of fish exposed to 143, 66 and 38 μg/liter lead (measured concentrations) as early as one week after exposure. It should be pointed out that concentrations used by Jackim were 1000 times greater than those observed to be effective in Hodson`s experiments. Possible reasons for these differences may be, according to Hodson, that Jackim used a static exposure system and salinic water so that lead concentrations declined with time. Some investigations have shown that the activity of ALAD also presumably depends upon the hardness of the water. It also seems likely that the high difference temperature can influence the response of the activity of the enzyme to the presence of lead (Ref. 14). Our in vitro and in vivo experiments confirmed the decrease of ALAD activity, and the lowest lead concentration which could cause inhibition was 3 ppm. The value obtained corresponds to 210 mg Pb for 70 kg which is the average body weight for the "standard man". The estimated lowest Pb concentration response for ALAD inhibition in man is 1.40 mg Pb orally/day for three days (Ref. 16 and 15). Comparing this value with that in dog-fish one may conclude that fish is much more resistent to lead than is man. It was also established that two-fold taking of blood samples from the same fish increases the activity of ALAD and this should certainly be taken into consideration while evaluating the results of experiments.

CONCLUSION

Although the experiments were carried out with relatively high concentrations, the results obtained in this work indicate that some simple and quick method such as the stippled cell determination and ALAD activity measurement in the peripheral blood of the fish could find some practical application as tests for the detection of lead pollution in the hydrosphere.

These preliminary results may also provide some initial data for more basic investigations on the influence of lead upon the biosynthesis of fish hemoglobin.

REFERENCES

(1) G. W. Klontz, Haemotological Techniques and the Immune Response in Rainbow Trout, Symposium of the Zoological Society of London 30, 89 (1972).

(2) T. Beritić and M. Vandekar, Some Observations on the Morphology of Erythropoietic Cells in Human Lead Poisoning, Blood 11, 114 (1956).

(3) T. Beritić, Siderotic Granules and the Granules of Punctate Basophilia, Brit. J. Haemat. 9, 185 (1963).

(4) T. Beritić, D. Prpić-Majić, V. Karačić, and S. Telišman, ALAD/EP Ratio as a Measure of Lead Toxicity, JOM 19, 551 (1977).

(5) A. Berlin and K. H. Schaller, European Standardized Method for the Determination of δ-aminolevulinic acid dehydratase Activity in Blood, Z. Klin. Chem. Klin. Biochem. 12, 389 (1974).

(6) J. Field, B. Elvehjem, and C. A. Juday, A Study of the Blood Constituents of Carp and Trout, J. Biol. Chem. 148, 261 (1943).

(7) T. Beritić, Toxicological Significance of Erythrocytic Inclusions, Arh. Hig. rada 8, 97 (1957).

(8) J. Zibar, Erythrocytic Inclusions in Carp, Cyprinus carpio, Induced by Lead, M. S. Thesis (1975).

(9) V. Chernikova, Seasonal Blood Characteristics of Fresh-water Fishes, in Report summaries of the All-Union Conference of the Ecology and Physiology of Fishes, Biol. Abstr. 48, 95053 (1967).

(10) A. Houston and H. De Wilde, Thermoacclimatory Variations in the Haematology of the Common Carp, Cyprinus carpio, J. Exp. Biol. 49, 71 (1968).

(11) A. Houston and H. De Wilde, Haematology and Blood Volume of Thermally Acclimated Brook Trout, Salvelinus fontinalis, Comp. Biochem. Physiol. 28, 877 (1969).

(12) E. Jackim, Influence of Lead and Other Metals on Fish δ-Aminolevulinate Dehydrase Activity, J. Fish. Res. Board Can. 30, 560 (1973).

(13) O. Hodson, δ-Aminolevulinic Acid Dehydratase Activity of Fish Blood as an Indicator of a Harmful Exposure to Lead, J. Fish. Res. Board, Can. 33, 268 (1976).

(14) P. Davies and H. Everhardt, Effects of Chemical Variations in Aquatic Environments, Lead Toxicity to Rainbow Trout and Testing Application Factor Concept., Ecol. Res. Serv. Environ. Prot-Agency, Washington, D. C., EPA-R3-73-011C 3, 80 (1973).

(15) E. Stuik, Biological Response of Male and Female Volunteers to Inorganic Lead, Int. Arch. Arbeutsmed. 33, 83 (1974).

(16) K. Mahaffey, Quantities of Lead Producing Health Effects in Humans: Sources and Bioavailability, Environ. Health Perspect. 19, 285 (1977).

Some Factors Influencing Lead Metabolism

K. Kostial

Institute for Medical Research and Occupational Health,
Zagreb, Yugoslavia

ABSTRACT

This paper represents a series of selected "highlights" from the author`s previous experiments performed with the purpose of evaluating factors which might influence lead absorption, retention and distribution. All data were obtained in the same laboratory. Lead-203 was used for studying lead pharmacokinetics in rats. The results show that pregnancy, lactation, age and nutrition greatly influence lead metabolism. They specially indicate that the early neonatal age represents a period of increased lead accumulation in the body. The neonates therefore seem to be at a higher risk than adults at the same level of environmental exposure. These findings support the assumption that standards for lead exposure in the youngest age group might need reconsideration.

INTRODUCTION

The increased amount of lead in the biosphere with its potential hazard to man has been the subject of a great number of investigations. However, only few investigations into the mechanism of intestinal lead absorption and into the factors which influence this process have been performed. On the other hand the available evidence yields strong implication that lead ingestion is a very important route of entry into the body especially in children (Ref. 1).

The purpose of this paper is to present some experimental evidence obtained in this laboratory which shows that several factors greatly influence lead absorption from the intestine. Our aim, in particular was to evaluate whether differences in lead absorption due to pregnancy, lactation, age and nutrition are a sufficient indication to reconsider standards for lead exposure.

METHODS

All experiments were performed on albino rats of different age. The radioactive tracer technique (^{203}Pb) was used for studying lead transport from mother to litter and lead absorption, retention and distribution in relation to age and diet. Dietary additives were used to decrease lead absorption from the intestine and chelating agents to enhance lead elimination from the body. The results are expressed as percentages of the oral or intraperitoneal dose of radioactive lead.

They are presented as arithmetic means and standard error of the means. The number of animals in each group is given in brackets.

RESULTS

Lead Transfer from Mother to Litter

Studies of lead transfer from mother to litter were performed to evaluate whether lead exposure in pregnancy and lactation represents a hazard for neonates.

Lead absorption was determined in rats on the 15th day of lactation by applying a single oral dose of radioactive lead (Fig. 1). The cumulative 5-day urinary excretion and whole body retention in lactating rats and in their litters was compared with the values obtained in control animals. The absorption of lead-203 was about 3.5 times higher than in controls. Since lactating animals consumed about twice as much food as controls the actual increase in the amount of stable or radioactive lead absorbed from the diet is supposed to be about seven times higher than in controls (Ref. 2).

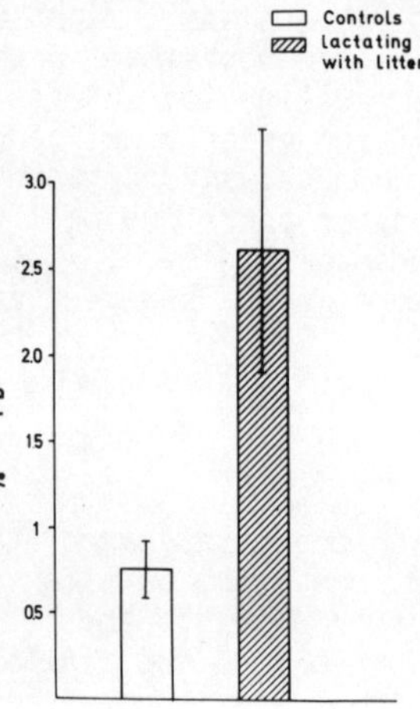

Fig. 1. The effect of lactation on intestinal lead absorption (from Ref. 2).

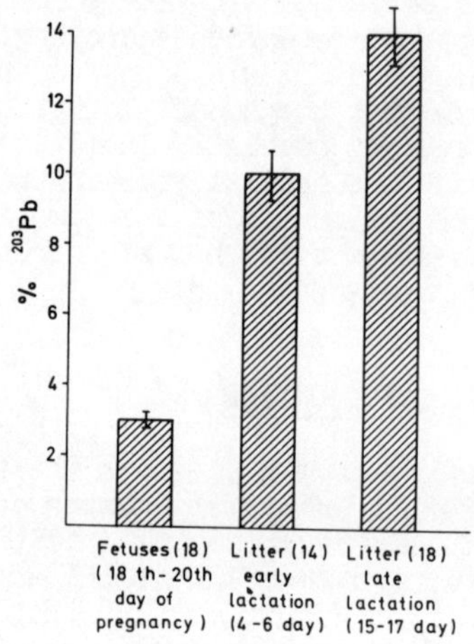

Fig. 2. Lead transport from mother to litter in pregnancy and lactation.

We also tried to obtain quantitative data on the transfer of lead during various stages of gestation and lactation in order to determine the period of maximum transfer from mother to litter. Tracer doses of radioactive lead were given to pregnant and lactating rats intravenously and the retention was determined in the whole body of their fetuses and litter 48 hours after lead-203 administration (Fig. 2). The results obtained indicate the period of late lactation as the time of the maximum transport of lead from mother to litter (Ref. 3).

Lead Absorption in Neonates

The purpose of our next experiment was to estimate lead absorption in the early neonatal period to find a possible explanation why the young of a given species are more prone to lead poisoning than the adult. We used the artificial feeding technique (Ref. 4) for administering radioactive lead to suckling rats. The absorption of lead-203 was very high in the youngest age group (Ref. 5). If we compare lead absorption in sucklings and in older animals (Ref. 6) a sharp decrease with increasing age was observed (Fig. 3).

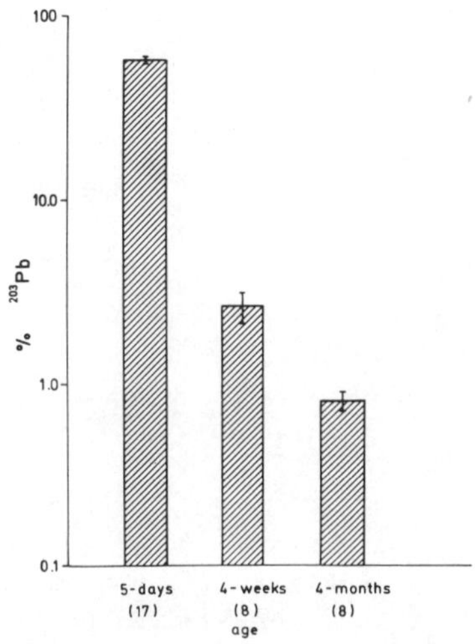

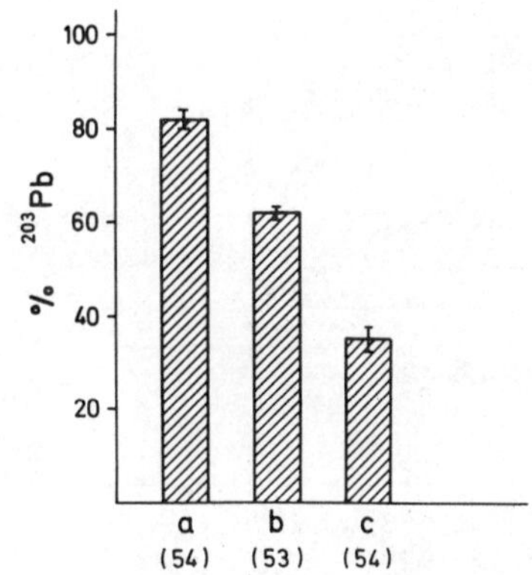

Fig. 3. The effect of age on intestinal lead absorption (first column from Ref. 5; second and third from Ref. 6).

Fig. 4. Influence of milk additives (mg/100 ml) on lead absorption in sucklings: a) 140 Ca, 95 P; b) 400 Ca, 230 P; c) 400 Ca, 230 P, 2000 sodium alginate OG.1. (Ref. 7).

Our purpose was further to evaluate whether the high absorption of lead from the intestine in newborns can be reduced by various additives to milk. The experiments were performed on 5-7-day-old baby rats which were artificially fed over a period of eight hours with cow`s milk and with milk to which calcium, phosphate and alginates were added (Fig. 4). The highest retention values were observed in the group of animals without additives. The retention of lead was about 1.4 times lower in animals which received calcium and phosphate milk additives and approximately 2.5 times lower in animals fed milk with calcium, phosphate and alginate additives (Ref. 7).

Influence of Milk Diet on Lead Absorption

The purpose of our further work was to evaluate the effect of milk diet on lead absorption from the intestine since milk represents the main source of food intake in the very young and might be one of the reasons for a high lead absorption in neonates. Rats were fed on different diets one week before radiolead administration. The body retention 6 days after administration was much higher in animals on milk diet (four times higher in rats on normal diet and cow`s milk; 57 in animal on cow`s milk; 33 times in rats on powdered milk) than in controls (Fig. 5). The milk diet had practically no effect on the whole body retention of radioactive lead after intraperitoneal administration (Ref. 8). This indicates that milk considerably enhances lead retention in the body mainly by greatly increasing its absorption from the intestinal tract.

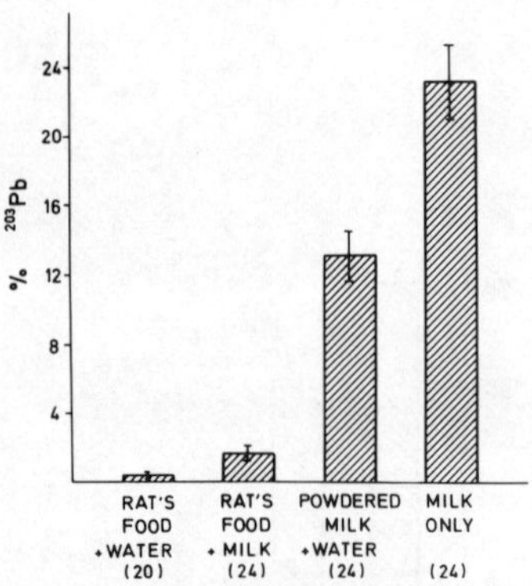

Fig. 5. Effect of milk diet on lead absorption (from Ref. 8).

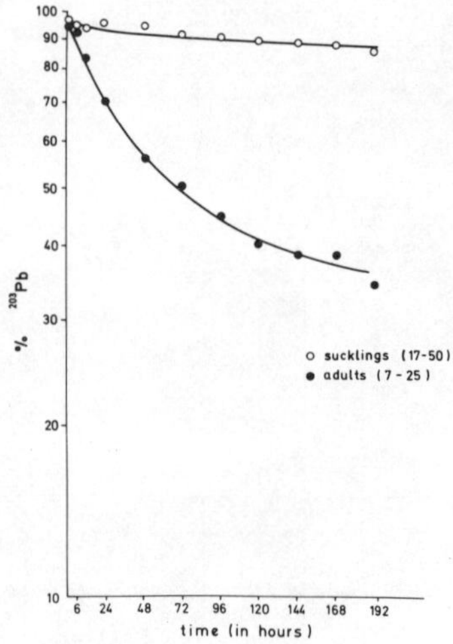

Fig. 6. Kinetics of lead retention in suckling and adult rats (from Ref. 9).

Lead Retention and Organ Distribution in Relation to Age

If we omit the gastrointestinal route of entry by administering lead intravenously or intraperitoneally other differences in lead metabolism between suckling and adult rats become apparent. The experiment presented on Fig. 6 was performed on 2-week and 18-week old female albino rats. The whole body retention determined at different time intervals after a single intraperitoneal administration was much higher in sucklings compared to adult rats and the elimination was lower. Marked differences were observed in the kinetics of lead retention and distribution in some other tissues. In blood, lead retention was 2.5 times higher and in the femur 1.5 times higher than in adults eight days after single intraperitoneal administration (Fig. 7). The most striking was the eight times higher concentration of lead in the brain of suckling rats (Ref. 9).

Effect of Chelating Agents on Lead Metabolism in Relation to Age

We also tried to estimate the efficiency of chalating agents in the treatment of lead poisoning in the very young. Chelating agents were found to be less efficient in removing lead from the body in suckling rats than in adults. This effect was independent of the route of lead administration, of the lead dose applied, of chelating agents, metal molar ratio and of the frequency of chelating treatment (Ref. 10). In the experiment presented in Fig. 8 rats received a single intraperitoneal dose of EDTA (75 mg per kg) immediately after radiolead administration. The effect on lead excretion was markedly lower in sucklings as indicated by whole body retention values at several intervals after lead and chelating agent administration. This effect was interpreted by a lower "free lead" level due to higher stability or inertness of lead linkages with body ligands in the immature (Ref. 10).

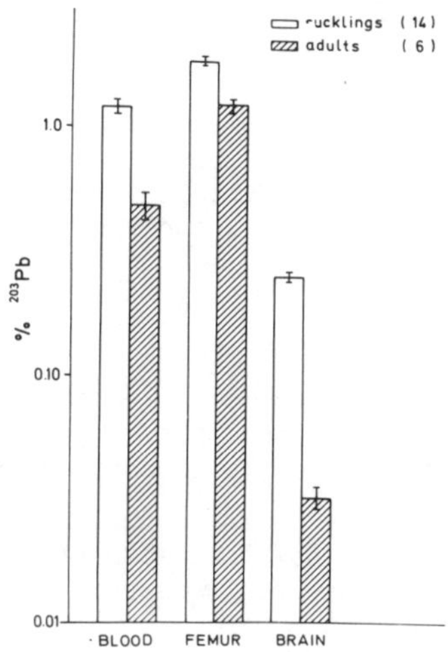

Fig. 7. Lead retention in tissues of suckling and adult rats (from Ref. 9).

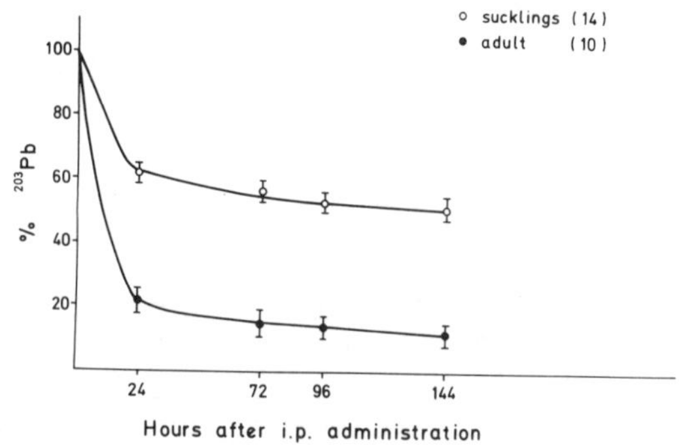

Fig. 8. Effect of EDTA on lead excretion in relation to age (from Ref. 10).

DISCUSSION

Changes in lead absorption, retention and distribution in pregnancy, lactation and neonatal age are most probably the results of morphological and functional characteristics of the body at these stages.

The increased intestinal absorption of lead during lactation is most probably due to the morphologic augmentation of the gastrointestinal tract (Ref. 11, Ref. 12). We found a similar increase in intestinal absorption of strontium and calcium in rats at the same phase of the lactation process (Ref. 13).

When comparing calcium and lead transport from mother to litter we noticed that the transmammary transport of lead was about four times lower and its transplacental transport eight times lower than that of calcium. This points at the placenta as a better barrier against lead than the mammary gland (Ref. 3).

A very high intestinal absorption in neonates is not specific for lead only, since we also found a very high absorption of cadmium, mercury and manganese in newborn rats (Ref. 14). Increased metal absorption in neonates was also reported by some other authors (reviewed in Ref. 15). This could be due to pinocytosis, specific for this age (Ref. 16, 17) to milk diet or other factors.

The mechanism by which milk increases lead absorption is still unknown. Lead probably forms ligands with milk constituents which are more easily absorbed (Ref. 18). Milk is also low in some trace elements especially in iron which might cause increased lead absorption (Ref. 19). This effect of milk is also not lead specific since we also found an increased absorption of some other metals in rats fed milk diet (Ref. 14).

A higher lead retention in the body and some organs is very likely the result of different binding of lead to body ligands at the phase of intense body growth and protein synthesis. The lower efficiency of chelating agents in enhancing lead elimination is most probably due to the same reason (Ref. 10).

However, the purpose of this paper is not to explain the mechanism by which differences in lead absorption, distribution and excretion occur but to demonstrate their existance and their magnitude and to indicate that they might cause considerable increase in the body burden of lead. Such differences should be taken into account when considering the biological consequences of lead occurrence in the environment.

CONCLUSIONS

On the basis of our experiments on rats it might be concluded that:

- lactation is a period of increased lead absorption from the intestine and considerable lead transfer from mother to litter;
- neonatal age is a period of greatly increased intestinal lead absorption (partly because of milk diet) and higher lead retention in the body, especially in the brain;
- chelating agents are less efficient in enhancing lead elimination from the body in sucklings than in adults.

Therefore neonates might be at a higher risk than adults at the same level of environmental exposure. This indicates that more attention should be paid to the early developmental stage in further research on lead toxicity.

ACKNOWLEDGEMENTS

My thanks are due to all members of the staff of the Laboratory of Mineral Metabolism and especially to Dr. B. Momčilović, Dr. D. Kello, Dr. S. Jugo, Mr. Tea Maljković, Marica Pišonić and Prof. I. Šimonović, who were coauthors in the experimental work presented. I also wish to thank Dr. J. F. Stara, HERL, EPA, Cincinnatti, Ohio, USA for his stimulating interest and help throughout this work.

This work was partially financed by the Environmental Protection Agency, USA.

REFERENCES

(1) Environmental Health Criteria 3 Lead 1977, WHO, Geneva, 126

(2) K. Kostial and B. Momčilović, The effect of lactation on the absorption of ^{203}Pb and ^{47}Ca in rats, Health Physics 23, 383 (1972).

(3) K. Kostial and B. Momčilović, Transport of ^{203}Pb and ^{47}Ca from mother to offspring, Arch. Environ. Health 29, 28 (1974).

(4) K. Kostial, I. Šimonović and M. Pišonić, Effect of calcium and phosphate on gastrointestinal absorption of strontium and calcium in newborn rats, Nature 215, 1181 (1967).

(5) K. Kostial, I. Šimonović and M. Pišonić, Lead absorption from the intestine in newborn rats, Nature 233, 564 (1971a).

(6) K. Kostial, D. Kello and G. M. Harrison, Comparative metabolism of lead and calcium in young and adult rats, Int. Arch. Arbeitsmed. 31, 159 (1973).

(7) K. Kostial, I. Šimonović and M. Pišonić, Reduction of lead absorption from the intestine in newborn rats, Environ. Res. 4, 360 (1971b).

(8) D. Kello and K. Kostial, The effect of milk diet on lead metabolism in rats, Environ. Res. 6, 355 (1973).

(9) B. Momčilović and K. Kostial, Distribution and kinetics of lead in suckling and adult rats, Environ. Res. 8, 214 (1974).

(10) S. Jugo, T. Maljković and K. Kostial, The effect of chelating agents on lead excretion in rats in relation to age, Environ. Res. 10, 271 (1975).

(11) R. N. Campbell and B. F. Fell, Gastrointestinal hypertrophy in the lactating rat, J. Physiol. Lond. 171, 90 (1964).

(12) R. Boyne, B. F. Fell and I. Robb, The surface area of the intestinal mucosa in the lactating rat, J. Physiol. 183, 570 (1966).

(13) K. Kostial, N. Gruden and A. Duraković, Intestinal absorption of calcium-47 and strontium-85 in lactating rats, Calcif. Tissue Res. 4, 13 (1969).

(14) K. Kostial, D. Kello, S. Jugo, I. Rabar and T. Maljković, Influence of age on mental metabolism and toxicity, Environm. Hlth. Perspec. (in print).

(15) S. Jugo, Metabolism of toxic heavy metals in growing organisms: A Review, Environ. Res. 13, 36 (1977).

(16) J. G. Lecce, Selective absorption of macromolecules into intestinal epithelium and blood by neonatal mice, J. Nutr. 102, 69 (1972).

(17) K. Martinsson and L. Jönsson, On the mechanism of intestinal absorption from the gut, Toxicol. Appl. Pharmacol. 40, 277 (1977).

(18) D. Kello and K. Kostial, Influence of age and milk diet on cadmium absorption from the gut, Toxicol. Appl. Pharmacol. 40, 277 (1977).

(19) M. K. Six and A. R. Goyer, The influence of iron deficiency on tissue content and toxicity of ingested lead in the rat, J. Lab. Clin. Med. 79, 128 (1972).

Acute and Chronic Effects of Lead on Mammalian Cells *In Vitro*

A. B. Fischer
Hygiene Institut, Zentrum für Ökologie, Universität Giessen,
German Democratic Republic

ABSTRACT

The acute effects of inorganic lead salts on mammalian cells in vitro were investigated using L-A cells, a subline of L 929 mouse fibroblasts, HeLa-cells, a human portio-carcinoma line, and human diploid fibroblasts. Lead salts caused a growth inhibition which was independant of the anion employed and could be prevented by the addition of Ca-EDTA. The LC_{50} was approximately 1 mM when exposing L-A cells for 2 days and 2×10^{-4} M when treating then for 7 days. The release of lactic acid dehydrogenase was not elevated indicating lack of gross membrane damage by lead. Concomitant with the dose-dependent inhibition of proliferation the fraction of cells synthesizing DNA and the mitotic rate were lowered; by cytophotometry a block in the G1 phase of the cell cycle was demonstrated. Energy metabolism appeared affected as evidenced by a rise in lactate production. In HeLa cells and human fibroblasts mitotic aberrations were observed.

Under continuous exposure L-A cells developed lead resistance, even to concentrations in the range of the LD_{50}. The kinetics of the acquirement of resistance are demonstrated. The resistant cells adjusted more quickly to higher lead doses than the controls. Low-resistance strains retained the growth characteristics of the parent cells but with increasing levels of tolerance multiplication rates declined significantly. After ca. 60 generations in heavy metal-free environment the cells had to a major degree lost their resistance, indicating that the induced lead tolerance is caused by adaptation rather than by mutation and selection.

INTRODUCTION

Concern about the environmental dangers of lead has stimulated a vast number of investigations on the pathological and subclinical effects of this heavy metal on human, animal, and plant life, supplemented by biochemical tests and tissue culture studies (Ref. 1-3). Functional, biochemical, and morphological effects have been investigated at various structural levels. Morphological changes may be observed at the macroscopical, light microscopical and at the ultrastructural level. Metals may give rise to gross structural organ damage and to alterations in cellular and subcellular morphology. Functional and biochemical changes have also been demonstrated at the subcellular, cellular and organ levels. Most of the biochemical knowledge of the effects of lead has come from in vitro experiments. It is obvious that changes in cellular functions must occur before changes in organ functions manifest themselves. Heavy metals interact with cells in many ways; reactions occur with membranes, organelles, and metabolic and mitotic systems. The cell is a complex entity with its own internal homeostatic mechanisms. The ideal way to elucidate the complex interactions of lead with cells would be to examine the cells in the living organism. Another more feasible approach is to use cells cultured in vitro. This test system has the following limitations: (I) lack of humoral, nervous and other integration mechanisms to which the cells are subjected in the whole organism, although some regulative and cooperative mechanisms do exist; (II) lack of metabolization (activation, detoxification) of a toxic substance; (III) absence of secondary effects caused by the accumulation of damage in the whole organism. On the other hand, the following advantages are found: (I) the possibility to study cellular events directly; (II) absence of secondary effects caused by primary changes in the whole organism at other sites; (III) the use of homogeneous cell populations; and (IV) well defined, quantifyable and reproducible experimental conditions (Ref. 4).

This communication deals with the acute and chronic effects of lead chloride on cells in tissue culture. In short-term experiments the following parameters were studied: viability, proliferation, DNA synthesis, mitotic index, chromosome morphology, lactate and pyruvate production, and lactic acid dehydrogenase release. In long-term experiments L-A cells developed resistance to lead concentrations which were effective during acute exposure. Data are furnished on the comparative behaviour of control and resistant cells with regard to growth rates and the acquirement and loss of resistance.

MATERIAL AND METHODS

Most of the methods applied have been described in detail (Ref. 5-7). In the majority of the experiments static suspension cultures of L-A cells, a subline of the mouse fibroblast line L 929 were used. L-A cells were propagated in Medium 199 supplemented with 10 % horse serum. In some cases also HeLa cells (ATCC) and human diploid fibroblasts were employed. These strains were grown in MEM (Eagle) supplemented with 10 % newborn calf serum. The cells were counted with a Coulter Counter Model ZB equipped with a 100 μm orifice. Viability was examined by adding trypan blue at a final concentration of 1.5 % to the cell suspensions and scoring 200 - 400 cells per sample. Mitotic indices were determined in at least 5000 cells per sample which were fixed with acetic acid: methanol and stained with hematoxilin-eosin. DNA synthesis was investigated by two methods. Cells in logarithmic phase were exposed to 0.15 μCi ^{3}H-thymidine (specific activity 5 Ci/mmole, Amersham, UK) for 24 h, washed with buffer, fixed with acetic acid: methanol and dripped on slides. The slides were air-dried, exposed to stripping film (Kodak AR 10) for 9 days and finally stained with hemalum. ^{3}H-thymidine uptake was also studied by exposing logarithmically growing L-A cells to 0.7 μCi/ml ^{3}H-thymidine at 0 - 2, 2 - 4, 4 - 6 or 24 - 26 h, following lead exposure. The cells were then pipetted into microtiter plates (approximately 33 000 cells/well) and immediately flushed on filter disks with a Multiple Cell-Culture Harvester (Skatron, Lierbyen, Norway). The disks were dried, put into vials together with a scintillation cocktail (5 g PPO and 0.5 g dimethyl-POPOP in toluol) and counted with a Packard Tricarb Liquid Scintillation Spectrometer. For biochemical tests Biochemical Test Combinations (Boehringer Corp., Mannheim, F. R. Germany) were used. Metal salt solutions were prepared from analytical grade compounds (E. Merck AG, Darmstadt, F. R. Germany).

For statistical evaluation simple analysis of variance and Scheffe`s test were used.

RESULTS

1. Acute Effects

Pilot studies with different lead salts showed that the biological effect on L-A cells was independent of the anion used. As can be seen in Fig. 1, there was no difference between the growth inhibition caused by lead chloride, acetate, and nitrate. Furthermore, this graph demonstrates that the action of lead can be prevented by the addition of a chelating agent EDTA.

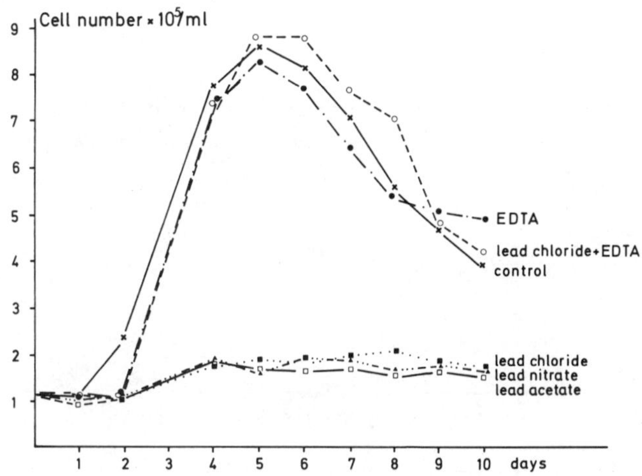

Fig. 1. Effect on cell growth of different lead salts administered alone or in combination with Ca-Na-ethylenediaminetetraacetate (EDTA). 4 milk dilution bottles per experimental group were inoculated with 1.2×10^{-5} L-A-cells/ml (18 ml/bottle). The lead salts were used in a concentration of 10^{-4} M and EDTA in a concentration of 2×10^{-4} M. Cell counts were done on 0.5 ml samples which were withdrawn from the bottles at one or two days' intervals.

Acute lead toxicity was determined using lethality and growth inhibition as the main parameters. Figure 2 demonstrates the effects on viability (right hand side, closed symbols). Lead is a slow poison; after 2 days' exposure (dashed line) the LC_{50} is approximately 1 mM, but after 7 days $3 - 5 \times 10^{-4}$ M, i.e., one half to one third. In the same figure the proliferation of the cells treated with lead is shown (open symbols) expressed as percentage of the increase of control cells. The 50 % growth inhibition dose (IC_{50}) lies in the range of $4 - 5 \times 10^{-5}$ M.

In comparison with inorganic mercury and cadmium lead possesses a smaller toxicity to L-A cells (Table 1). The three heavy metals can be ranked according to increasing toxicity in the following order: lead, mercury, and cadmium. Of special interest is the finding that the ratio between growth inhibition (IC_{50}) is much higher in the case of lead than in the case of mercury and cadmium.

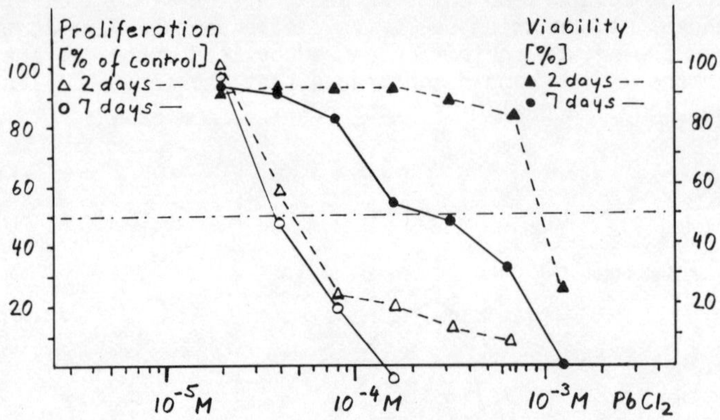

Fig. 2. Effect of lead on viability and proliferation of L-A-cells cultured for 2 or 7 days in the presence of the heavy metal.

TABLE 1 Comparison of the LC_{50}, IC_{50} and Quotients of LC_{50} and IC_{50} of the Heavy Metals Lead, Mercury and Cadmium.

	LC_{50} : IC_{50}	
Pb	$\frac{3 \times 10^{-4}\ M}{4 \times 10^{-5}\ M}$ =	7.5
Hg	$\frac{5 \times 10^{-5}\ M}{1.3 \times 10^{-5}\ M}$ =	3.84
Cd	$\frac{1.3 \times 10^{-5}\ M}{7.5 \times 10^{-6}\ M}$ =	1.9

Figure 3 shows the main parameters of cell multiplication: cell number, mitotic rate, DNA synthesis, and in addition viability. The values were determined after 2 days of exposure with 0, 2, 8 and 32 x 10^{-5} M $PbCl_2$. The difference between the controls and the cells treated with 2 x 10^{-5} M was not significant in any of the parameters listed ("no effect level"). Cells exposed to 8 and 32 x 10^{-5} M exhibited markedly reduced cell numbers and mitotic rates, though viability was little affected. Concentrations inhibiting proliferation caused a marked depression of the fraction of ^3H-thymidine labelled cells. While approximately 94 % of the controls and of the cells treated with the "no effect" dose metabolized ^3H-thymidine, only 83.3 of 39.9 % of the cells exposed to the elevated lead concentration took

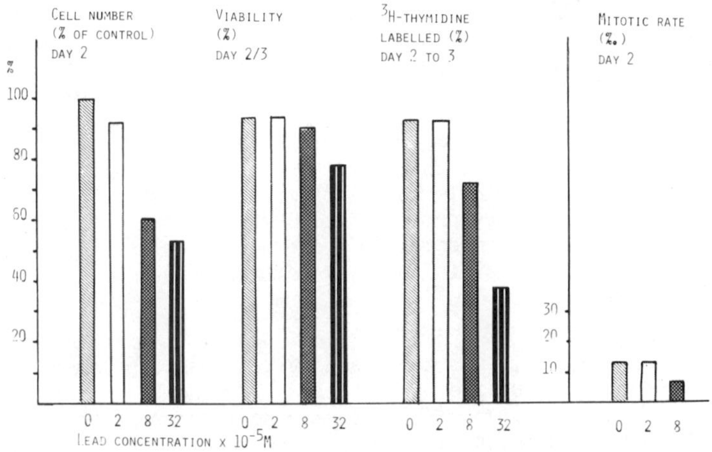

Fig. 3. Effect of lead on proliferation. L-A-cells were cultured for 2 or 3 days, respectively, as described in Fig. 1, but with 0, 2, 8, or 32 x 10^{-5} M $PbCl_2$.

part in DNA synthesis; this is a much smaller number than that corresponding to the fraction of viable cells. From these data it can be concluded that the observed growth inhibition is essentialy caused by the existence of a large subpopulation which either does not fully participate in proliferation or is retarded somewhere along the cell cycle. As this inhibition takes place before DNA synthesis starts, it seems that the cells are blocked in the G_1 phase of the cycle. We tried to confirm this hypothesis by measuring the DNA content of proliferating controls and lead-exposed cells by cytophotometry and found that the lead cultures had indeed a greater fraction of G_1 cells (Table 2). However, this question is not yet comple-

TABLE 2 Effect of Lead on the Distribution of Cells in Cell Cycle Phases.

	G_1 %	S %	G_2 + M %
Control	78.7	14.8	6.3
5x10^{-5} M $PbCl_2$	89.1	7.5	3.4
10^{-4} M $PbCl_2$	91.8	5.1	3.9

Logarithmically growing L-A-cells were exposed to 0, 5 or 10 x 10^{-5}M $PbCl_2$ for 24 h. Cells were fixed in 70 % and absolute ethanol, digested with pepsin (5 % in 0.05 M HCl; 5 min), centrifuged in physiological saline + tris buffer (pH = 7.5) and stained with 4', 6-diamidino-2-phenylindol (Serva, Heidelberg; 3 μM in tris buffer + 0.02 M $MgCl_2$). Cellular DNA content was measured by cytophotometry.

tely solved. When conducting comparative studies on the effect of lead chloride on HeLa cells and human diploid fibroblasts, we noted changes in the morphology of the mitotic chromosomes. These changes consisted mainly of spiralization defects, leading to "fuzziness" and clumping of the chromatin and to disturbances in chromosome distribution. In HeLa cultures these features appeared 24 and 48 h after incubation with 10^{-4} M $PbCl_2$ (Fig. 4). At the same time the mitotic rate rose markedly and there was a concomitant increase in the early phases of mitosis, notably metaphases. The ratio between pro- + metaphases and ana- + telophases was significantly elevated (Table 3). In human fibroblasts no increase in mitotic index was found, but on the third day of exposure to 1.2 and 1.6 x 10^{-4} M $PbCl_2$ chromosome clumping and an increased ratio of the early to the late phases of mitosis were also observed (Fig. 5, Table 4). These findings indicate that cells are held up during the early and middle stages of mitosis, possibly by the observed spiralization defects, which may prolong the formation of the metaphase plate and impede the distribution of chromosomes to the daughter cells. In conclusion it can be stated the lead is capable of influencing several phases of the cell cycle, particularly G_1 and mitosis.

The finding that DNA synthesis was depressed by lead was confirmed by kinetic studies on the ^{3}H-thymidine uptake by L-A cells. During the first two hours of incubation with 0.5 and 1 x 10^{-4} M PbCl, thymidine incorporation was still unaltered. However, during the next four hours there was a dose-dependent decrease to 73.7 and 67.1 % of control values. After 24 h the lead-treated cells incorporated 52.1 and 18.5 % of the controls (Fig. 6).

As a further parameter of acute toxicity lactic acid dehydrogenase (LDH) release was studied. Release of intracellular enzymes gives an indication of membrane permeability. The amounts of LDH in the supernatant of cells exposed to 4 and 8 x 10^{-5} M were significantly lower than those of the controls on days 4 to 8 of the experiment. The LDH curves paralleled the growth curves, which means that the controls and the lead-treated cells gave off approximately the same amounts of the enzyme per cell (Fig. 7).

Finally, the effects on lead on lactate and pyruvate metabolism were studied; these are important parameters of carbohydrate metabolism. The complex kinetics of lactate production or uptake per cell and measuring interval of 2 days are presented in Fig. 8. At the beginning of their growth period, in the lag-log phase, L-A cells release lactate into the nutrient medium. This is later on again taken up, but in the stationary phase of growth lactate is set free once more. Under lead lactate synthesis was already stimulated during the first 2 days of exposure; the differences between the controls and cells treated with 8 x 10^{-5} M were significant on days 2, 6, and 8. As expected, the increase of lactate in the supernatant is accompanied by a decrease of pyruvate. Table 5 shows the results of an experiment, where logarithmically growing L-A cells were exposed to 0, 8, 12, and 16 x 10^{-5} M $PbCl_2$ for 24 hours. A dose-dependent stimulation of lactate and depression of pyruvate was observed.

2. Chronic Exposure

During long-term exposure L-A cells developed resistance to increasing lead levels. Figures 9 and 10 illustrate the kinetics of the development of resistance. In these graphs cumulated cell numbers are plotted logarithmically and the time of cultivation or number of cell passages, respectively, are plotted linearly. It can be seen how, after an initial depression of proliferation, the lead-exposed cells normalized their growth rates in the course of maximally 8 weeks. The preexposed cells adjusted more quickly to the elevated lead levels than the "naive" controls. It is remarkable, however, that even the controls managed to normalize their growth rate under different lead concentrations, i.e., also acquired resistance.

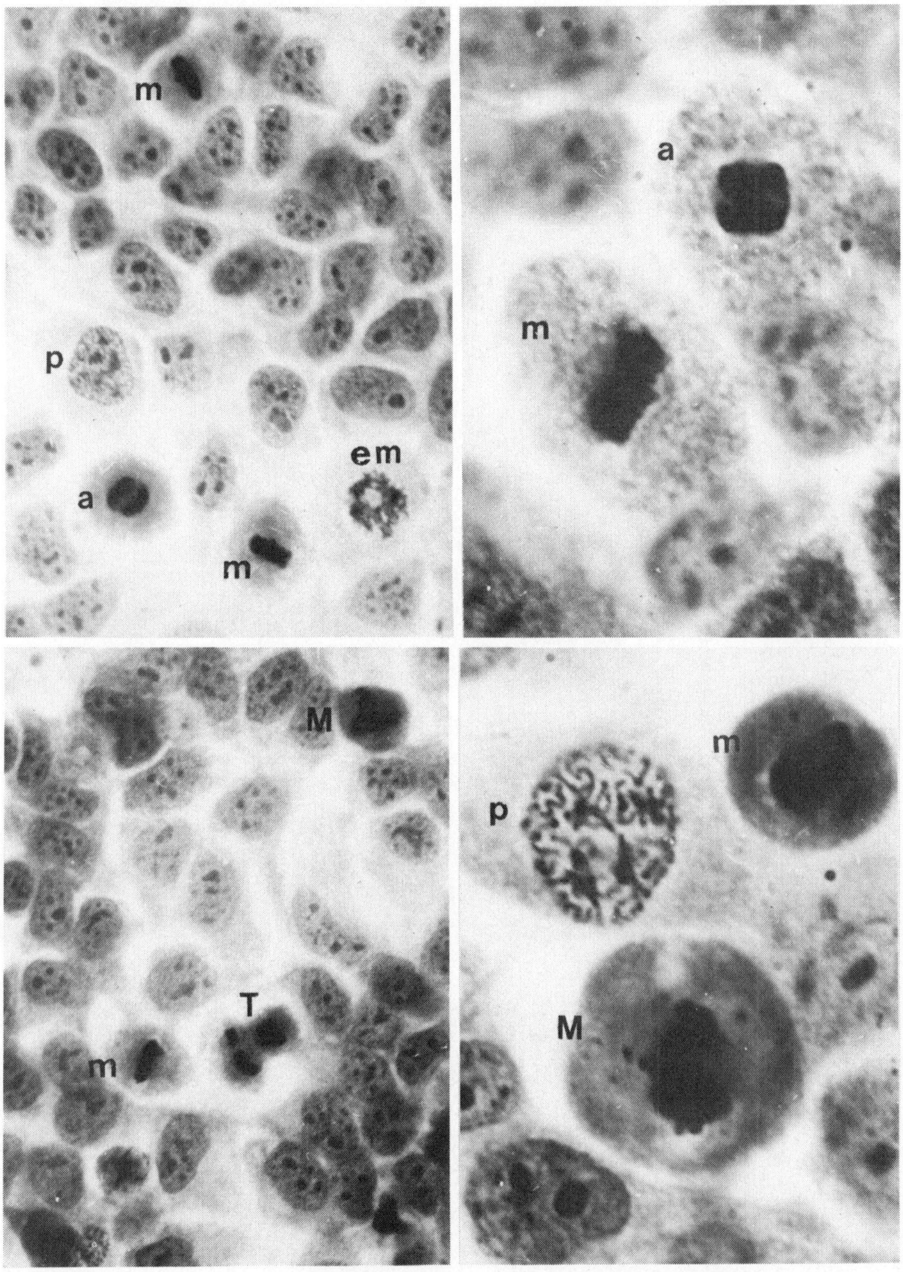

Fig. 4. Effect of lead on chromosome morphology. HeLa cells were cultured for 48 h with or without 10^{-4} M PbCl. Upper row: controls, lower row: lead-exposed cells; p = prophase, em = early metaphase, m = metaphase, a = anaphase, M = abnormal metaphase, T = abnormal telophase.

TABLE 3 Effect of Lead on Cell Division of HeLa Cells which were Cultured for 24 and 48 H in the Presence or Absence of 10^{-4} M $PbCl_2$.

Sample	No.of cells counted	Prophase (‰)	Metaphase (‰)	Anaphase (‰)	Telophase (‰)	Mitotic rate (‰)	$\frac{\text{Pro + Meta}}{\text{Ana + Telo}}$
Control 24 hours	3000	2.3	23.3	1.7	24.3	53	0.99
10^{-4} M $PbCl_2$ 24 hours	5000	2.8	52.4	1.0	19.2	72	2.73
Control 48 hours	4000	.0	22.8	1.8	13.3	43	1.85
10^{-4} M $PbCl_2$ 48 hours	4000	7.5	53.3	3.5	22.8	88	2.31

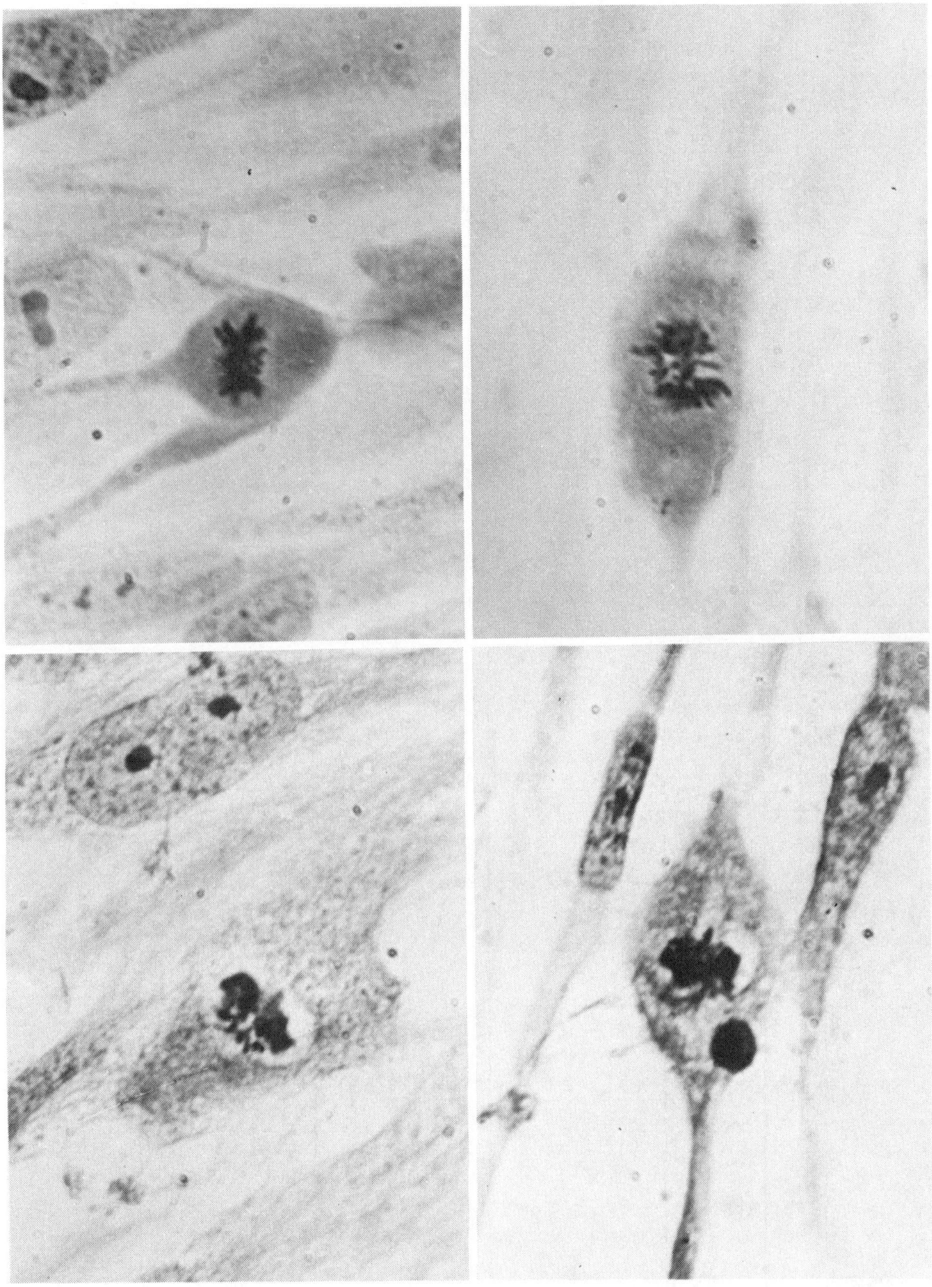

Fig. 5. Effect of lead on chromosome morphology. Human diploid fibroblasts were cultured for 72 h with or without 1.6 x 10^{-4} M $PbCl_2$. Upper row: controls, lower row: lead-exposed cells. Only metaphases are shown.

TABLE 4 Effect of Lead on Cell Division and Chromosome Morphology of Human Diploid Fibroblasts Cultured for 72 H in the Presence of 0, 1.2, and 1.6 x 10^{-4} M $PbCl_2$.

Sample	Mitotic phase	Chromosome morphology "good"	"medium"	"bad"	Sum	Mitotic rate ca. %	Pro- + metaphases / ana- + telophases
Controls 72 hours	prophase	9	-	-	9	3.4	52 : 45 = 1.13
	metaphases	40	3	0	43		
	anaphase	5	-	-	5		
	telophase	31	2	8	41		
	all phases	85	5	8	98		
$1.2x10^{-4}$ M $PbCl_2$ 72 hours	prophase	2	1	-	3	2.0	31 : 15 = 2.07
	metaphase	12	7	9	28		
	anaphase	2	1	-	3		
	telophase	5	-	8	12		
	all phases	21	9	17	47		
$1.6x10^{-4}$ M $PbCl_2$ 72 hours	prophase	6	-	-	6	1.7	41 : 15 = 2.73
	metaphase	17	3	15	35		
	anaphase	1	-	1	1		
	telophase	11	2	1	14		
	all phases	35	5	16	56		

Over 20 000 cells per sample were scanned. Mitoses were graded as "good", when chromosomes were clean and showed the characteristic mitotic configuration, "bad" when chromosome clumping or fuzziness occurred and when the pattern of chromosome distribution was disturbed, and "medium" when the morphology was intermediate.

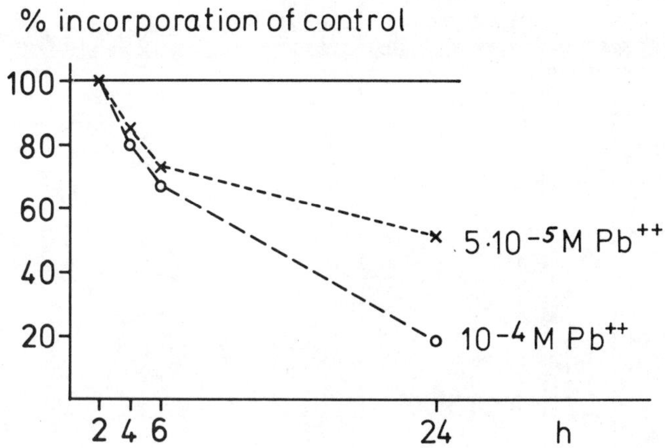

Fig. 6. Effect of lead on ^{3}H-thymidine incorporation. L-A-cells in logarithmic growth were exposed to 5 and 10 x 10^{-4} M $PbCl_2$ and pulsed at 0-2, 2-4, and 24-26 h following lead exposure.

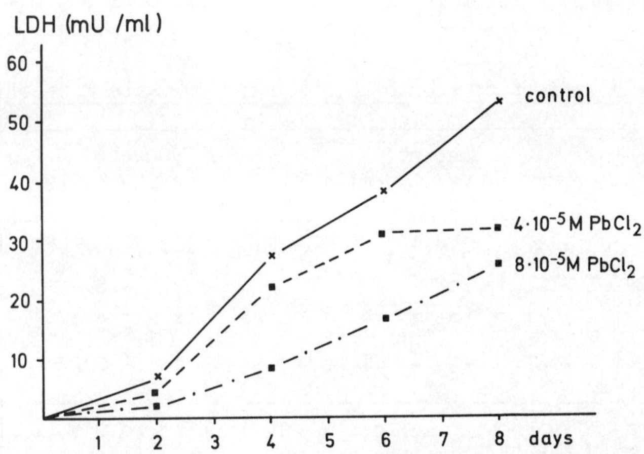

Fig. 7. Effect of lead on LDH release. 5 milk dilution bottles per experimental group were inoculated with 20 ml of a cell suspension containing 1.2 x 10^5 L-A-cells/ml. The cultures were grown in lead-free medium or exposed to 4 and 8 x 10^{-5} M $PbCl_2$. Every other day the enzyme activity was determined in the culture medium.

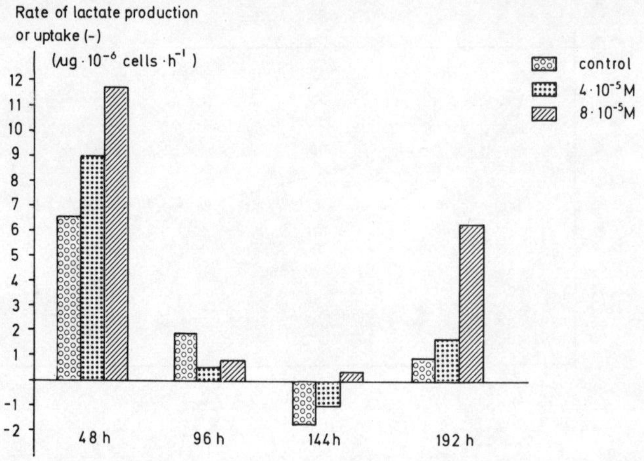

Fig. 8. Influence of lead on the rate of lactate production or uptake. L-A-cells were cultured as described in Fig. 7 and the lactate content of the supernatant was assayed every other day. The amounts of lactate newly released into or taken up from the culture medium in a measuring interval were related to the average cell number present in, and the number of hours of, that interval.

TABLE 5 Effect of Lead on Lactate and Pyruvate Metabolism.

	Lactate $\mu g/10^6$ cells	Pyruvate $\mu g/10^6$ cells
Control	179 ± 8	12.9 ± 0.1
0.6×10^{-4} M $PbCl_2$	266 ± 23	14.4 ± 0.2
1.2×10^{-4} M $PbCl_2$	272 ± 8	6.5 ± 0.2
1.6×10^{-4} M $PbCl_2$	299 ± 7	6.6 ± 0.2

Logarithmically growing L-A-cells were exposed to 0.8, 1.2, and 1.6×10^{-4} M for 24 h and the concentrations of lactate and pyruvate determined in the culture supernatant were related to cell numbers.

When comparing the growth rates of controls and cells with different degrees of lead resistance in parallel passages, a dose-dependent decrease in proliferation was observed (Fig. 11). The differences in replication rates between non-exposed and lead-tolerant cells were not significant, when $PbCl_2$ concentrations up to 7×10^{-5} M were employed, but became significant with the 11, 14, and 20×10^{-5}M concentrations.

It was of interest to see whether reversal of resistance occurred. Figure 12 shows the results of a typical experiment. The following cells were employed: controls (0-0), cells cultured continuously with 14×10^{-5} M lead cultured for 6 passages (46 days) in the absence of lead (14-0). These cells were exposed in parallel to lead-free medium or 7 and 14×10^{-5} M $PbCl_2$. Two findings can be gathered from this experiment. The first one, confirming the results of the previous calculations, is that the XIV-cells proliferated more slowly with than without lead attaining cell numbers 60 - 70 % of the non-exposed controls. The second finding is that the growth of the 14-0 cells was clearly depressed by the reintroduced lead to about one half of the cell number of the 14-14 cells. On the other hand, the previously untreated controls were still more affected; their proliferation remained below 10 % that of the lead tolerant cells. The resistant cells had to a major degree - though not completely - lost their resistance following 46 days cultivation in lead-free environment.

DISCUSSION

In short-term experiments the lethal and growth inhibiting contrations of lead for L-A cells have been determined as follows: minimal growth inhibiting concentration (MGIC) $2\text{-}4 \times 10^{-5}$ M; IC_{50} $4\text{-}5 \times 10^{-5}$ M; and LC_{50} after 2 days and 7 days ca. 1 mM and ca. 2×10^{-4} M, respectively. The range between "no effect" level and lethal action covers approximately one power of 10. In this respect lead differs from the two more toxic heavy metals cadmium and mercury, where the growth inhibiting concentrations approach the lethal concentrations. It was demonstrated that the inhibition of proliferation by inorganic lead salts occurred independently of the anion used and could be reversed by Ca-EDTA. Lead concentrations depressing cell multiplication did not increase lactic acid dehydrogenase release. This finding together with the results of the viability staining tests, which were also in the control range, indicate that lead does not cause gross membrane damage in the sense of elevated permeability.

The question whether the proliferation inhibition is caused by a block or a prolongation of specific phases of the cell cycle or whether all phases are influenced by lead is still incompletely answered. It seems that the G_1 phase in L-A cells is mainly affected. However, in HeLa cells and in human diploid fibroblasts the early phases of mitosis are prolonged following lead exposure as evidenced by a raised ratio between pro- + metaphases and ana- + telophases. Morphological changes of the chromosomes suggest spiralization defects due to lead. The same phenomena were previously also observed in human lung fibroblasts treated with particulate PbO ($40 \mu g/10^5$ cells, Ref. 8). Obe and Sperling (Ref. 9) reported a "woolly" structure of chromatin in human lymphocytes cultured in the presence of high concentrations of lead acetate (10^{-3} M, 24 h). The morphological chromosome changes may be caused by an antagonism of lead to iron which is needed for the proper folding of the chromosomes, or by the binding of lead to chromosomal proteins (Ref. 10-12). As the problem of lead effects on cell cycle phases is of fundamental interest, it should be pursued using synchronized cells.

It was shown in autoradiographic studies that the fraction of cells synthesizing DNA during a period of 24 h was lowered. A similar decrease of ^3H-thymidine incorporation was found in kinetic studies lasting for 24 h. These results confirm

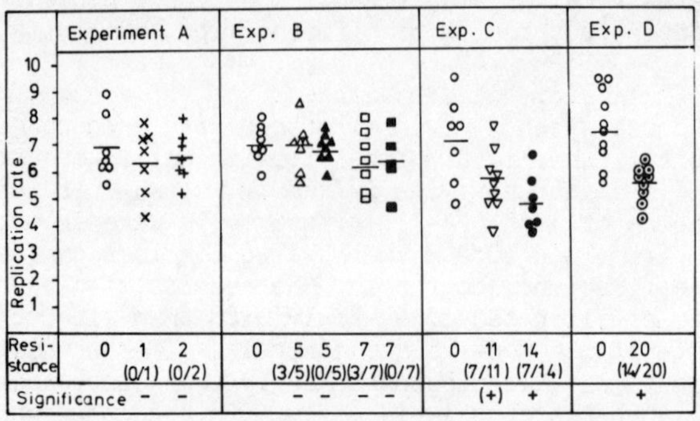

Fig. 11. Comparative growth rates of controls and resistant cells tolerating various lead concentrations. The proliferation rates of each experiment were determined in parallel passages. In experiments A and B the cells were passaged at 5 days', in experiments C and D at 7 days' intervals. The numbers on the abscissa denote the degree of resistance expressed as multiplicities of 10^{-5} M and the numbers in parentheses denote the origin of those cells (e.g., 5 (3/5) = cells resistant to 5 x 10^{-5} M $PbCl_2$ and developed from cells tolerating 3 x 10^{-5} M).

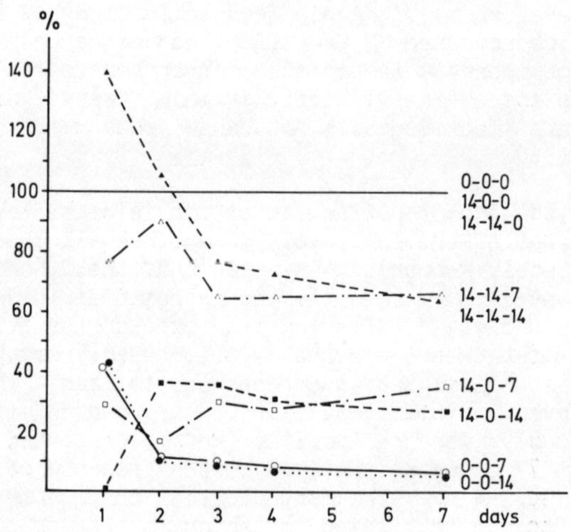

Fig. 12. Comparative effect of lead on controls (0-0), XIV-cells cultured continuously with lead (14-14), and XIV-cells cultured for 6 passages (46 days) without lead (14-0). The cells were exposed to lead-free medium and to 7 and 14 x 10^{-5} M $PbCl_2$. There were 4 replicates per experimental group with an inoculation number of 1.2 x 10^{-5} cells/ml.

the more extensive investigations by Škreb and Habazin-Novak (Ref. 13) who noted a reduction in DNA synthesis in HeLa cells under the influence of 2.5×10^{-4} M $PbCl_2$. Their finding that the effect of a lead intoxication was reversible agrees with our observation on the reversibility of growth inhibition obtained with lead-resistant cells.

In spite of reduced proliferation lactic acid production is relatively and absolutely increased after an initial lag. Enhanced lactate production has also been demonstrated in nonproliferating mouse peritoneal macrophages exposed to $PbCl_2$ _in vitro_ (Ref. 14). These findings are in keeping with an _in vivo_ observation by Muhle and co-workers (Ref. 15) who noted increased glycolytic activity in the lung macrophages of rats chronically inhaling low levels of $PbCl_2$ aerosols. Since several authors have shown that lead is bound to mitochondria, where it causes structural alterations (Ref. 12 and 16) and leads to impairment of cellular oxidation processes (Ref. 3 and 17), it appears plausible that the affected cells switch over from oxidation carbohydrate metabolism to glycolysis.

In the context of lead resistance the following questions were of interest: (1) what are the kinetics of the development of resistance; (2) does tolerance to certain lead levels protect the cells against higher levels of exposure; (3) is there a limit to resistance; (4) are there differences in proliferation behaviour between controls and resistant cells; (5) is resistance developed through selection or adaption; and (6) what is the mechanism of resistance?

The kinetics of the acquirement of resistance has been described in detail and it has been demonstrated that resistant cells tolerate higher lead doses better than controls. This was especially obvious for highly resistant cells, which needed less time for adjustment and showed better growth rates and viability. The absolute limit to resistance has not been determined so far. Tolerance to 2 and 2.5×10^{-4}M is still possible; these are concentrations approaching the LD_{50}. Under this heavy metal burden, however, growth capacity diminishes markedly, indicating that tolerance cannot be infinitesimally increased.

Several results point to the acquirement of resistance in L-A cells by adaptation rather than mutation and selection. After a lead-free interval of 46 days (in another experiment 39 days) the resistant cells lost to a major degree their ability to cope with the previously tolerated lead concentrations. Under our experimental conditions the average population doubling time was 20 h. Accordingly maximally 60 generations could have been produced in that interval. However, backward mutation of a genetically fixed property within so few generations is highly improbable. Another indication for the modification of the tolerant cells by adaptation is the finding that they multiply more slowly in the presence than in the absence of the heavy metal. If the new growth characteristics of the resistant cells had arisen by selection, they should not be influenced by the removal of the selective agent. Furthermore, the dose-dependent and at low concentrations very rapid time-course of the development of resistance by the controls (to 3 and 5×10^{-5} M less than 2 weeks; to 7×10^{-5} M or 10.5 and 14×10^{-5} M 3 and over 8 weeks, respectively) and the accelerated adjustment by resistant cells (to 7×10^{-5} M 2 weeks; to 10.5 and 14×10^{-5} M 5 - 6 or 7 - 8 weeks, respectively) argue against selective forces. Škreb (unpublished results) has confirmed these results with HeLa cells. She demonstrated that lead tolerance developed within 2 weeks and concluded that it must be brought about by a process of adaptation.

The mechanism of adaptation is still unclear. We analysed the lead content of lead-exposed controls and resistant cells and found comparable Pb amounts in the cells (Ref. 7). Thus, resistance is apparently not achieved by lead exclusion. Three other mechanisms are conceivable: (1) Alteration of metabolic pathways in the cells, like e.g., the switch from respirative to glycolytic energy production.

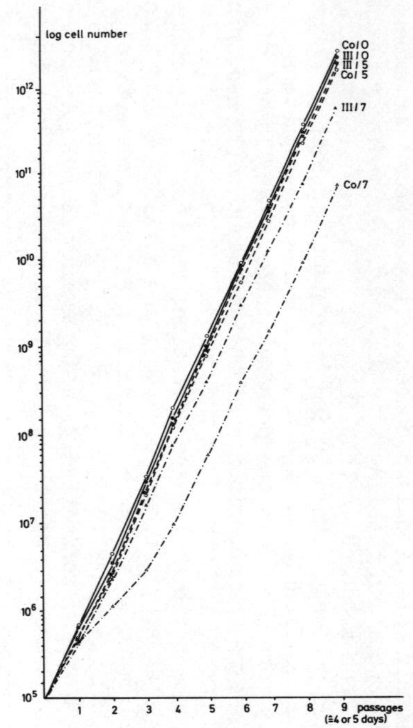

Fig. 9. Long-time effect of lead on L-A-cell growth. Control cells (Co) and cells tolerating $3x10^{-5}$ M $PbCl_2$ (III) were continuously cultured in the absence or presence of 5 and $7x10^{-5}$ M $PbCl_2$. Cumulated cell numbers are logarithmically plotted on the ordinate and the number of passages (4 or 5 days/passage) linearly on the abscissa.

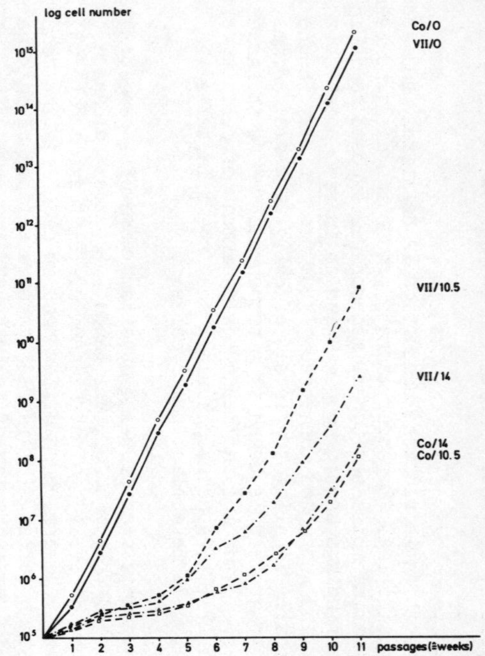

Fig. 10. Long-time effect of lead on L-A-cell growth. Control cells (Co) and cells tolerating $7x10^{-5}$M $PbCl_2$ (VII) were continuously cultured in the absence or presence of 10.5 and $14x10^{-5}$ M $PbCl_2$. Cumulated cell numbers are logarithmically plotted on the ordinate and the number of passages (1 week/passage) linearly on the abscissa.

(2) Detoxification of the heavy metal by binding it in an insoluble comples. It has been proposed that the lead-induced nuclear inclusion bodies, such as demonstrated in liver and kidney cells and osteoclasts which are composed of lead-protein complexes serve to spare toxic injury to cytoorganelles (Ref. 11). (3) Recently a factor has been isolated from liver cytosol and in smaller quantities from erythrocytes, which protects against lead inhibition of uroporphyrin I synthetase activity. The factor was not identical to metallothionein, a protein protecting against cadmium and mercury. The factor was heat stable, but was destroyed by acid, base, ashing or preincubation with protease. These findings suggest that the factor is probably protein in nature (Ref. 18).

The biological capacity for adaptation is not only of theoretical interest but also of practical relevance. Lead tolerance *in vivo* has been observed in occupational medicine (Ref. 19 and 20). Experimental investigations with laboratory and domestic animals have shown that pretreatment with lead protects against lethal doses and clinical manifestations following acute exposure (Ref. 21-23). In the course of a long-term exposure stabilazation of blood lead levels, excretion of d-aminolevulinic acid in urine, d-aminolevulinic acid dehydratase in blood and free erythrocyte porphyrin have been noted (Ref. 24-28). Further evidence for regulatory and compensatory mechanisms at the cellular level was obtained in studies with rats chronically exposed to lead chloride aerosols, whose alveolar macrophages showed enlarged cellular volume, increased phagocytic capacity, and elevated glycolysis. The physiological alterations were interpreted to be possible adaptations to the heavy metal (Ref. 15).

ACKNOWLEDGEMENT

I thank Mrs. U. Krämer for the statistical analysis and Miss U. Conrad for her skillful technical assistance.

This work was in part supported by the European Communities under Contract No. 071-74-1 ENVD and by the European Communities together with the German Umweltbundesamt under Contract No. 137-76-10-ENVD.

REFERENCES

(1) R. E. Engel, D. I. Hammer, R. J. M. Horton, N. M. Lane, and L. A. Plumlee, Environmental lead and public health, Air Pollution Control Office Publication No. AP-90, 1-34 (1971).

(2) R. A. Goyer and B. C. Rhine, Pathological effects of lead, *Int. Rev. exp. Path.* 12, 1-77 (1973).

(3) B. L. Vallee and D. D. Ulmer, Biochemical effects of mercury, cadmium and lead, *Ann. Rev. Biochem.* 41, 91-128 (1972).

(4) E. G. Beck, N. Manojlović, and A. B. Fischer, The cell culture as a test system in applied environmental hygiene, *Symposium Recent Advances in the Assessment of Health Effects on Environmental Pollution*, Paris, June 24-28, 1974.

(5) A. B. Fischer, The effect of lead on cells cultivated *in vitro*. I. Acute Effects, *Zbl. Bakt. Hyg.*, I. Abt. Orig. B 161, 26-37 (1975).

(6) A. B. Fischer, The effects of lead on cells cultivated *in vitro*. II Chronic exposure and development of resistance, *Zbl. Bakt. Hyg.*, I. Abt. Orig.

B 161, 317-330 (1976).

(7) A. B. Fischer, Die Toxizität von Schwermentallen in der Zellkultur, Zbl. Bakt. Hyg., I. Abt. Orig. B 162, 77-84 (1976).

(8) A. B. Fischer, E. G. Beck, and N. Manojlović, Die Wirkung von partikelförmigem Blei auf die Vermehrung von in vitro gezüchteten Zellen, Zbl. Bakt. Hyg., I. Abt. Ref. 36, 1973.

(9) G. Obe and K. Sperling, Zur Wirkung von Bleiacetat auf menschliche Leukozytenchromosomen in vitro, Blei und Umwelt, Komm. f. Umweltgefahren des Umweltbundesamtes, Arbeitsgr. Blei 2.12.1970.

(10) E. Robbins and T. Pederson, Iron: Its intracelullar localization and possible role in cell division, Proc. Nat. Acad. Sci. 66, 1244-1251 (1970).

(11) J. F. Moore, R. A. Goyer, and M. Wilson, Lead-induced inclusion bodies, Lab. Invest. 29, 488-494 (1973).

(12) E. Sabbioni and E. Marafante, Identification of lead-binding components in rats liver: in vivo study, Chem. Biol. Interactions 15, 1-20 (1976).

(13) Y. Škreb and V. Habazin-novak, Reversible inhibition of DNA, RNA and protein synthesis in human cells by lead chloride, Toxicology 5, 167-174 (1975).

(14) N. De Ruiter, N. Seemayer, and N. Manojlović, Einfluss von Zink-Ionen auf die toxische Wirkung von Bleichlorid ($PbCl_2$) untersucht an Mäusemakrophagen in vitro, Zbl. Bakt. Hyg., I. Abt. Orig. B 164, 90-98 (1977).

(15) H. Muhle, F. Otto and D. Stüer, The effect of chronic inhalation of lead aerosols on rat alveolar macrophages, 10th FEBS Meeting, Paris, 1975.

(16) J. Bruch, A. Brockhaus, and W. Dehnene, Elektronenmikroskopische Beobachtungen an Rattenlungen nach Exposition mit partikelförmigem Blei, Proc. Int. Symp. Environmental Health Aspects of Lead, Luxembourg, 221-229 (1973).

(17) R. A. Goyer and R. Krall, Untrastructural transformation in mitochondria isolated from kidneys of normal and lead-intoxicated rats, J. Cell. Biol. 41, 393-400 (1969).

(18) W. N. Piper and R. B. L. Van Lier, Isolation of a factor that protects against lead inhibition of hepatic and erythrocytic uroporphyrinogen I synthetase activity, Life Science 19, 1225-1234 (1976).

(19) F. Kölsch, Handbuch der Berufskrankheiten, VEB G. Fischer Verlag, Jena, 1959, 2. Aufl., p. 240.

(20) H. Valentin, W. Klosterkötter, G. Lehnert, H. Petry, J. Rutenfranz, and H. H. Wittgens, Arbeitsmedizin, G. Thieme Verlag, Stuttgart, 1971, p. 184.

(21) W. Garber and E. Wei, Adaptation to the toxic effect of lead, Amer. Industr. Hyg. Ass. J. 33, 756-760 (1972).

(22) H.-J. Haple, Wirkungen und Schäden durch Blei, Cadmium und Zink bei Nutztieren, Staub-Reinhalt. Luft 34, 8-10 (1974).

(23) C. P. Fassbender, Experimentelle akute Bleivergiftung bei subklinisch belasteten Schafen, Berl. Münch. Tieräarztl. Wschr. 87, 270-273 (1974).

(24) H.-J. Hapke and E. Prigge, Interactions of lead and glutathione with delta-aminolevulinic acid dehydratase, Arch. Toxicol. 31, 153-161 (1973).

(25) S. Tola, S. Hernberg, S. Asp, and J. Nikkanen, Parameters indicative of absorption and biological effect in new lead exposure: a prospective study, Brit. J. Industr. Med. 30, 134-141 (1973).

(26) T. B. Griffin, F.Coulston, L. Golberg, H. Wills, and J. C. Russell, Biologic effects of airborne particulate lead on continuously exposed rats and rhesus monkeys, in Lead (F.Coulston et al., eds.), Georg Thieme Publishers, Stuttgart, Academic Press, New York, 1975, pp. 202-220.

(27) T. B. Griffin, F. Coulston, L. Golberg, H. Wills, J. C. Russell, and J. H. Knelson, Clinical studies on men continuously exposed to airborne particulate lead, in Lead (F. Coulston et al., eds.) Georg Thieme Publishers, Stuttgart, Academic Press, New York, 1975, pp. 221-240.

(28) T. J. Kneip, N. Cohen, V. Rulon, D. Goldstein, and N. Nelson, Dose-effect relationships for lead in infant baboons, in Effects of dose-response relationships of toxic metals, (G. F. Nordberg, ed.), Elsevier, Amsterdam, 1976, 434-445.

Lead in the Marine Environment: Some Biochemical Considerations

J. M. Wood

Freshwater Biological Institute, College of Biological Sciences, University of Minnesota, Navarre, Minnesota, U.S.A.

ABSTRACT

Recent studies in our laboratory have shown that there is a correlation between the standard reduction potential for inorganic complexes and the mechanism for biomethylation of representative heavy metals.

In the case of inorganic lead salts we have shown that electrophilic attack on the cobalt-carbon bond of methyl-B_{12} occurs, facilitating methyl-transfer from cobalt to lead. This reaction is important in that methyl-B_{12} has been shown to be the only methyl-transfer coenzyme capable of transferring methyl-groups as CH_3^-, S-adenosylmethionine and N^5-methyltetrahydrofolate transfer methyl-groups as CH_3^+, and this rules out chemical or biochemical transfer to lead by these alternative methyl-donors.

At this point in our research effort we have been unable to synthesize a monomethyl-lead complex which is stable in aqueous solution, although we have demonstrated methyl-transfer to diethyl-lead dichloride by the mechanism shown in equation (1).

$$(Et)_2Pb^{IV}Cl_2 + CH_3Co^{III}B_{12} + H_2O \longrightarrow (Et)_2Pb^{IV}(CH_3)Cl + H_2O\ Co^{III}B_{12}$$

INTRODUCTION

In 1975 three publications appeared in Nature (Ref. 1, 2, 3) addressing the subject of biomethylation of organo-lead and inorganic-lead compounds in the aqueouse environment. Since that time no attempt has been made to put this observation on a firm biochemical or chemical basis. This is unfortunate because until we have hard evidence for biological methyl-transfer we have an important gap in the biogeochemical cycle for lead.

In this review I will attempt to bring together what is known on the cycling of organo-lead compounds; in addition I will discuss the special problems which need to be overcome in order to biomethylate inorganic-lead in the aqueous environment.

Organo-lead Compounds in the Marine Environment

Tetra-ethyl and tetra-methyl lead have been shown to be extremely unstable in the marine environment (Ref. 4), with dealkylation leading to the formation of the corresponding trialkyl and dialkyl products. These trialkyl and dialkyl lead compounds appear to be quite stable in seawater, presumably due to the ionic nature of these species, which do not bioaccumulate to any appreciable extent in marine biota. However, tetra-ethyl lead and to a lesser extent tetra-methyl lead are very non-polar compounds which may be taken up rapidly by marine organisms, probably by diffusion controlled processes. The uptake of tetra-alkyl lead compounds can be compared with the uptake of dimethyl-mercury or methyl-mercuric chloride which are similarly non-polar in the marine environment. Methyl-mercuric chloride has been shown to diffuse through a 40A° (angström) membrane in 20 x 10^{-9} seconds (Ref.5). This is close to the diffusion of methylmercuric chloride through an equivalent thickness of an organic solvent such as toluene. Therefore, it seems likely that if organo-lead compounds are to be taken up by marine organisms, then the concentration of tetra-alkyl lead in seawater will determine the extent of contamination. Once the tetra-alkyl lead has been taken up by a living organism then it must be dealkylated to a charged chemical species before it can react at the molecular level inside the cell. Since diffusion is a two-way process it is to be expected that diffusion and partition of tetra-alkyl lead compounds, through and into cell membranes, will represent important processes prior to conversion to toxic reactive molecules such as triethyl-lead and trimethyl-lead salts (Ref. 6).

Alkyl-transfer reactions between alkyl-metals and inorganic metal ions in aqueous solution occur by a number of facile disproportionation reactions. For example, alkyl-lead compounds have been shown to disproportionate with mercury, gold, palladium and platinum salts to give the corresponding alkylated products (Ref. 7, 8). Trimethyl-lead readily disproportionates with the principle species of mercury in sea-water $HgCl_4^{2-}$ to give methylmercuric chloride as the product (Table 1).

TABLE 1 Transmethylation Reactions Attempted in Aqueous Solutions**

Methylator	$HgCl_2$	$NaAuCl_4$	Na_2PdCl_4	K_2PtCl_4	$PtCl_4$ [c]	Me_2TlCl	Me_2SnCl_2	$CdCl_2$
Me_3Sn^+	+ [a]	+	+	+	0			
Me_3Pb^+	+ [a]	+	+	+		0	0	0
$MeHg^+$		+	+	+			0	
Me_2Tl^+	0		+					
Me_2Sn^{2+}	0		0					
Me_3S^+	0 [b]							

** Reactions were carried out in 5 mm o.d. NMR tubes at -10^{-2} m for each reactant. Plus signs indicate cases where reaction occurred and zeros indicate cases in which no reaction was observed over a 100 h period.

a Rate of transmethylation retarded by presence of CN^- or SCN^-.

b No reaction up to 100°C.

c Conducted in degassed water.

Since triethyl and trimethyl-lead salts appear to be quite stable in the marine environment, it is important that ethyl and methylmercury levels be checked routinely in marine organisms which live in, or close to, an area polluted with organo-lead compounds.

Recently, a biological and biochemical mechanism has been found for the biomethylation of dialkyl and trialkyl-lead compounds to the tetra-alkyl lead level. Methylation of stable organo-lead compounds has been demonstrated in sediments (Ref.1), by mixed microbial populations (Ref. 9) and by a B_{12}-dependent pathway (Ref. 10, 11, 12). This reaction can be shown to occur under anaerobic conditions, though for the B_{12}-mechanisms aerobic conditions are favored. Methyl-transfer occurs by electrophilic attack on the Co-C bond of methylcobalamin (Fig. 1). The discovery of this reaction together with field observations on alkylation of alkyl-lead compounds provides us with a rudimentary biological cycle for organo-lead compounds (Fig. 2).

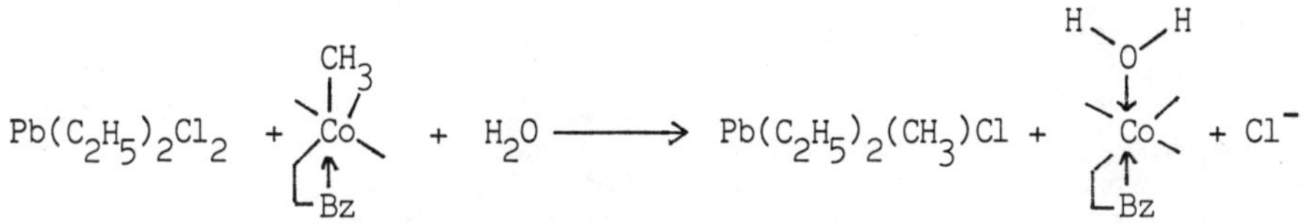

Fig. 1. Proposed mechanism for the methylation of alkyl-lead compounds.

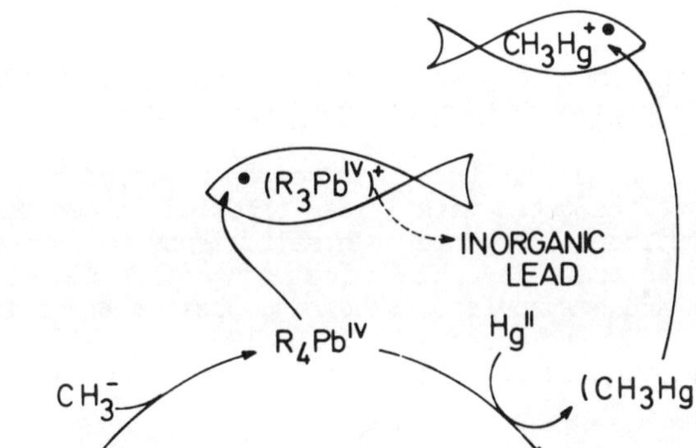

Fig. 2. Proposed biogeochemical cycle for organo-lead compounds in the marine environment.

In this cycle dealkylation occurs by either oxidation or disproportionation, and biomethylation re-establishes low steady-state concentrations of methyl lead compounds. In this cycle it is to be anticipated that methyl-groups will replace ethyl-groups with each turn of the cycle.

Inorganic Lead Compounds in the Marine Environment

The buffering capacity of seawater, together with the fact that the marine environment is oxygen rich, provides conditions for the precipitation of insoluble Pb^{IV} complexes. Therefore, it is to be expected that much of the Pb^{IV} in seawater exists in particulate matter. Polarographic studies predict that the major soluble species of lead in seawater are Pb^{II} salts in approximately the following proportions (Ref. 13): $PbCO_3$ (38%), $PbOH^+$ (33%), $PbCl_2$ (19%), $PbCl^+$ (8%), and Pb^{2+} (2%). It should be borne in mind that this picture is an oversimplification because it does not take into account the formation of coordination complexes of lead with strong chelating agents such as the "so called" marine humic and fulvic acids.

Recent studies in our laboratory have shown that there is a correlation between the standard reduction potential for inorganic complexes and the mechanism for biomethylation of representative heavy metals (Ref. 12). In the case of inorganic-lead salts we have shown that electrophilic attack on the Co-C bond of methyl-B_{12} occurs to displace the methyl group as CH_3^-, but the resulting monomethyl-lead species is very unstable and is hydrolyzed in milliseconds to regenerate the inorganic-lead salt (Ref. 12). Taylor and Hanna (Ref. 14) have shown that prolonged incubation of ^{14}C-methyl-B_{12} with fine suspensions of Pb^{IV} oxides results in demethylation of B_{12}, but the resulting ^{14}C lead complex is unstable. In these experiments the ^{14}C was volatilized from reaction mixtures. In Taylor's laboratory, and in our own, we have shown that there is no reaction with Pb^{II} salts (Ref. 14, 15).

So far we have been unable to isolate or synthesize a monoalkyl lead complex which is stable in water, and therefore we can find no hard scientific evidence to support the published work on biomethylation of inorganic lead (Ref. 1, 2, 3).

It is of crucial importance that these studies be repeated by using a stable lead isotope to ensure that the tetra-methyl lead produced in sediment experiments (Ref. 1, 3) originates from added inorganic-lead salt. From an environmental standpoint we need to know whether there is a biological connection between the biogeochemical cycles for inorganic and organo-lead compounds, just as there is between inorganic and organometallic mercury compounds (Ref. 16, 17, 18). At the present time we are trying to isolate compounds of biological origin which contain a single water stable Pb-C bond.

ACKNOWLEDGEMENTS

Some of the research in this paper was supported by grants from the Public Health Services (AM 18101) and the International Lead Zinc Research Organization.

REFERENCES

(1) P. T. S. Wong, Y. K. Chau and P. L. Laxon, Nature (London) 253, 263 (1975).

(2) A. W. P. Jarvie, R. N. Markall and H. R. Potter, Nature (London) 255, 217 (1975).

(3) U. Schmidt and F. Huber, Nature (London) 259, 157 (1975).

(4) F. G. Noden, Proceedings of the International Experts Discussion on "Lead-Occurrence, Fate and Pollution in the Marine Environment", Rovinj, October 1977, M. Branica and Z. Konrad (eds.), Pergamon Press, 1979, 83-91.

(5) J. M. Wood, A. Cheh, L. J. Dizikes, W. P. Ridley, S. Rakow, and J. R. Lakowicz, Federation Proceedings 37,16-21 (1978).

(6) B. G. Maddock and D. Taylor, Proceedings of the International Experts Discussion on "Lead-Occurence, Fate and Pollution in the Marine Environment", Rovinj, October 1977, M. Branica and Z. Konrad (eds.), Pergamon Press, 1979, 233-261.

(7) C. Huey, F. E. Brinkman, S. Grim and W. P. Iverson, Proceedings of an International Conference on Transport of Persistent Chemicals in Aquatic Ecosystems, ed. Q. N. Lettam, National Research Council of Canada, Ottawa, 73-78 (1974).

(8) J. M. Wood, Biochemical and Biophysical Perspectives in Marine Biology, D. C. Malins and J. R. Sargent, eds., Academic Press (London) 3, 407-431.

(9) Y. K. Chau and P. R. S. Wong, Proceedings of the International Expert Discussion on "Lead-Occurrence, Fate and Pollution in the Marine Environment, Rovinj, October 1977, M. Branica and Z. Konrad (eds.), Pergamon Press 1979, 225-231.

(10) J. M. Wood, J. D. Lipscomb, L. Que, R. S. Stephens, W. H. Orme-Johnson, E. Münck, W. P. Ridley, L. H. Dizikes, A. Cheh, M. Francis, T. Frick, and R. Zimmermann, Proceedings of an International Conference on Bio-inorganic Chemistry, Ed. D. H. Dolphin, Wiley Interscience, Chapter 7, 261-288 (1977).

(11) J. M. Wood, Chemical and Engineering News, April 25th issue, 23 (1977).

(12) W. P. Ridley, L. J. Dizikes and J. M. Wood, Science 197, 329-332 (1977).

(13) L. Sipos, P. Valenta, H. W. Nürnberg, and M. Branica, Proceedings of the International Experts Discussion on "Lead-Occurrence, Fate and Pollution in the Marine Environment, Rovinj, October 1977, M. Branica and Z. Konrad (eds.), Pergamon Press, 1979, 61-76.

(14) R. T. Taylor and M. L. Hanna, J. Environ. Sci. Health Environ. Sci. Eng. 3, 201 (1976).

(15) J. M. Wood, Science 183, 1049-1052 (1974).

(16) J. M. Wood, Die Naturwissenschaften 62, 1 (1975).

(17) J. M. Wood, La Recherche 70, 711 (1976).

(18) J. M. Wood, W. P. Ridley, A. Cheh, H. J. Segall, W. Chudyk and J. S. Thayer, Proceedings of an International Conference on Heavy Metals in the Environment, ed. T. C. Hutchinson, Univ. of Toronto Press, Toronto, 49-68 (1977).

The Cavtat Incident

G. F. Harrison
Associated Octel Company Limited, London, U.K.

ABSTRACT

On July 14th, 1974 the Yugoslavian vessel, m.v. "Cavtat" (2090 tons) was in collision with the Panamian vessel "Lady Rita" (18,000 tons) and after about three hours sank in 95 m of water at a point some 6.4 km south east of Capo d'Otranto. The cargo of the m.v. "Cavtat" included 900 50-gallon steel drums containing Antiknock Compound based on Tetra-ethyl and Tetra-methyl lead.

The paper reviews information regarding the solubility, rate of solution, decomposition and dispersal of lead alkyls in sea water. Information is also given relating to the mammalian and aquatic toxicity of those materials, and the conclusion is reached that only very minor local environmental effects would be caused. At the time of writing, a salvage operation has recovered 495 drums (from the original deck stow of 496) and of these 20 were empty following mechanical damage which occurred during the sinking. In spite of the presence of over 1000 gallons (approx. 7200 kg) of Compound on and around the wreck, all the drums and the wreck are visibly supporting a vigorous sea life. Analyses of water and of biota taken near the wreck have revealed no abnormal concentration of lead nor have any ill effects due to lead been revealed.

INTRODUCTION

Anti-knock Compound containing tetra-alkyl lead, manufactured in the U.K. plants of the Associated Octel Co. Ltd., is shipped to oil refineries all over the world with the exception of North America.

The containers used for this traffic include 50 gallon (230 litre) drums, 5, 10 and 20 ton demountable tanks and specially constructed bulk chemical carriers.

Occasionally one or two deck stowed drums have been lost overboard during storms at sea but in all cases these losses have been in deep water. No losses have occurred during transport using demountable tanks or from the bulk carriers.

Two incidents are on record involving the loss of the whole cargo of drums following the total loss of the vessel by mishap.

The first of these incidents occurred in July, 1966 when the s.s. Seafarer, a dry cargo vessel, ran aground in fog on rocks in 6 m of water within 300 m of the promenade of a populous seaside resort near Cape Town, South Africa. The cargo included 190 drums of Anti-knock Compound but due to heavy seas the ship broke up before salvage could be effected and only 33 drums were recovered. Since that time no evidence of the presence of lead or of any ill effects due to lead have been reported. At the time of the incident many dead fish were washed ashore but the local Government Health Authority diagnosed death as being due to pentachlorophenol - a large consignment of which was being carried on the ship.

The second incident involving the loss of a cargo of Anti-knock Compound occurred on 14th July, 1974 at 04.00 hrs when there was a collision between the m.v. Cavtat (2090 tons, Yugoslavian) and the m.v. Lady Rita (18,000 tons, Panaminian), at a point some 6.4 km south-east of Capo d'Otranto. The m.v. Cavtat was badly damaged and after 3 hours sank in 95 m of water. The cargo of the m.v. Cavtat included 900 drums of Anti-knock Compound stowed as follows:

1. Location of Anti-knock Compound Cargo

1.1 Original drum stow

900 drums comprising:

496 TML-CB	(Light Blue)
274 TEL-CB	(Orange)
130 CR50-CB	(Orange)

The contents of these drums may be represented as follows:

Drums	Anti-knock Cpd. (Kg)	Lead Alkyl (Kg)	Lead (Kg)
496 TML	178,064	90,490	70,139
274 TEL	100,284	61,663	39,502
130 CR 50	47,060	26,429	18,537
	325,408	178,582	128,178

The drums were stowed on m.v. Cavtat as follows:

496 drums TML Compound

On deck, in three blocks, both port and starboard.
Single tier of drums on end.

404 drums TEL & CR

Under deck.

350 drums aft end No. 2 hold in two tiers
54 drums fow'd end No. 3 hold in one tier.

2. Present situation of Compound drums

(a) Drums in hold

The 404 drums stowed in Nos. 2 and 3 holds are presumably still there. In view of the extensive damage to the starboard side of m.v. Cavtat in line with the fow'd end of No. 3 hold, the possibility of damage to drums in that hold cannot be ruled out.

(b) Drums on deck

The deck cargo (originally 496 drums) was surveyed during October 1975 by divers operating under the control of the salvage company "Brodospas". Their report indicated that six dives were made, five with a television camera.

On the basis of this inspection Brodospas concluded:

(i) that the total number of drums on deck was about 250;
(ii) that a number of the drums on deck were damaged but not perforated, except possibly for one drum very badly damaged.

Furthermore, the video tape recordings made during five of the six dives showed:

(i) drums apparently trapped in and by wreckage;
(ii) drums piled up on top of each other;
(iii) many drums on the 'roll', not, as originally stowed, 'on end'.

Brodospas reported that no drums were found on the sea bed anywhere within 10 to 15 m of the ship.

3. Drums

The drums used for carrying Anti-knock Compound are specially made, using high grade steel plate 2.5 mm thick. They are zinc sprayed and painted. A drum can be damaged by: (a) Pressure; (b) Impact and (c) Corrosion.

(a) Pressure. Anti-knock Compound drums have been tested for resistance to external pressure by two methods:

(i) In the laboratory
Anti-knock Compound drums containing either water, or a mixture of dichloroethane, dibromoethane and toluene, have been subjected to external hydraulic pressure. At a pressure of 115 psig (8 bar) equivalent to the pressure at a depth of water of 76 m, one test drum containing water showed a longitudinal collapse of the body shell and inward deformation of the dished ends.

The same drum was then subjected to a pressure of 185 psig (12.8 bar), equivalent to that of water at a depth of 122 m.

The maximum depth of shell collapse was 168 mm (6 5/8") measured from the inside of the rolling hoop.

A third test was made during which the pressure applied was 305 psig (21 bar) equivalent to that at a water depth of 204 m.

The maximum depth of collapse was 171 mm (6 3/4"). There was no leakage from the drum in spite of the deformation.

Tests on other drums containing organic liquids showed only minimal shell collapse and end deformation following exposure to a pressure of 135 psig (9.3 bar) equivalent to that found at a depth of 92 m of water.

(ii) In the field
During the diving operations in October 1975 a Compound drum was filled with a sand-water mixture to give the same ullage and weight as if containing Compound. This drum was then dropped to the sea bed in the vicinity of the m.v. Cavtat and subsequently recovered. Visual inspection indicated that no damage or even distortion had occurred. (Verbal report, Mr. W. A. Hurst, Octel.)

From these tests it can be concluded that Anti-knock Compound drums will withstand, without leaking, the external water pressure encountered at a depth of 100 m.

(b) Impact. The drum manufacturing specification requires an impact test on one drum per 4,000. For this test the drum is filled to 98% of minimum capacity (e.g. 51.9 IG, 236 l) with water and dropped 4 ft (1220 mm) on to concrete so as to strike diagonally on the chimb. No leakage must occur. Other tests have been made when such drums filled with water to 98% of minimum capacity have been dropped 7 ft (2200 mm) on to the chimb, or on to an 80 mm square steel projection. Although in these tests the drum shell was locally flattened and on occasion the rolling hoops fractured, no leakage occurred.

Nonetheless, it is known from field experience that the impact of a full drum of Anti-knock Compound with a sharp projection, e.g. a piece of angle iron or a stanchion, can result in a puncture of the body shell.

It must be realised that even allowing for the buoyancy of the water, a full drum of Compound on the sea bed weighs about 500 lbs. (227 kg) whereas on land a drum full of water weighs about 680 lbs (309 kg).

From the diving report and the video tape recording, it is clear that many drums broke away from the deck stow and rolled towards the stern of the vessel, or got tangled up in wreckage.

(c) Corrosion. Corrosion of drums can occur internally, due to the action of Anti-knock Compound, or externally due to the action of sea water.

Anti-knock Compound is mildly corrosive and prolonged storage in drums can and does result in the occurrence of pin-hole leaks. Since such drums are removed from service we have no experience to indicate the time span required for internal corrosion to result in a major failure or collapse of the drum.

Very little information is available regarding external corrosion due to the prolonged action of sea water.

Occasionally deck stowed drums have been lost overboard during storms at sea but in all cases these losses have been in deep water. There is, however, one well documented incident involving the recovery of drums of Anti-knock Compound after prolonged immersion in the sea.

In brief, in November 1961 two lighters, carrying 81 drums and 62 drums respectively of TEL-B Compound, broke away from their towing tug during a storm whilst in transit between Port of Spain and Pointe-a-Pierre, Trinidad. The lighter with 81 drums was found aground two days later but extensive search by helicopters and ships failed to locate the other. This lighter was eventually located some 14 months later, in January 1963, under 20 ft of water.

During a salvage operation on January 31st, 1963 61 drums were recovered, the other drum was never found. The drums were heavily encrusted with marine growths, barnacles etc., but apart from two with pin-hole leaks, were all sound - as indicated by a test with a seven pound hammer! The drums involved in this incident had been in service between 1 and 8 years.

The age of the drums on m.v. Cavtat vary between 1 year and 10 years.

In a previous document ('The Cavtat Incident', April 1975) it was stated that "after a long period in the sea, corrosion will result in leaks and Anti-knock Compound will escape into the sea. It is not possible to predict with any accuracy the time required for all 900 drums to empty but it could be as long as ten years". No reason is seen for any change in this conclusion.

4. Properties of Lead Alkyls

Anti-knock Compounds are heavy, oily liquids, S.G. 1.58 - 1.60. Material released from drums will remain as a pool, but whether or not it penetrates the silt layer depends on the nature and density of that layer.

The behaviour of Anti-knock Compound in contact with silt, mud, sand, etc., has been investigated in the laboratory. The density of Anti-knock Compound and typical estuary deposits is as follows:

Material	Density (g/cm^3)
TML-CB	1.59
TEL-CB	1.60
CR50-CB	1.58
Wet sand	2.2
Compact mud	2.0
Silt	1.2

When placed on a bed of wet sand or estuary mud, Anti-knock Compound showed no tendency to penetrate into the bed.

Silt was not penetrated by Compound for a time, but eventually, apparently quite spontaneously, penetration occurred and the silt then formed a layer on the surface of the Compound.

Without more detailed knowledge of the sedimentation layer near the m.v. Cavtat no prediction can be made regarding this aspect of the behaviour of Compound.

Until recently it was thought that the saturation solubility of TEL in sea water was about 2.0 mg/l (Pb) but there is now evidence that the true solubility is much lower than this, probably below 100 μg/l Pb. TML is rather more soluble than is TEL, about 15 mg/l (Pb) in sea water. Laboratory studies have shown that the rate of solution of lead alkyls is low and that consequently the concentration of lead found in water flowing over the surface of a pool of such material is approximately 4 μg/l (Pb) from TEL and 20 μg/l from TML (Ref. 1).

At ordinary temperatures Anti-knock Compounds are stable but in dilute solution in water, both TEL and TML decompose to give tri-alkyl salts, then di-alkyl salts and finally inorganic lead compounds. In this respect TML is more stable than is TEL and the tri-alkyl salts are more stable than either the tetra or di-alkyl compounds (Ref. 1).

Lead alkyls released from drums on or near the wreck of the m.v. Cavtat will dissolve in sea water and will be dispersed by ocean currents and by diffusion processes. There is an extensive literature reporting both experimental and theoretical studies of the dispersal of material in water but much of this work is concerned with flow from estuaries, sewage outfalls and from pipes, etc. Very much less is known about the dispersal of material on the sea bed at depths, for example, of 100 m. Calculation of the dispersal characteristics in the present situation is further complicated by the lack of any data relating to thermal or salinity stratification and the velocity profile of the ocean currents.

Notwithstanding these problems, and neglecting the large turbulence due to the presence of the wreck itself, J. W. Cleaver (Ref. 2) has examined the situation mathematically and has concluded that dissolved material would be diluted to 'safe' concentrations (e.g. 1 μg/l Pb) within a comparatively short distance of the wreck (e.g. 500 m).

V. D'Arrigo, (Ref. 3) using a much simplified mathematical approach showed that the total cargo of lead alkyl could be absorbed by a small fraction of the total Adriatic water, namely that enclosed by the Otranto channel, to give a concentration much below that of the existing lead in solution. He concluded that "all the alarm sounded relative to the dangers of pollution of the Strait of Otranto and its beaches, and in fact the Adriatic itself... is not justified, and the attempts to recover the cargo of the ship can be regarded as a purely psychological action".

5. Toxicology of Lead Alkyls

5.1 Terrestrial animals. Tetra-ethyl and tetra-methyl lead can gain access to the body of land animals, including man, by three routes - through the gastro-intestinal tract following ingestion, via the lung, after inhalation and through the skin by contact with the liquids.

In the case of tri-alkyl lead compounds, which are nonvolatile and water soluble, the only route of access is by ingestion.

During the course of manufacture, distribution and use of Anti-knock Compounds containing lead alkyls, the major hazard is that of inhalation. By the application of rigorous safety measures at all stages, exposure to lead alkyl vapour can be virtually eliminated with the result that, apart from isolated accidents, no cases of even mild intoxication are encountered.

Laboratory studies using rats, rabbits, monkeys, dogs, etc. have clearly shown that:

(i) TML is considerably less toxic than is TEL and that,

(ii) marked differences occur in the relative toxicity of these two materials, depending on the animal species being tested.

The work of Cremer (Ref. 4), Cremer and Callaway (Ref. 5), Bingham (Ref. 6), strongly suggest that the tetra-alkyl lead compounds are non-toxic in the mammalian system and that the observed effects of TEL and TML are due to the tri-alkyl lead compounds, formed by de-alkylation of the tetra compound by enzymatic action in liver. The lower toxicity fo TML is probably due to the much greater stability of TML compared with TEL and, hence its greater resistance to de-alkylation.

The observed inter-species variation in the relative toxicity of the two tetra-alkyl compounds can be explained as being due to differing activities of the de-alkylating enzymes in the liver in different animals.

Quantitatively, acute toxicity is usually indicated by the LD_{50} value that is, the quantity of substance, administered as a single dose, required to kill half an exposed population within a period of 14 days. Examples of the results reported for lead alkyl compounds are as follows, all values being in mg lead/kg body weight.

Rat Cremer (Ref. 7)	Intravenous injection
Tetra methyl lead	80 mg Pb/kg
Tetra ethyl lead	10 " "
Tri methyl lead	20-25 " "
Tri ethyl lead	8 " "

Rat Ethyl Corporation (Ref. 8)	Ingestion
Tetra methyl lead	108 mg Pb/kg
Tetra ethyl lead	12 " "

Rat Buck & Kumro (Ref. 9) Intra Peritoneal

Tetra methyl lead	70-100 mg Pb/kg
Tetra ethyl lead	10 " "
Tri methyl lead	17 " "
Tri ethyl lead	5 " "

There are differences, not unexpected, between these results but it clearly emerges that TML is about 8 times less toxic to rats than is TEL.

Similarly, again using rats, Cremer (Ref. 5) has shown that by inhalation the LC_{50} 1 hour value for TML is about ten times that for TEL. (LC_{50} is the concentration of a toxicant, measured in mg/l of air, exposure to which for one hour will result in the death of half an exposed population.)

In considering the possibility of any hazard to human beings resulting from the presence of Anti-knock Compound in the wreck of the m.v. Cavtat, it is clear that inhalation can be eliminated. Dermal contact could only be a factor if substantial quantities of Compound were washed ashore. Since the materials are heavy liquids (S.G. 1.58 - 1.60) and the sea currents are away from the land it is thougt to be impossible for this to happen.

Intoxication of drinking sea water containing lead alkyls in solution is not considered to be a threat to health. The LD_{50} for TEL for rats is equivalent to about 10 mg Pb/kg. In the unlikely event that human beings are ten times more susceptible than the rats, this would give LD_{50} dose of TEL of 70 mg Pb. The ingestion of 70 mg will require an input of 70 l of sea water containing 1 ppm Pb as TEL. For TML the quantity would be about 60 l of a saturated solution.

Little is known from animal experiments about the effects of the ingestion of small quantities of lead alkyls on a continuous basis. In the case of human beings it is known that exposure to atmospheres containing 100 μg Pb/m^3 as TEL for 8 hours per day, 5 days per week for life, does not lead to any impairment of health. Assuming a respiratory volume of 10 m^3 for an 8 hr working day this gives an exposure of 10 x 100 μg Pb equal to 1.0 mg Pb/day. Assuming only 50% absorption in the lung, the daily dose would be 0.50 mg Pb. To ingest such a quantity would require the drinking of 5 l of a saturated solution per day. By analogy with respiratory exposure, such ingestion every day, five days per week for life would have no deleterious effects on health.

The chance of attaining a saturated solution in the open sea is so remote - as discussed in Section 9 below - that it is felt not unreasonable to discount completely any possibility of hazard from the ingestion of sea water.

5.2 Aquatic animals. Although there is extensive literature regarding the effects of heavy metals, including lead, on aquatic animals, little is known concerning the effects of lead alkyl compounds. The acute toxicity of substances with respect to aquatic animals is usually expressed as the LC_{50} value, that is, the concentration of the substances, in mg/l, that will result in the death of half an exposed population of test animals during a stated period of exposure. Values for the LC_{50} for lead alkyls are reported to be as follows:

Fish Blue gill sun fish (Lepomis macrochirus) (Ref. 10)

TEL	LC_{50}	24 hr	2.0 ppm Pb
		48 hr	1.4 ppm Pb

Rainbow Trout (Salmo irideus) (Ref. 11)

Tri methyl lead chloride	LC_{50}	96 hr	32 ppm Pb
Tri ethyl lead chloride	LC_{50}	96 hr	9 ppm Pb

Maddock & Taylor (Ref. 12) have reported the results of an extensive study of the toxicity of tetra, tri and dialkyl lead compounds with respect to fish, mussels, shrimps and marine algae.

Their results may be summarised as follows:

TABLE 1 LC_{50} (96 hr) mg/l Pb

Compound Species	Shrimp	Mussel	Plaice	Dab	Algae[3]
Tetra methyl lead	0.11	0.27	0.05	-	1.3
Tetra ethyl lead	0.02	0.10	0.23	-	0.1
Tri methyl lead chloride	8.8	0.50	24.6	20.0	0.8
Tri ethyl lead chloride	5.8	1.1	1.7	2.0	0.1
Di methyl lead chloride	-	-	300	-	-
Di ethyl lead chloride	-	-	75	-	-
Inorganic lead	(375)[1]	(7,500)[1]	(180)[2]	-	< 5.0

Figures in brackets refer to related species:

1 Ref. 13

2 Ref. 14

3 Parameter measured is reduction of photosynthetic activity (6 hr EC_{50}).

From these data it will be seen that:

(a) Both TEL and TML are toxic with respect to marine animals.
(b) The di-alkyl and tri-alkyl compounds are less toxic than are the corresponding tetra alkyl compounds. In this respect fish differ markedly from the mammals where the tri-alkyl compounds are somewhat more toxic than are the tetra compounds.

Ingress of toxic agents into marine animals is almost always through the respiratory system (the gills), not through the gastro-intestinal tract. This may account for the different toxicity of the tri and tetra compounds although the details of the mechanism are quite unknown at present.

The data presented above shows that a saturated solution of both TEL and TML would be fatal to fish. However, the concentrations expected in sea water following release of Compound from drums on the m.v. Cavtat are likely to be much lower than given by the solubility data, as shown in the section discussing properties of lead alkyls.

6. Bioaccumulation

In the preceding section, the acute or short-term toxicity of lead alkyl compounds has been discussed. It is now necessary to examine the possibility of the occurrence of long-term effect on marine animals due to exposure to low concentrations of lead alkyl compounds. One such effect to be considered is bio-accumulation, that is, the concentration by an aquatic organism, of a pollutant present in quantities very much lower than those leading to physiological impairment or death.

Classic examples of bio-accumulation are found in shell fish, which concentrate lead, arsenic, copper, etc. from dilute solution. Inorganic lead, unlike mercury, is reported not to be accumulated by fish (Ref. 15).

Until recently, no information was available regarding the accumulation of lead by marine organisms exposed to low concentrations of lead alkyl compounds. Hence, Maddock & Taylor (Ref. 12) have reported results of such accumulation by mussels and by fish exposed to dilute solutions of tri alkyl lead chloride. Their findings are summarised in Tables 2 and 3.

TABLE 2 Accumulation of Lead by Mussels (Mytilus edulis)
Summary of Results

	Control	Trimethyl lead chloride			Triethyl lead chloride		
Exposure concentration (nominal) ppm/Pb	0	0.01	0.05	0.10	0.01	0.05	0.10
Fraction of LC_{50} 96 hr value		1/50	1/10	1/5	1/100	1/20	1/10
Exposure time (days)	35	9	35	21	9	35	21
Time to reach steady state (days)	-	-	9	7	-	6	5
Steady state conc. (mean) ppm/Pb Dry Weight	6	10	27	68	9	29	48
Concentration factor:							
Tissue/water	-	400	420	620	300	460	420
Exposed/control	1	1.6	4.5	11	1.5	4.8	8
Depuration time (days)	27	-	27	27	-	27	27
Half-life (mean)	-	-	3	3	-	4	3

As will be seen, very little accumulation occurred following exposure of mussels to the lowest concentration (0.01 ppm Pb). Exposure to concentrations of 0.05 ppm and 0.1 ppm Pb resulted in rapid accumulation during the first 5 to 9 days following which no further increases in tissue concentration occurred even after prolonged exposure. When placed in lead-free sea water exposed mussels rapidly lost the lead they had accumulated.

The behaviour of fish exposed to low concentrations of alkyl lead compounds differed from that of mussels in that, the increase of tissue concentration of lead was (with one exception) linear with respect to time. The exception was in the tissue of fish exposed to tri methyl lead chloride, where a steady concentration was reached after approximately 20 days.

TABLE 3 Accumulation of Lead by Fish (Limanda limanda)
Summary of Results

	Control	Trimethyl lead chloride		Triethyl lead chloride	
Exposure concentration (nominal) ppm Pb	-	1.0	2.0	0.1	0.2
Fraction of LC_{50} 96 hr value		1/20	1/20	1/20	1/10
Exposure time (days)	41	41	41	41	41
Time to reach steady state: liver	-	20	20	> 41	> 41
muscle	-	> 41	> 41	> 41	> 41
Tissue concentration ppm Dry wt.: liver	1.9	17*	32*	9	21
muscle	1.4	22	33	10	18
mean	1.6	20	32	10	20
Concentration factor Tissue/water	-	18	15	84	92
Exposed/control	-	12	20	6	12
Depuration time (days)	56	49	49	56	56
Half-life (days): liver	-	70	62	> 56	> 56
muscle	-	> 49	41	56	> 56

* Steady state value

When placed in lead-free sea water exposed fish lost the lead they had accumulated (in muscles and liver tissues) but the rate of loss was low, the half-life being in excess of 50 days.

The data given previously show that flowing water in contact with either TEL or TML dissolves only sufficient material to give concentrations of approximately 4 μg/l Pb from TEL or 20 μg/l Pb from TML. Furthermore, the experimental evidence suggests that such concentrations would only be found in a layer of water a few centimetres thick above the lead alkyl surface. It is almost inconceivable that a fish could live in such a layer for a time sufficient for its tissues to attain a significant lead concentration. Giaccio (Ref. 16) has described experimental work which indicates that a number of species of fish can detect the presence of very low concentrations of TEL/TML and will move away.

The primary lead solution derived by contact of sea water with lead alkyls will be diluted by diffusion and dispersion. These processes will result in a very much larger volume of water being contaminated with lead at very much lower concentrations. Fish could remain in such diluted solution for protracted periods but the tissue concentration that would result would be insignificantly increased above the natural level.

7. Analyses of Marine Samples

Attempts have been made to determine the extent of pollution, caused by leakage of Anti-knock Compound from drums on m.v. Cavtat, by the analyses of samples of water, sediment and marine biota. However, this work has been hampered by difficulties of sampling, by the use of a variety of methods of analyses and lack of reference material of known lead content.

7.1 Sea Water. Patterson et al. (Ref. 17 and 18) has directed attention at the problems encountered in trace lead analysis and it is certain that without sophisticated methods of sampling and its exercise of extreme care during subsequent handling and analysis, erroneously high values will be obtained.

There is, indeed, some uncertainty regarding the base line concentrations of lead in the Adriatic sea water. For example, D'Arrigo (Ref. 3) quotes 0.02 - 0.03 ppb Pb, Sipos et al. (Ref. 19) gives 0.02 ppb as a "realistic value", Nürnber et al. (Ref. 20) suggest 0.1 - 0.2 ppb for "unpolluted coastal water", and Stoeppler (Ref. 21) refers to 5.0 ppb as being typical of sea water.

The lead content of various smaples taken since the m.v. Cavtat sank show similar large variations, ranging from about 1 μg/l (reported by Central Hydrobiological Laboratory, Rome, Ref. 22) to about 10 μg/l (reported by Instituto di Ricerca Sulle Acque, Ref. 23).

A limited number of water samples have been analysed for organo lead compounds. Although the possibilities of contamination are much less than in the case of inorganic lead, such analyses pose other problems. Tetra alkyl lead compounds are very readily lost from dilute aqueous solutions unless stored in completely filled, stoppered vessels. Furthermore, these compounds, especially tetra ethyl lead, decompose quite quickly to give water soluble non-volatile tri alkyl derivates. There appears to be little merit in quoting the sparse data reported, in view of the lack of standardised methods of analysis and paucity of information regarding sampling techniques, etc. In passing it may be of interest to note that CNR Bari laboratory (Ref. 23) found about twice the concentration of "organic lead" in the surface water off Otranto, as they found near Anti-knock Compound drums on the deck of the wreck. This lead was attributed to atmospheric contamination and to petrol-engined boats.

7.2 Fauna. Samples of marine fauna were collected by CNR Laboratory, Bari (IRSA) on 15.6.76, and the lead content determined by that laboratory and by the Instituto Superior de Sanità Rome (ISS). The interpretation of these results is made difficult because:

(a) the methods of analysis used by IRSA differed from those used by ISS;

(b) IRSA analysed the whole sample, including skin, etc., whereas ISS analysed only specific organs (e.g. liver, brain, kidney). It is known that the concentration of lead in fish skin is very much higher than that of the muscle, tissue, etc;

(c) Only a limited number of species were collected both near the wreck of m.v. Cavtat and distant from it.

Samples of seven species collected near the wreck and also at a distance of 12 miles north of the wreck were analysed for total lead by ISS. In every case the latter showed a higher lead concentration than the former.

A further series of samples of marine fauna was collected during the extensive survey co-ordinated by the Central Hydrobiological Laboratory, Rome in October 1976. The results show that no relationship could be found between the lead content of the tissues and the distance from the wreck or the main flow of the ocean currents.

7.3 Sediments. Marine sediments collected up to 6 km from the wreck of m.v. Cavtat have been found (with certain exceptions) to contain 'normal' quantities of lead. For example, the Central Hydrobiological Laboratory, Rome (Ref. 22) reported values between 11.8 and 24.0 ppm with a mean value of 16.7 ppm. Similarly, CNR (Bari) (Ref. 23) found values ranging from 5.1 ppm to 28.7 ppm (dry weight) in samples collected between 0.25 miles and 12 miles from the wreck. The exceptions mentioned above refer to samples of sediment taken from the deck of the wreck, where it is known that lead alkyls from damaged drums had collected. Such samples show lead contents of up to 6,000 ppm (wet weight).

A limited number of samples collected by Octel during October, 1976 show concentrations ranging from 2.3 ppm to 19.4 ppm total lead (wet weight basis).

Except in the immediate vicinity of the wreck, it appears that the lead alkyl cargo has had no measurable effect on the lead content of the marine sediments.

8. Present Situation Regarding m.v. Cavtat

Since March, 1977 the Italian company, SAIPEM, has been engaged in an operation to recover drums of Anti-knock Compound from the wreck of the m.v. Cavtat. At the time of writing (December 1977), 495 drums have been recovered from the original deck stow of 496. Of the 495 drums recovered 20 were emtpy following mechanical damage, almost certainly caused during the sinking and 80 were leaking, mainly from pin-holes in longitudinal body welds. Neglecting any consideration of the 404 drums still inside the ship, it appears that about 1000 gallons of Anti-knock Compound have been lying about on the wreck or on the sea bed near the wreck since July, 1974. In spite of this, the drums and all parts of the wreck are visibly supporting a flourishing marine life. Furthermore, analyses of water and of biota taken in the vicinity of the wreck have not revealed any abnormal concentrations of lead nor have any ill effects due to lead been revealed.

This minimalecological impact can be explained - and was, in fact, predicted - by a knowledge of the extremely low rate of solution of lead alkyls, by the rapid dispersal due to diffusion and to ocean currents and by the relatively low level of accumulation of organo lead by aquatic fauna.

REFERENCES

(1) J. R. Grove, Proceedings of the International Experts Discussion on "Lead-Occurrence, Fate and Pollution in the Marine Environment", Rovinj, October 1977, M. Branica and Z. Konrad (eds.), Pergamon Press, 1979, 45-52.

(2) J. Cleaver, Proceedings of the International Experts Discussion on "Lead-Occurrence, Fate and Pollution in the Marine Environment", Rovinj, October 1977, M. Branica and Z. Konrad (eds.), Pergamon Press, 1979, 325-343.

(3) V. D'Arrigo, Quad Merceol. 16, 47 (1977).

(4) J. E. Cremer, Brit. J. Ind. Med. 16, 191 (1959).

(5) J. E. Cremer and S. Callaway, Brit. J. Ind. Med. 18, 277 (1961).

(6) F. Springman, E. Bingham and K. L. Stemmer, Arch. Env. Hth. 6, 369 (1963).

(7) J. E. Cremer, Ann. Occup. Hyg. 3, 226 (1961).

(8) Ethyl Corporation, Private Communication.

(9) J. S. Buck and K. Kumro, J. Pharm. Exp. Therap. 38, 161 (1930).

(10) H. Turnbull, J. G. De Mann and R. F. Weston, Ind. Eng. Chem. 46, 324 (1954).

(11) Dept. Scientific & Ind. Research, Water Pollution Research Lab.(1964).

(12) B. G. Maddock and D. Taylor, Proceedings of the International Expert Discussion on "Lead-Occurrence, Fate and Pollution in the Marine Environment", Rovinj, October 1977, M. Branica and Z. Konrad (eds.), Pergamon Press, 1979, 233-261.

(13) J. E. Portmann, K. W. Wilson, Shellfish Information Leaflet 22, Min. of Ag. & Fish., UK. (1971).

(14) E. Jackim et al., J. Fish Res. Bd., Canada 27, 383 (1970).

(15) Survey of lead in food, First Supplementary Report, Min. of Ag. & Fish., UK. (1975).

(16) M. Graccio, Quad. Merceol. 16, (1), 55 (1977).

(17) C. C. Patterson, D. Settle and B. Glover, Marine Chemistry 4, 305 (1976).

(18) T. J. Chow, C. C. Patterson and D. Settle, Nature 251, 159 (1974).

(19) L. Sipos, P. Valenta, H. W. Nürnberg and M. Branica, Proceedings of the International Expert Discussion on "Lead-Occurrence, Fate and Pollution in the Marine Environment", Rovinj, October 1977, M. Branica and Z.Konrad (eds.), Pergamon Press, 1979, 61-76.

(20) L. Mart, H. W. Nürnberg and P. Valenta, Proceedings of the International Expert Discussion on "Lead-Occurrence, Fate and Pollution in the Marine Environment", Rovinj, October 1977, M. Branica and Z. Konrad (eds.), Pergamon Press, 1979, 155-179.

(21) M. Stoeppler, Proceedings of the International Expert Discussion on "Lead-Occurrence, Fate and Pollution in the Marine Environment", Rovinj, October 1977, M. Branica and Z. Konrad (eds.), Pergamon Press, 1979, 207-224.

(22) Central Hydrobiological Laboratory, Rome, Private Communication.

(23) CNR Laboratory, Bari, Private Communication.

Lead and the Historical Sedimentary Record

K. K. Bertine
Department of Geological Sciences, San Diego State University, San Diego, California, U.S.A.

In recent years there has been ever increasing concern about the extent to which the environment has been affected by man's activities. Lead in particular is worrisome since it has been extensively released to the atmosphere from the burning of gasoline containing lead alkyls. Unfortunately, environmental samples have usually not been taken prior to the widespread dissemination of lead to the environment. Therefore, it is difficult to know the extent to which the environment has been altered with respect to the concentration of this element.

The sediments of glaciers, lakes, estuaries, lagoons and other coastal marine areas can be used to obtain the historical record of lead inputs (both anthropogenic and natural) to an area. To do this, an undisturbed sedimentary core is necessary. Box coring techniques or using divers to ensure a slow entry of a coring device are best. Once the sediment is recovered it can be subsampled at appropriate intervals depending upon either the anticipated sedimentation rate or the layering present. An x-radiograph is useful for revealing subtle differences in sediment properties not readily visible to the naked eye. It can also serve as a permanent record for later use. If the water content of the sediment is high, the core should be frozen prior to extraction from the coring device. Finally, the lead contents of the sediments are measured at intervals down the lengths of the core. Generally, a pollutant lead input to the sediments is suspected if the Pb contents are higher in the upper sediments. However, the absolute concentrations of lead with depth in the sediment column do not give an accurate measurement of the degree to which the lead content has been altered by man's activities. For instance, Fig. 1 (Ref. 1, 2) shows the concentrations of Pb with depth in four cores from nearshore Southern California Basins. If just the Pb concentrations are looked at, there is a leveling off and even decrease in the Pb content in the tops of the San Pedro and Santa Barbara Basin cores, suggestive of a lessening of the anthropogenic Pb input in recent times. However, when the Pb/Al (Al concentration is assumed to be a measure of the detrital mineral content) ratio is plotted, it can be seen that there has been no decrease in the recent anthropogenic Pb influx in recent times. The reason for the apparent decrease was due to a greater dilution of the upper (more recent) sediment with biogenous phases and sea salt. Al was chosen as the normalizing element because of its high concentrations in detrital minerals that are presumed to have had a uniform input over the past century. A check must be made, however, to ascertain that the detrital sediment sources have not changed over the time period investigated. Bertine et al. (Ref. 3) have found that in semiarid environments where stream flow is episotic, the detrital mineralogy and hence the Al content varies over a year's time span. In such areas some normalizer other

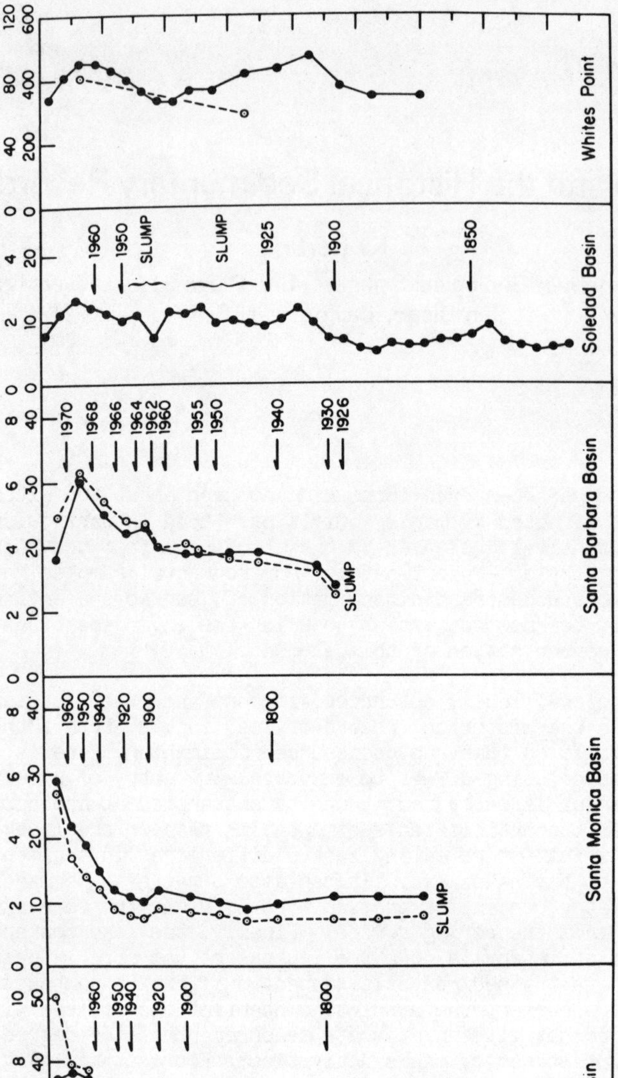

Fig. 1. Pb content (●) and Pb/Al ratios (⊙) in Southern California Basin sediments, and Whites Point sewer outfall sediments (Ref. 2 and 3).

than Al content must be used.

If the flux of detrital minerals to the basins also varies from area to area, both the measured content of Pb in the sediments and the Pb/Al ratios do not give accurate indications of the extent to which different areas have been affected by anthropogenic Pb inputs. Regarding Fig. 1 again, it would seem that San Pedro sediments have received the greatest anthropogenic Pb inputs since the Pb content is higher in the upper sediments of that core than in the other two cores. However, when anthropogenic Pb fluxes are calculated, Santa Barbara Basin sediments have had a greater anthropogenic Pb flux (2.1 $\mu g/cm^2$/year) than San Pedro (1.7 $\mu g/cm^2$/year) or Santa Monica (0.9 $\mu g/cm^2$/year) Basins. This is because Santa Barbara Basin has a considerably higher sedimentation rate (0.4 cm/year) compared to 0.09 cm/year for the San Pedro sediments and 0.07 cm/year for the Santa Monica Basin sediments. Hence, the anthropogenic Pb contribution is more diluted in Santa Barbara Basin sediments. Thus, in order to compare the degree of Pb contamination from one spot to another, fluxes rather than concentrations must be used. This is particularly applicable to closed basins such as lakes where sedimentation rates usually decrease away from shore (river input usually being the primary source of sediment).

Other areas where Pb fluxes rather than concentrations are essential in order to interpret the historical record are where the sedimentation rate has changed over the length of the core studied. That portion with the higher sedimentation rate would have proportionately lower Pb contents due to greater dilution by the detrital fraction. Areas with changing sedimentation rates are quite common since man's activities (in particular building construction) have increased sedimentation rates in recent times (e.g. Ref. 4). Sedimentation rates can be determined by a variety of methods, the most common of which is Pb-210 dating (e.g. Ref. 5). Th-228/Th-232 (Ref. 5), age horizons, varving, C-14 dating, artificial radionuclide geochronologies (Ref. 6) and other methods have been used. For best results, two different methods should be used since all of the above have been found to have given erroneous results.

Another advantage of having determined anthropogenic Pb fluxes is that it is then possible to compare the anthropogenic Pb fluxes measured in the sediment with the Pb fluxes emitted from known sources. For instance, in the study on the California Basins the anthropogenic Pb fluxes in the sediments were compared to waste water (sewage) and atmospheric (primarily automobile exhaust) Pb fluxes (Table 1).

TABLE 1 Fluxes of Lead to Southern California Coastal Region in Units of Metric Tons/Year/12,000 Km^2 (from Bruland *et al.*, 1974)

Element	Flux of anthropogenic components to sediments	Waste waters (SCCWRP, 1973)	Storm water plus dry weather flow (SCCWRP, 1973)	Washout fluxes
Pb	190	213	90	156

The fluxes of all three were similar indicating that both were possible sources.

One problem in comparing the anthropogenic fluxes in the sedimentary record with fluxes from known sources is that many sediments may have magnified concentrations of elements such as Pb which had a major atmospheric transport mechanism. Atmospheric fallout occurs over wide areas of land surface. This deposited atmospheric fallout can be removed by storm runoff from the large area and concentrated into the rivers and then the lakes or basins. Thus the sediments depositing in these lakes or basins may have magnified contents of those elements (in particular Pb) concentrated in atmospheric particulates over those calculated from direct atmospheric fallout. The degree of magnification can best be obtained by integrating the Pu-239+240 fluxes to the sediments (determined by measuring the Pu-239+240 activities in the sediments) and comparing the total flux found in the sediments to the total flux of atmospheric fallout of Pu-239+240 at that latitude (see for example Ref. 8).

Pu-239+240 is used since its sole source is production in the atmosphere from weapons testing. (A side benefit is that Pu-239+240 can serve as the confirmatory dating technique discussed above.) Magnification factors so obtained have ranged from 2 (California coastal sediments; Ref. 8), to 4-5 (Palace Moat, Tokyo, Japan; Ref. 8), to a high of 8 (Barrett Reservoir, San Diego, California; Ref. 3). The highest magnification factors are found in lakes. Thus, in order to compare anthropogenic fluxes in sediments to those from known sources with a primary atmospheric transport path, corrections for the magnification factors must be applied.

Further identification of the source of anthropogenic Pb to the sediments can be made by using the isotopic ratios of Pb. Table 2 (Ref. 4) illustrates this approach for Lake Whitney, New Haven, Connecticut, USA sediments. Prior to 1877, the Pb isotopic ratios were constant and very radiogenic. The source of the Pb is the glacial till of the area. By the early 1900's the Pb isotopic ratios had been lowered due to the influx of lead from the heavy coal burning industries in the area.

TABLE 2 Lead Isotope Ratios in Lake Whitney Sediments (from Ref. 4)

Depth (cm)	Year	Pb $\frac{206}{204}$	Pb $\frac{206}{207}$	Pb $\frac{206}{208}$	Pb concentration (μg/g)
1-2	1974	18.91	1.2021	0.4901	707
10-11	1965	18.77	1.191	0.4869	1100
18-19	1956	18.64	1.179	0.4826	470
26-27	1948	18.78	1.190	0.4822	270
34-35	1912	18.93	1.202	0.4880	170
38-39	1896	19.22	1.216	0.4880	170
43-44	1877	19.77	1.241	0.4926	100
48-49	1858	20.79	1.302	0.5103	100
53-54	1839	20.90	1.305	0.5076	44
56-57	1827	20.50	1.291	0.5057	40
			Gasoline[a]		
Boston, Mass.	1964	18.45	1.179	0.4819	-
			Coal[a]		
Ashley, Pa.		18.64	1.189	0.4824	34
Jeddo, Pa.		18.82	1.197	0.4850	11
Trevorton, Pa.		18.82	1.200	0.4849	10
St. Nicholas, Pa.		18.79	1.199	0.4849	14

[a] Ref. 9

In the middle 1900's, gasoline burning of lead alkyls took over as the principal source and resulted in a further lowering of Pb isotopic ratios (below the levels of Pb isotopes in coal). The increase in the Pb isotopic ratios in the most recent samples is most probably attributable to a change in the source of Pb (each Pb mine may have a different isotopic ratio due to the various ages of the deposits) added to gasolines. Such an increase has also been found in California coastal sediments (Patterson, personal communication). The very high Pb concentrations and also fluxes in the recent Lake Whitney sediments result from automobile exhaust Pb from a major through street paralleling Lake Whitney for most of its length. The decrease in Pb content at the 1-2 cm depth is significant and is due to the opening of a highway out of Lake Whitney's drainage area which reduced the volume of traffic on Whitney Avenue and hence the anthropogenic Pb input.

In some areas the historical record in the sediments can be disturbed. This is most apparent in oxic environments where burrowing organisms are present. Figure 2 (Ref. 10) illustrates that burrowing organisms can homogenize sediment and therefore Pb contents to depths of tens of centimeters. In San Diego Bay sediments, homogenization has been found down to a depth of 60 cm. In general, areas in which the overlying waters are anoxic are to be preferred since the sediments rarely contain burrowing organisms.

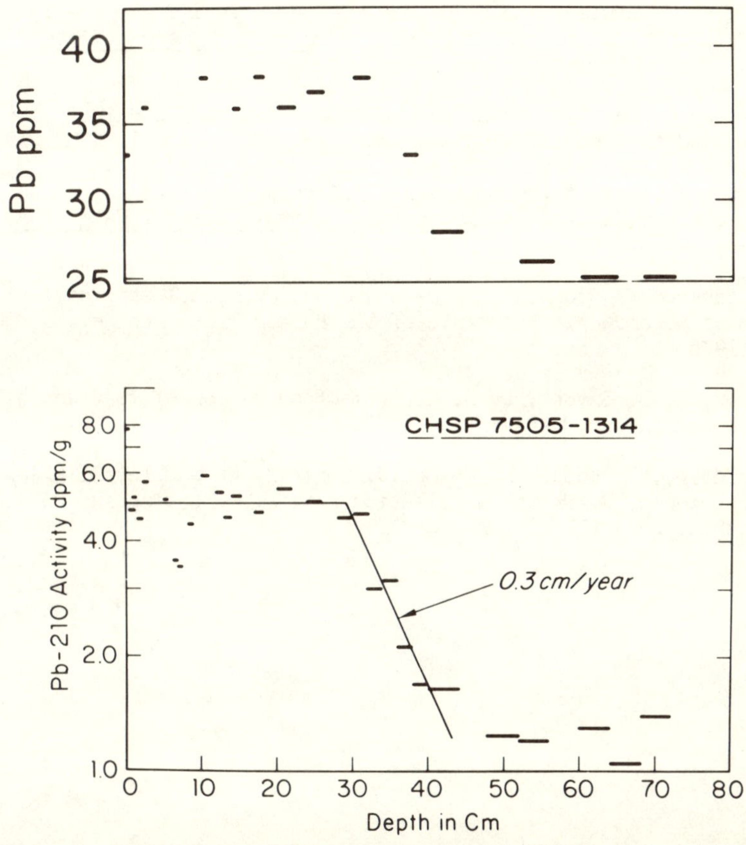

Fig. 2. Pb and Pb-210 in Chesapeake Bay sediments showing mixing of the surface sediments.

In conclusion, sediments maintain a record of the fluxes of Pb, both natural and anthropogenic, to a region. By means of careful analysis of this sediment, the historical record can be revealed.

REFERENCES

(1) T. J. Chow, K. W. Bruland, K. K. Bertine, A. Soutar, M. Koide and E. Goldberg, Lead Pollution: Records in Southern California Coastal Sediments, Science 181, 551 (1973).

(2) K. W. Bruland, K. Bertine, M. Koide and E. Goldberg, History of Metal Pollution in Southern California Coastal Zone, Environ. Sci. Technol. 8, 425 (1974).

(3) K. K. Bertine, S. J. Walawender and M. Koide, Historical record of metal fluxes and transport paths to a semiarid lake (submitted to Geochim. Cosmochim. Acta) (1978).

(4) K. K. Bertine, and M. F. Mendeck, Industrialization of New Haven, Conn., as Recorded in Reservoir Sediments, Environ. Sci. Technol. 12, 201 (1978).

(5) M. Koide, K. W. Bruland and E. Goldberg, Th-228/Th-232 and Pb-210 geochronologies in marine and lake sediments, Geochim. Cosmochim. Acta 37, 1171 (1973).

(6) M. Koide, J. J. Griffin and E. Goldberg, Records of plutonium fallout in marine and terrestrial samples, Jour. Geophys. Res. 80, 4153 (1975).

(7) SCCWRP, the Ecology of the Southern California Bight: Implications for Water Quality Management, Southern California Coastal Water Research Project (1973).

(8) E. D. Goldberg, V. Hodge, M. Koide and J. J. Griffin, Metal pollution in Tokyo as recorded in sediments of the Palace Moat, Geochem. J. 10, 165 (1976).

(9) T. J. Chow, C. B. Snyder, J. L. Earl, Isotope ratios of lead as pollutant source indicators, IAEA-SM-191/4 95 (1975).

(10) E. D. Goldberg, V. Hodge, M. Koide, J. Griffin, E. Gamble, G. Bricker, G. R. Holdren, R. Broun, A pollution history of Chesapeake Bay, Geochim. Cosmochim. Acta (1978) (in press).

Dispersion of Lead Alkyls from Pools Located on the Sea-Bed

J. W. Cleaver

Department of Mechanical Engineering, University of Liverpool, U.K.

ABSTRACT

A survey is made of methods available for calculating the dispersion of material continually released from a source on the sea bed. All the techniques discussed have been developed for the corresponding atmospheric problem and found to be satisfactory. The lack of field data for sources in the sea makes it difficult to assess whether the methods still retain their usefulness for the corresponding sea pollution problem. With this limitation the methods are applied to the problem of assessing the dispersion of lead alkyls lying in a pool on the sea bed. In particular for conditions likely to exist in the Straits of Otranto it is estimated that safe concentrations of TML and TEL will be reached well within a few hundred metres from a source located on the sea bed.

INTRODUCTION

Lead alkyls have a specific gravity significantly in excess of unity and in consequence any bulk leakage of material at the sea will rapidly sink and rest in pools on the sea bed. The material will diffuse into the surrounding sea water and local currents will tend to enhance the uptake of lead alkyl from the pool and transport it with the flow. Dilution will then occur through the stretching of the fluid elements arising from large scale regular patterns of water movement. In addition dilution will also arise through molecular diffusion and by small scale irregular movements of fluid, usually referred to as turbulence. At this stage it may be surmised that molecular diffusion is relatively insignificant and that the dispersion of material will be governed by the prevailing current and nature of the turbulence. General review articles concerned with Turbulence and Dispersion at sea are given by Bowden (Ref. 1-4).

Because of the random nature of a turbulent flow field, calculation of the resulting dispersion of a passive containment relies heavily on experimental data. In the vicinity of the sea surface such data exists. Many observations have been made of the horizontal and lateral dispersion of material arising from an instantaneous or continuous point source discharge and empirical formulae are available for making order of magnitude predictions of concentration profiles and plume widths downstream of a given source, see for example Okubo (Ref. 5) and Talbot (Ref. 6). Corresponding studies for sources located on the sea bed appear to be more restricted and in consequence the methods adopted in the report for estimating concentration distributions must remain speculative in that it has been impossible to make comparisons with full scale measurements.

At very large distances from the source of lead alkyl it may be anticipated that the shape of the pool will not be important and a sensible approximation is to consider the mass transfer from a point source. Near the source this will not be satisfactory and its finite size and shape should be allowed for. Because the diffusion of the lead alkyl is relatively slow the source may be considered steady. With such approximations pessimistic estimates of the spread and concentration of lead alkyls downstream of a pool may be obtained by first estimating the source strength and then assessing the influence of the prevailing currents on the resulting turbulent dispersion.

Similar problems have been considered in relation to the pollution by ground level sources in the lower atmosphere and evidence (Ref. 7 and 8) suggests that sensible predictions of the dispersion can be made. Most of the methods rely heavily on local measurements of the wind profiles near the ground and some measure of the turbulence characteristics. Even in the absence of local data estimates can be made by making use of the considerable amount of accumulated statistical evidence of 'typical' wind conditions likely to exist near the ground (Ref. 9). Corresponding sea bed data appears to be scarce and application of methods developed for the atmosphere will depend mainly on the ability to assess local flow conditions.

The present note reviews methods used in atmospheric problems and attempts to relate them to the sea bed situation relevant to the dispersion of lead alkyls from static pools lying on the sea bed.

Diffusion from a Continuous Point Source

The dispersal of a passive contaminant has been studied from two fundamental view points. In one the behaviour of an identifiable particle is studied by fixing the co-ordinate system within the moving particle of fluid (Lagrangian system) and in the other the co-ordinates are fixed in space and properties of many particles studied at given points (Eulerian system). Since diffusion is primarily concerned with particle displacements the most natural approach to the problem is through a Lagrangian framework. The relative spread and concentration of a passive contaminant from a fixed source may then be obtained by determining the probability that particles emanating from the fixed point in space will pass through other given positions later in time. Such treatments of turbulent dispersion have led to a better understanding of the problem, cf, (Ref. 10). A summary of the statistical approach is given in Csanady (Ref. 11). Despite the attraction of working in a Lagrangian co-ordinate system the governing equations are more difficult to handle and from a practical point of view it is more convenient to discuss the problem in relation to a co-ordinate framework fixed in space. Working in an Eulerian Framework also has the added advantage that some quantities, such as the velocity correlation function, are more easily obtained experimentally than the corresponding Lagrangian data.

In a Eulerian frame of relevance the mean concentration c in a turbulent flow field is given by

$$\frac{\partial c}{\partial t} + u_i \frac{\partial c}{\partial x_i} = - \frac{\partial}{\partial x_i} (\overline{u_i' c'})$$

where u_i are the mean velocities in the x_i directions, and $\overline{u_i'c'}$ represents the convariance of the velocity fluctuation with the concentration fluctuation. Terms representing molecular diffusion have been ommitted since they are invariably small in Oceanic and Atmospheric problems. The difficulty associated with determining the three covariances has given rise to semi-empirical approaches to the problem which relate them to the mean concentration gradients by

$$\overline{u_i' c'} = - K_{ij} \frac{\partial c}{\partial x_j}$$

where the tensor K_{ij} is called the 'eddy diffusivity'. It is generally assumed that it is only the diagonal components of K_{ij} which are important and it is taken that the mean concentration is governed by

$$\frac{\partial c}{\partial t} + u_i \frac{\partial c}{\partial x_i} = \frac{\partial}{\partial x_i} (K_i \frac{\partial c}{\partial x_i}).$$

For steady parallel flow in the x direction*

$$U \frac{\partial c}{\partial x} = \frac{\partial}{\partial y} (K_y \frac{\partial c}{\partial y}) + \frac{\partial}{\partial z} (K_z \frac{\partial c}{\partial z}) \tag{1}$$

where it has been assumed that diffusion in the direction of flow is small in comparison to the spread through convection. The main difficulty with such an approach to the problem lies in specifying the values of K_y and K_z. It is well known that turbulent diffusion, unlike molecular diffusion, is not a property of the fluid but varies with the scale of the diffusing process.

A wide spectrum of turbulence exists in the ocean with very rapid changes occurring near the sea bed, in consequence the effective eddy diffusivities will vary rapidly with distance from the wall, as will the mean velocity u. The variation of u, K_z, K_y will depend on local flow conditions. A relatively simple, but highly effective approximation can be made by assuming that K_z is proportional to the coefficient of vertical momentum transfer. It must be noted however, that this has only been found true for neutrally stable flows. For more realistic situations in which density gradients are present this approximation may well be inadequate. For example, in positive density gradients turbulence is suppressed and there is evidence to suggest that vertical dispersion can be quite small (Ref. 1 and 11).

In the corresponding atmospheric problem observation has shown that the wind profile follows closely the logarithmic law in the first 10 to 100 m from the surface (Ref. 7), this corresponds to a region in which the shearing stress is essentially constant. Although data is limited there is some evidence that the logarithmic profile also exists near the sea bed, Bowles et al. (Ref. 12), see also Bowden (Ref. 1). Unfortunately the solution of eq. (1) for such a profile is not available and this has led various authors to consider power law variations in the mean velocity. Smith (Ref. 13) assumes that the mean velocity may be represented by

$$\frac{u}{u_o} = (\frac{z}{z_o})^{\alpha} \tag{2}$$

where u_o and z_o are convenient reference parameters, α will depend on the stability of the flow, but for neutrally stable flow in the atmosphere the classical 1/7 power law for turbulent flows has been found to hold. There is no reason to suppose that great departures from this observation will hold for the sea bed situation, provided the flow is neutrally stable. Assuming that the mass and momentum based eddy diffusivities are proportional to each other this suggests that within the constant stress region the eddy diffusivity will vary as

* See Fig. 1 for definition of co-ordinate system.

$$\frac{K_z}{K_{z_o}} = \left(\frac{z}{z_o} \right)^{1-\alpha} \tag{3}$$

where K_{z_o} is a reference value at z_o. For the atmospheric case Smith (Ref. 13) uses evidence by Sutton (Ref. 14) to argue that the transverse diffusivity will vary in the same way as K_z. However, he recognised that this is somewhat speculative and generalises his analysis to let K_y vary as

$$\frac{K_y}{K_{y_o}} = \left(\frac{z}{z_o} \right)^{1-\alpha+\mu} \tag{4}$$

where μ is a free parameter to be fitted to available experimental data. Making the single assumption that the lateral profile of concentration will always be gaussian in character, Smith is able to obtain an analytic solution of eq. 1 for a point source located at the origin on the ground subject to the following boundary conditions.

(i) $c = Q\,\delta(0)$

where Q is the strength of the source (mass/time) and $\delta(0)$ is the delta function.

(ii) The ground is impervious to contaminant so that $\frac{\partial c}{\partial z} = 0$ when $z = 0$.

(iii) $c \to 0$ as $z \to \infty$

and that (iv) $\int_{-\infty}^{\infty} \int_{0}^{\infty} cu \, dzdy = Q, \quad 0 < x \leq \infty$

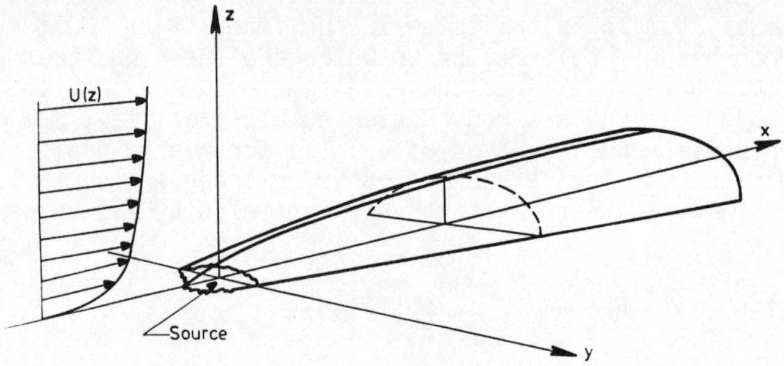

Fig. 1. Co-ordinates of plume.

The general solution is found to be

$$\bar{c} = \bar{c}_o \sqrt{\frac{\bar{c}_o}{2\pi\bar{c}_2}} \exp - \frac{\bar{y}^2}{2\frac{\bar{c}_2}{\bar{c}_o}} \tag{5}$$

where $\bar{c} = c/c_*$, c_* some reference value and $\bar{c}_o$ and $\bar{c}_2$ are given by

$$\bar{c}_o = \frac{\bar{Q}\ \bar{x}^{-(\frac{1+\alpha}{1+2\alpha})}}{(1+2\alpha)^{\frac{1}{1+2\alpha}}\left[-\frac{\alpha}{1+2\alpha}\right]!} \exp - \frac{\bar{z}^{1+2\alpha}}{(1+2\alpha)^2\ \bar{x}}$$

$$\bar{c}_2 = \frac{2\bar{Q}\left[1+2\alpha\right]^{2b-2}\ \bar{x}^{b-a}\ G(\eta)\ \exp - \eta}{\left[1+2\alpha\right]^{\frac{1}{1+2\alpha}}\left[-\frac{\alpha}{1+2\alpha}\right]!}$$

$$G(\eta) = \frac{(b-1)!\ (b+a-2)!}{(2b-1)!}\left[\frac{(b-1)!}{(a-1)!}\ {}_1F_1(b;a;\eta) - \eta^b\ V(b;a;\eta)\right]$$

$$V(b;a;\eta) = \sum_{r=0}^{\infty} \frac{(2b+r-1)!\ \eta^r}{(b+r)!\ (b+a+r-1)!}$$

${}_1F_1(b;a;\eta)$ = Kummers Function

$$\eta = \frac{\bar{z}^{1+2\alpha}}{(1+2\alpha)^2\ \bar{x}} \quad ; \quad a = \frac{1+\alpha}{1+2\alpha} \quad ; \quad b = \frac{2+\mu}{1+2\alpha}$$

The non-dimensional quantities $\bar{Q}$, $\bar{x}$, $\bar{y}$, $\bar{z}$ are given by

$$\bar{Q} = \frac{Q}{z_o^2\ c_*\ u_o}\sqrt{\frac{K_{z_o}}{K_{y_o}}}$$

$$\bar{x} = \frac{K_{z_o}}{u_o z_o^2}\ x, \quad \bar{y} = \frac{1}{z_o}\sqrt{\frac{K_{z_o}}{K_{y_o}}}\ y$$

$$\bar{z} = z/z_o$$

Of particular interest in the present context is the variation of maximum concentration, which will occur on the sea bed and along the axis of the plume, i.e. $y = 0$, $z = 0$. With $\mu = 0$

$$\frac{c}{c_*} = \frac{\bar{Q}\,\bar{x}^{-\left(\frac{1+\alpha}{1+2\alpha}\right)}}{2\sqrt{\pi}(1+2\alpha)^{\frac{2-2\alpha}{1+2\alpha}}\,\Gamma\left(\frac{1+\alpha}{1+2\alpha}\right)\sqrt{\dfrac{\Gamma\left(\frac{2}{1+2\alpha}\right)\Gamma\left(\frac{2-\alpha}{1+2\alpha}\right)\Gamma\left(\frac{2}{1+2\alpha}\right)}{\Gamma\left(\frac{4}{1+2\alpha}\right)\Gamma\left(\frac{1+\alpha}{1+2\alpha}\right)}}} \tag{6}$$

where $\mu = 0$ and Γ is the gamma function.

In the atmosphere the value of α depends on the stability of the atmosphere. However, for neutrally stable conditions, $\alpha = 1/7$, is found to represent the velocity profile very closely and measurements of concentration downstream of a ground level source are found to decay as $x^{-1.76}$ (Ref. 7) which is in close agreement with $x^{-5/3}$ in eq.(6). In the absence of detailed measurements it may conjectured that 1/7 power law would also be a fair representation of conditions near the sea bed.

An alternative approach to the problem is to make the assumption that at any station downstream of the source the distribution of concentration in the plume is Gaussian in both the transverse and vertical directions. If diffusion in the direction of flow is again ignored then the mean concentration is given by

$$c = \frac{Q}{2\pi\,\bar{u}\,\sigma_y\,\sigma_z} \exp -\left\{\frac{y^2}{2\,\sigma_y^{\,2}} + \frac{z^2}{2\,\sigma_z^{\,2}}\right\} \tag{7}$$

σ_y and σ_z are functions of x only and are the dispersions or standard deviations of the distribution about the centre line, and Q is the source strength. Provided the eddy diffusivities are defined as

$$K_y = \frac{\bar{u}}{2}\,\frac{d\,\sigma_y^{\,2}}{d\,x}\,,\quad K_z = \frac{\bar{u}}{z}\,\frac{d\,\sigma_z^{\,2}}{d\,x}$$

then it is readily verified that eq. (7) is a solution of the basic diffusion eq.(1).

The assumptions lying behind eq.(7) are that the plume is elongated in the x-direction, the turbulence intensity is small and that $x \gg y, z$. Frenkiel (Ref.10) demonstrated that this expression is valid for both small and large distances downstream of the source.

Two difficulties associated with the use of such approximations lies in determining a suitable mean speed when the flow is sheared, e.g. near the sea bed, and having access to reliable data for σ_y and σ_z as a function of x. In the atmosphere sufficient empirical data now exists to allow sensible predictions of σ_y and σ_z to be made, but in the sea bed case this is not the case and in consequence the usefulness of eq. (7) may be limited.

An attempt to remove some of the empiricism in eq. (7) was made by Sutton (Ref. 7). By assuming a power law variation of the velocity profile, eq. (3), and postulating a specific variation of the Lagrangian correlation function Sutton was able to relate the standard deviation of the plume width to the turbulence intensity in the following way.

$$\sigma_y^{\,2} = \frac{1}{2} \; \frac{4\,\nu^n}{(1-n)(2-n)\,\bar{u}^{\,n}} \left(\frac{\overline{v'^2}}{\bar{u}^{\,2}}\right)^{1-n} x^{\,n-2}$$

$$\sigma_z^{\,2} = \frac{1}{2} \; \frac{4\,\nu^n}{(1-n)\,(2-n)\,\bar{u}^{n}} \left(\frac{\overline{w'^2}}{\bar{u}^{\,2}}\right)^{1-n} x^{\,n-2}$$

$$n = \frac{2\,\alpha}{1+\alpha}\,, \quad \frac{\sqrt{\overline{v'^2}}}{\bar{u}}\,, \text{ and } \frac{\sqrt{\overline{w'^2}}}{\bar{u}}$$

are the turbulence intensities in the y and z directions and $\bar{u}$ is some suitable mean value of the velocity in the stream direction.

Again Sutton has shown that in neutrally stable atmospheres a value of $\alpha = 1/7$ describes the spread and concentration of a dispersing plume very well. However, it must be noted that success of the method depends on the ability to predict or measure local wind profiles and turbulent intensities. Such data is not often available in the sea bed situation.

Equation (7) applies to a point source situated in a infinite environment. The case of a source being situated at ground or sea bed level can readily be obtained by the method of images which gives

$$c = \frac{Q}{\pi\,\bar{u}\,\sigma_y\,\sigma_z} \exp - \left\{ \frac{y^2}{2\sigma_y^{\,2}} + \frac{z^2}{2\,\sigma_z^{\,2}} \right\} \tag{8}$$

Provided the sea bed is deep eq. (8) should be adequate. However, if data is needed at distances downstream of the source where the vertical spread of the plume begins to interact with the sea surface the above equation will not tell the whole story. In the case of the sea the thermocline near the sea surface will give rise to very high values of local density gradients which will provide very effective damping of the turbulence, and in practice will almost act as an impervious surface to the diffusing material. For such cases the problem is equivalent to a diffusing point source between two impervious boundaries distance H apart. Again a solution can be obtained by the method of images (Ref. 11) and the concentration distribution is given by

$$c = \frac{Q}{\pi\,\sigma_y\,\sigma_z} \exp - \frac{y^2}{2\,\sigma_y^{\,2}} \sum_{n=-\infty}^{n=\infty} \exp - \frac{(z+2nH)^2}{2\,\sigma_z^{\,2}} \tag{9}$$

Again the case of particular interest is the maximum concentration and this is found along $y = 0$, $z = 0$. This is

$$c = \frac{Q}{\pi\,\sigma_y\,\sigma_z} \sum_{n=-\infty}^{n=\infty} \exp - \frac{(2nH)^2}{2\,\sigma_z^{\,2}}$$

For small distances from the sources

$$c \sim \frac{Q}{\pi \sigma_y \sigma_z}$$

which is exactly the same as eq. (8). However, at large distances from the source the dispersion is restricted in the z direction and it is readily shown that for $x \to \infty$.

$$c \sim \frac{Q}{\sqrt{2}\pi \sigma_y \bar{u}H} \quad \exp - \frac{y^2}{2 \sigma_y^2} .$$

Thus at large distances downstream the decay of C becomes significantly slower.

Whether eq. (9) has any strict justification is open to question. The upper boundary will significantly influence the velocity profiles and there is no reason to expect an infinite array of sources, with σ_y and σ_z calculated for single boundary flows, to describe the distribution of the concentration in bounded situations. In consequence eq. (9) is highly speculative and its justification can only be assessed by comparison with measurements. If this data were available suitable adjustments to σ_y, σ_z might be made to make it a useful empirical formula.

Concentration Distribution Near the Source

The concentration in the near vicinity of an idealised point source will be unrealistically large. If concentrations just downstream of a source is required it is necessary to make some allowance for its finite size and shape. Davies (Ref. 15) has presented a general theory for diffusion from an arbitrary shaped source, with the assumption that the turbulent velocity profile and eddy diffusivities can be represented by power laws (eqs. (2) and (3)). Unfortunately, this solution is not in a readily accessible form and for the present study it is simpler to use the theory of Calder (Ref. 16) which deals with diffusion downstream of a two-dimensional source of finite extent in the stream direction. For two-dimensions the corresponding diffusion equation is

$$u(z) \frac{\partial c}{\partial x} = \frac{\partial}{\partial z} \left(K_z(z) \frac{\partial c}{\partial z} \right)$$

which must be solved subject to the following boundary conditions

$$\begin{array}{lll} C(x,0) = C_* & \text{(Const.)} & 0 < x \leqslant x_o \\ C(x,\infty) = 0 & & 0 < x \leqslant x_o \\ C(0,Z) = 0 & & 0 < Z \\ \frac{\partial c}{\partial z}(x,0) = 0 & & x > x_o \end{array}$$

Where x_o is the extent of the source. With $U(z) = U_o(z/z_o)^\alpha$ and $K(z) = K_o(z/z_o)^{1-\alpha}$. Calder shows that downstream of the source (x x_o) that the concentration is given by

$$\frac{c}{c_x} = \frac{\sin \sigma\pi}{\pi} \int_o^{x_o} \theta^{\sigma-1}(x-\theta)^{-\sigma} \exp\left(-\frac{\gamma}{x-\theta}\right) d\theta \qquad (10)$$

where $$\sigma = \frac{\alpha + 1}{2\alpha + 1}$$

and $$\gamma = \frac{u_o \, z_o^{1-2\alpha} \, z^{2\alpha+1}}{K_{z_o} \, (2\alpha+1)^2 \, x}$$

He also shows that the total uptake from the strip per unit area is given by

$$q = c_x \frac{\sin \sigma\pi}{\sigma \Pi} \, x_o^{\sigma-1} \, N \tag{11}$$

where $$N = (2\alpha+1)^{\frac{1}{1+2\alpha}} \, \Gamma\left(\frac{\alpha+1}{2\alpha+1}\right) \, K_{z_o}^{\frac{\alpha+1}{2\alpha+1}} \, u_o^{\frac{\alpha}{2\alpha+1}} \, z_o^{-\frac{1}{2\alpha+1}}$$

In addition to the power law approximations to u and K_z, and the assumption that the flow is two-dimensional, Calder has assumed that over the diffusing strip $0 \leqslant x \leqslant x_o$, that the concentration remains constant. Whether the equilibrium concentration of TEL and TML compounds can be maintained at the surface under flow condition might be queried, however the assumption that the equilibrium concentration is maintained at the surface will give rise to pessimistic approximations to the resulting dispersion.

Application to the Cavtat Incidence

The Cavtat lies 6.4 km South East of Capo d`Otranto in 95 m of water (Ref. 17). "Anti-Knock" compounds are stored in drums and it is conjectured that any leakage through corrosion or damage will rapidly form pools on the sea bed because of their high specific gravities. For reasonably smooth sea bed conditions Harrison (Ref.17) has estimated that the contents of a typical drum might form a shallow pool (1 cm in depth) covering an area of 2.2810^5 cm^2. Since leakage will probably occur through corrosion the orifice through which the compound will flow will be small and hence the relatively slow leaks will give rise to pools of lesser extent.

In order to calculate the concentrations of Lead Alkyls downstream of such a pool from eqs. (6) and (8) some knowledge of the velocity distribution near the sea bed is required and some indication is needed of the rate of uptake of the Lead Alkyls from the pool. From the limited amount of data available (Appendix 1) it would appear that near the sea bed a reasonable representation of the velocity profile* is given by $u/u_o = (z/z_o)^{1/7}$ where u_o = 12.5 cm/s and z_o = 100 cm. Although this may be inadequate away from the sea bed, as will be shown later, this may not be significant since the concentration of the diffusing material becomes insignificant above a few metres from the sea bed. It must be emphasised that this velocity distribution can only be thought of as some typical average condition that might exist in the region. Naturally, the currents will be continually changing but the estimation of dispersion under such conditions would be a formidable task.

Estimates of the uptake Q (gram/sec) have been made by McManamey (Ref. 18) and these are summarised in Appendix II. Maximum rates of mass transfer likely to be realised from a pool 2.28×10^5 cm^2 in area are estimated to be 0.000328 gram/s for T.E.L. and 0.00179 gram/s for T.M.L. An order of magnitude check on these values can be made by means of eq. (11). The equivalent mass transfer coefficient would

---

+ See Fig. 2.

then be

$$\frac{\sin \sigma \pi}{\sigma \pi} \; x_o^{\;\sigma - 1} \quad N$$

With α = 1/7 and x_o = 200 cm (i.e., a pool 2m x 10m is assumed) this would give rise to uptake of TEL and TML as follows 0.01 gram/sec and 0.049 gram/sec. If the pool is lengthened to 400 cm the answer is hardly changed. These estimates are an order of magnitude higher than those predicted by above. In view of the agreement obtained in laboratory tests this may indicate that the equivalent surface concentrations may be less than C* in eqs. (10) and (11), and that the 1/7 power law is not an adequate assumption of the profile when attempting to predict uptake rates.

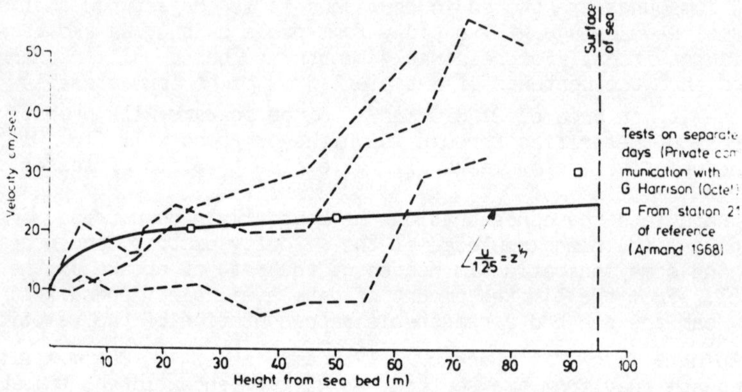

Fig. 2. Velocity distributions in the Otranto Straits.

With α = 1/7 eq. (6) may be used to assess the maximum concentrations downstream of the source. For this particular case eq. (6) becomes

$$\begin{aligned} c_{TEL} &= 6910 \;\; x^{-5/3} \text{ ppm} \\ c_{TML} &= 37600 \;\; x^{-5/3} \text{ ppm} \end{aligned} \tag{12}$$

This is shown in Fig. 3 and it can be seen that by about 100 m downstream of the source the concentration for TEL and TML has been reduced to about 0.001 ppm, which according to Maddock and Taylor (Ref. 19) is within safe limits. Within 10 m of the pool eq. (12) suggests that maximum concentrations of almost 1 ppm may exist. However, it should be noted that the theory will over-estimate the concentrations in this region and that the two-dimensional solution given by eq. (10) will probably be more representative.*

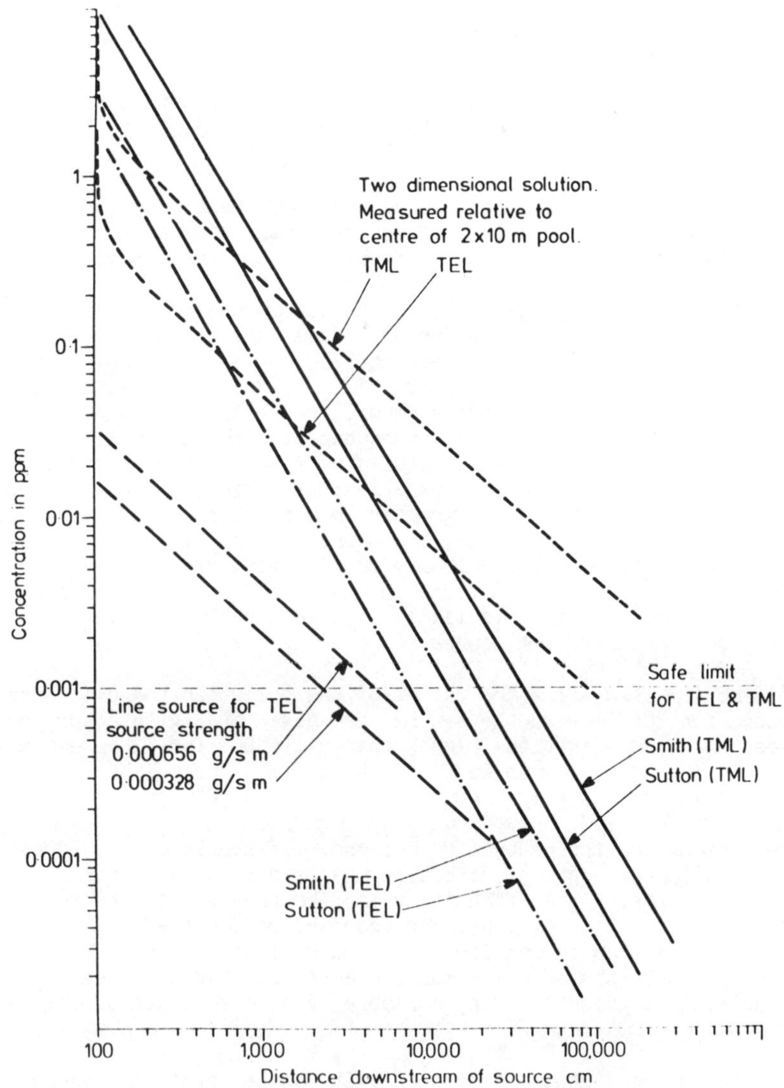

Fig. 3. Maximum concentration variation downstream source of TEL and TML.

* Calder's theory may also be used to determine the maximum concentration downstream of a two-dimensional line source. This is shown in Fig. 3 for TEL for sources of length 5 and 10 m.

Some comparison can be made by utilising the theory of Sutton (Ref. 7) and summarised in eq. (8). This again assumes a power law for the velocity profile but in addition incorporates information about the mean turbulence structure. In a turbulent flow near a surface the mean turbulent intensity normal to the surface $\sqrt{\bar{w}'^2}/u$ is usually of the same order as the friction velocity u^* (see Appendix 1). For example, in the atmospheric boundary layer $\sqrt{\bar{w}'^2} \sim 1.25\ u^*$, Couniham (Ref. 9). In addition the lateral turbulent intensity $\sqrt{\bar{v}'^2}/\bar{u}$ is generally given by $\sqrt{\bar{v}'^2} \sim 1.5 \sqrt{\bar{w}'^2}$. This is confirmed for the atmospheric case, Counihan (Ref. 9) and also from measurements near the sea bed by Bowden (Ref. 1).

Thus with $\sqrt{\bar{w}'^2} = u^*$ and $\sqrt{\bar{v}'^2} = 1.5 \sqrt{\bar{w}'^2}$ equation (8) is used to predict the maximum concentration. For the present problem this gives $c_{TEL} = 5800\ x^{-1.75}$ and $c_{TML} = 31600\ x^{-1.75}$, cf. Fig. 3. As can be seen the two quite different approaches agree modestly well. Confirmation of the accuracy of these methods for sea bed conditions is not possible due to lack of measurements. In the atmosphere they are known to be sensible approximations to available measurements (Ref. 8).

Because of the finite width of the pool, the three-dimensional solution with a point source might well be inadequate for distances less than about 20,000 cm. If the plume edge is defined as that approaching 1/10 the maximum concentration on the centre-line then this is almost given by $y = 2\sigma_y$. Thus for a source of dimensions 2m x 10m, the influence of the outside edge will not be significant until about 10,000 - 20,000 cm, see Fig. 4, and in consequence the two-dimensional solution will probably give the most pessimistic estimate of the peak concentrations for $0 < X < 20{,}000$ cm. The concentration downstream of a finite width source, given by equation 10, is an order of magnitude greater than the corresponding 2 D line source. However, as indicated earlier, the boundary conditions together with the 1/7 power law approximation to the velocity profile give rise to overall source strength greater in excess of those found in practice. In consequence the 2D line source solution present in Fig. 3 is likely to give a better estimate of the concentration just downstream of the source.

Equations (6) and (8) also allow the extent of the pollution to be assessed. Plots of the concentration profiles on the sea bed are given in Fig. 5. For the worst case it is seen that concentrations higher than c = 0.001 for TEL are confined to a narrow plume which has a maximum width of about 2 m. The height of the plume will be less than 1m, as shown in Fig. 6 for TEL at x = 6000 cm.

The above predictions are for neutral conditions. If stable density gradients are present dispersion in the vertical direction may be seriously hampered, due to the suppression of turbulence. This effectively implies that K_z will be much smaller than predicted for neutral conditions. For example, Bowden (Ref. 20) has shown that in stable conditions K_z may be smaller by an order of magnitude. In relation to eq. (6) this would imply that the maximum concentration might increase by a factor of 10. However, in the absence of any suitable data no estimate of this effect can be made.

Whether in neutral or very stable conditions, it can be noted that the plume lies close to the sea bed for steady flow conditions and unless concentrations significantly below 0.0001 are important the effect of the sea surface or the thermocline on the concentration in the plume may be neglected.

As shown by Neuman (Ref. 21) the effect of the shape of the source influences the resulting concentrations. For a given surface area Neuman shows that elliptically shaped sources result in concentrations less than those arising from rectangular sources and that triangular sources give rise to even lower values.

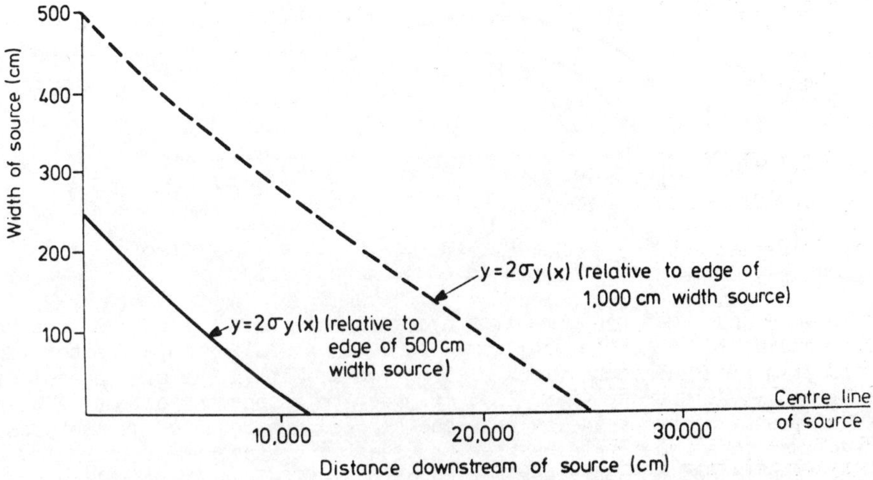

Fig. 4. Influence of finite width of source on plume developments for TEL z=0.

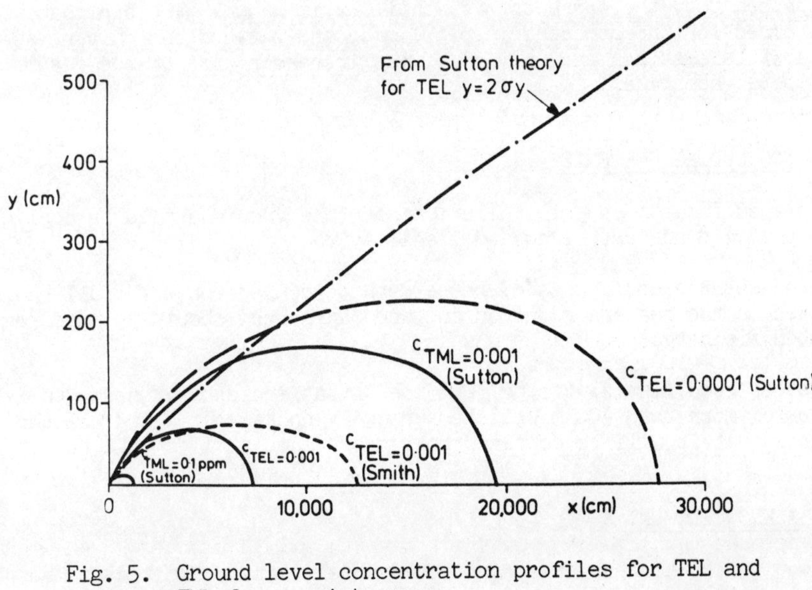

Fig. 5. Ground level concentration profiles for TEL and TML for a point source.

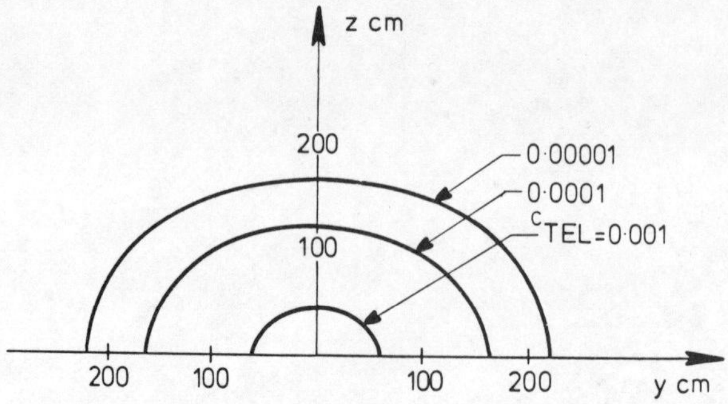

Fig. 6. Concentration profile for TEL at a distance of 6000 cm from source (Sutton Analysis).

The calculations have been performed for hydraulically smooth flow. If the bottom surface is considered rough the additional turbulence will promote greater mixing and the plumes maximum concentrations will be less. Within the atmosphere this is well documented, e.g. Csanady (Ref. 11). In order to assess the effect of this, detailed measurements of the velocity near the sea bed are required and methods developed for the atmosphere could be used to assess the approximate power law and surface friction needed to apply eqs. (6) and (8), cf Calder (Ref. 16) and Appendix 3.

Bearing in mind these additional problems, and that the plume may meander through 360 degrees it seems highly unlikely that the contaminated area will have a radius much in excess of 100 m from the source. All the evidence suggests that safe average concentrations will be well within this zone.

On the general question of assessing dispersion from pools of lead alkyl on the sea, there is enough theoretical work existing to make sensible predictions provided the nature of the sea bed, the currents and stability is known. The applicability of methods developed for the atmospheric problem to the sea bed can only be assessed by full scale trials at sea and in view of the increasing use of the sea bed such work needs to be undertaken.

CONCLUDING COMMENTS

Methods developed for atmospheric pollution near the ground should be applicable to the sea bed provided adequate flow and stability data is available.

In order to assess the usefulness of these methods comparison with full scale measurements taken on the sea are required and oceanographers should be encouraged to undertake such field tests.

For spillages of Lead Alkyls arising from the Cavtat incident it seems unlikely that a radius of more than 500 m will be seriously contaminated and all the evidence suggests that it will be confined to regions much smaller than this.

ACKNOWLEDGMENT

The author would like to thank the Associated Octel Company Limited for supporting this work.

APPENDIX 1

ESTIMATION OF VELOCITY PROFILES

Three velocity profiles taken in the Otranto Straits are shown in Fig. 3 (private communication with Mr. Harrison). Large variations are seen to exist and even if the conditions are considered steady the solution of eq. (1) with the given velocity profile would have to be found numerically. However, over the lower layer a mean approximation of the velocity can be obtained by assuming that

$$u = 12.5 \left(\frac{z}{100}\right)^{1/7} \quad \text{cm/sec}$$

where z is measured in cm. This is only marginally satisfactory but should give some order of magnitude of the concentration when used with eq. (6). Ideally measurements are needed within the first few metres above the sea bed.

In order to estimate the effective eddy diffusivity further assumptions must be made. For the constant stress region

$$K_z = K_{z_o} \left(\frac{z}{z_o}\right)^{1-1/7}$$

See Sutton (Ref. 14). For flow near a smooth surface it has been verified (cf. Sutton, Ref. 14) that

$$K_z (z) \frac{\partial u}{\partial z} = 0.020 \; \nu^{1/4} \, u_o^{7/4} \, z_o^{-1/4}$$

and therefore

$$K_z (z) = 0.14 \; \nu^{1/4} \, u_o^{3/4} \, z_o^{3/4} \left(\frac{z}{z_o}\right)^{6/7}$$

Thus it may be noted that with z_o =100 cm, u_o = 12.5 cm/sec that K_o = 9.33 cm^2/sec. The use of a power law variation for u and K_z is usually confined to a height of about 10^3 ν/u^*, which for the present case is less than 100 cm. Thus some caution must be exercised in interpreting the final answer, but in the absence of detailed measurements such an approximation must suffice.

Again assuming the sea bed is smooth a 1/7 power law implies that

$$u^{*2} = \frac{\tau\omega}{\rho} = 0.02 \; \nu^{1/4} \, u_o^{7/4} \, z_o^{-1/4} = 0.166 \text{ cm}^2/\text{sec}^2$$

for the present problem, i.e., u^* = 0.41 cm/sec. More detailed measurements closer to the sea bed would allow more precision to be obtained.

APPENDIX II

ESTIMATION OF THE SOURCE STRENGTH

According to Davies (Ref. 22), the mass transfer coefficient can be written as

$$K = 0.075 \;\; u^* \left(\frac{D}{\nu}\right)^{2/3}$$

where $u^* = \sqrt{\tau\omega/\rho}$ is the friction velocity, D the molecular diffusivity and ν the kinematic viscosity. In consequence the rate of uptake is given approximately by

$$Q = 0.075\, u^* \left(\frac{D}{\nu}\right)^{2/3} Ac^*$$

where c^* is the equilibrium solubility and A the area of the source. McManamey (Ref. 18) estimates the diffusion coefficients of TEL and TML in water and quotes

$$\left(\frac{\nu}{D}\right)_{TEL} = 1820 \quad \text{and} \quad \left(\frac{\nu}{D}\right)_{TML} = 1370.$$

Using values of $c^*_{TEL} = 0.8$ ppm and $c^*_{TML} = 15$ ppm for pure water modestly good agreement was obtained with experiment provided equation (2) was factored by 3.5. This can be accounted for by noting that in experimental conditions the surface of the pool was wavy and from the point of view of mass transfer could be considered rough.

Bearing this in mind Q for the pools lying on the sea bed was calculated from
$Q = 3.5 \times 0.075\, u^* \left(\frac{D}{\nu}\right)^{2/3} A\, c^*$.

With $A = 2.28\ 10^5\ cm^2$, $c^*_{TEL} = 2$ ppm, $u^* = 0.41$ cm/sec and $(\nu/D)_{TEL} = 1820$, $Q_{TEL} = 0.000328$ gr/sec and with $A = 2.28\ 10^5\ cm^2$, $c^*_{TML} = 9$ ppm, $u^* = 0.41$ cm/sec and $(\nu/D) = 1370$, this gave $Q_{TML} = 0.00179$ gr/sec.

APPENDIX III

IDEAL FLOW NEAR THE SEA BED

For steady flow past a smooth boundary the velocity distribution can be written, to a high degree of accuracy by

$$\frac{u(z)}{u^*} = 2.5 \log_{\ell} \left(\frac{u^* z}{\nu}\right) + 5.5$$

where $u^* = \sqrt{\tau\omega/\rho}$ = friction velocity
ν = kinematic viscosity
$u(z)$ = mean velocity at distance z from surface.

This equation holds except very close to the surface. In the atmosphere this represents the wind profile under steady conditions very well and Bowden (Ref. 1) indicates that it may also be true near the sea bed.

If the flow is rough

$$\frac{u}{u^*} \sim 2.5 \log_{\ell} \left(\frac{z}{z_r}\right)$$

where z_r is the mean roughness height and depends on the size, shape and spacing of the roughness elements.

Experiments have shown that the flow may be considered smooth if $\dfrac{u^* z_r}{\nu} < 5.5$

and rough if $\frac{u^* z_r}{\nu} > 75$.

As pointed out in Section 2 such relations for the velocity lead to difficulties in the subsequent solution of the diffusion equation and it is more convenient to approximate the velocity as a power law.

For smooth flow, one such approximation is given by

$$\frac{u}{u^*} = 8.74 \left(\frac{u^* z}{\nu}\right)^{1/7}$$

which agrees well up to $\frac{u^* z}{\nu} \sim 10^3$. For higher values of $\frac{u^* z}{\nu}$ the 1/7 power law has to be modified to 1/8, 1/9 etc. Thus when the flow is smooth Calder argues that the velocity can be written as

$$\frac{u}{u^*} = q \left(\frac{u^* z}{\nu}\right)^{\alpha}$$

where q , α depend on the range of $\frac{u^* z}{\nu}$.

For rough flow

$$\frac{u}{u^*} = q' \left(\frac{z - d}{z_r}\right)^{\alpha}$$

where q', α will depend on range of $\frac{z - d}{z_r}$ and d is a zero point displacement. From a practical point of view if a velocity is measured near the surface u^* could be estimated from the logarithmic law and other velocities away from the surface computed and a power law fitted.

Thus assuming

$$\frac{u}{u_o} = \left(\frac{z}{z_o}\right)^{\alpha}$$

Calder shows that

$$\frac{K_z}{K_{z_o}} = \left(\frac{z}{z_o}\right)^{1-\alpha}$$

where for smooth flow

$$K_{z_o} = \left(\frac{\varepsilon}{\alpha}\right) \frac{\delta^{2\alpha\beta} \, u_o^{2\beta-1}}{z_o^{2\alpha\beta-1}}$$

$$\varepsilon = \left(\frac{1}{q}\right)^{\frac{2}{1+\alpha}} \quad ; \quad \delta = \nu \, , \quad \beta = \frac{1}{1 + \alpha}$$

and for rough flow

$$K_{z_o} = \left(\frac{\varepsilon'}{\alpha}\right) \frac{\sigma'^{2\alpha\beta} \, u_o^{2\beta-1}}{z_o^{2\alpha\beta-1}}$$

With $\varepsilon' = \frac{1}{q'^2}$; $\delta' = Z_r$; $\beta = 1$

Thus the theories of Calder (Ref. 16) and Smith (Ref. 13) are applicable to rough and smooth flow provided field measurements can be taken to enable q, q', z_r, α, to be calculated.

For rough flow velocities measurements at three vertical stations are required to estimate u^*, z_r, α in conjunction with the assumed log-law variation of the velocity.

REFERENCES

(1) K. F. Bowden, Turbulence, Ocean. Mar. Biol. Ann. Rev. 2, 11-30 (1964).

(2) K. F. Bowden, Turbulence II, Ocean. Mar. Biol. Ann. Rev. 8, 11-32 (1970).

(3) K. F. Bowden, Turbulent Diffusion, Mem. Soc. Roy. des Sc. de Liege 6, 79-89 (1972).

(4) K. F. Bowden, Measurements of Diffusion in the Sea, Mem. Soc. Roy. des Sc. de Liege 6, 91-97 (1972).

(5) A. Okubo, Oceanic Diffusion Diagrams, Deep Sea Research 18, 787-802 (1971).

(6) J. W. Talbot, Interpretation of Diffusion Data, Proc. Int. Symp. on Discharge of Sewerage from Sea Outfalls, 1974.

(7) O. G. Sutton, The Problem of Diffusion in the Lower Atmosphere, Quart. J. R. Met. Soc. 73, 257 (1947).

(8) F. Pasquil, Atmospheric Dispersion of Pollution, Quart. J. R. Met. Soc. 97 369-395 (1971).

(9) J. Counihan, Adiabatic Atmospheric Boundary Layer, Atm. Env. 9, 871-905 (1975).

(10) F. N. Frenkiel, Turbulent Diffusion, Adv. Appl. Mech. 3, 61 (1953).

(11) G. T. Csanady, Turbulent Diffusion in the Environment, D. Riedal Publishing Co., 1973.

(12) Bowles, et al., Proc. 2nd Int. Conf. Peaceful Uses of Atomic Energy U. N., Geneva 18, 376-389 (1968).

(13) F. B. Smith, Diffusion of smoke from a continuous elevated point source into a turbulent atmosphere, J. F. M. 2, 49 (1957).

(14) O. G. Sutton, Micrometeorology, McGraw Hill, New York, 1953.

(15) D. R. Davies, Three-dimensional turbulence and evaporation in the lower atmosphere, Quart. J. Mech. & Appl. Maths. 3, 64 (1950).

(16) K. L. Calder, Eddy Diffusion and Evaporation in Flow over Aerodynamically Smooth and Rough Surfaces, Quart. J. Mech. & Appl. Maths. 2, 153 (1948).

(17) G. K. Harrison, The Cavtat Incident.{Internal Report, Octel Co. Ltd. Also Proceedings of the International Experts Discussion on the "Lead - Occurrence, Fate and Pollution in the Marine Environment", Rovinj, October 1977, M. Branica and Z. Konrad (eds.), Pergamon Press, 1979, 305-317.

(18) W. J. McManamey, cf. J. R. Grove, Investigations into the formation and behaviour of acqueous solutions of lead alkyls. (Also, Proceedings of the International Experts Discussion on the "Lead - Occurrence, Fate and Pollution in the Marine Environment", Rovinj, October 1977, M. Branica and Z. Konrad (eds.), Pergamon Press, 1979, 45-52.

(19) B. G. Maddock and D. Taylor, The acute toxicity and bioaccumulation of some lead alkyl compounds in marine animals. Also Proceedings of the International Experts Discussion on the "Lead -Occurrence, Fate and Pollution in the Marine Environment", Rovinj, October 1977, M. Branica and Z. Konrad (eds.), Pergamon Press, 1979, 233-261.

(20) K. F. Bowden, Horizontal Mixing in the Sea due to Shearing Current, J. F. M. 21, 83-95 (1965)

(21) J. Neumann, Turbulent Diffusion of Pollutants From Some Plane Area Sources, Atm. Env. 9, 785-792 (1975).

(22) J. T. Davies, Turbulent Phenomena, Academic Press, 1972.

The Relative Importance of Lead as a Marine Pollutant

M. Bernhard

Divisione Protezione Ambiente, Comitato Nazionale per l'Energia Nucleare, I-19030 Fiascherino, La Spezia, Italy

ABSTRACT

Estimations based on the recently determined low total lead levels in the marine environment have shown that the nation with the highest average sea fish consumption (Portugal) derives only about 0.2% of its total lead intake through the marine pathway. Also, heavy fish eaters such as fishermen, fish vendors, etc., consume only insignificant amounts of lead (about 2.5%) through sea foods. Total lead concentrations observed in highly polluted areas are 10 times lower than the lowest concentrations showing effects on marine organisms. The environmental concentrations of several other heavy metals exceed toxic levels. Organic lead levels are practically unknown. It is concluded that, contrary to the considerable hazard of industrial lead to man from non-marine pathways, inorganic lead at least seems to be a pollutant of minor importance in the marine context.

INTRODUCTION

Ample evidence is now available that the majority of lead concentrations determined in environmental samples are actually several order of magnitude lower than so far assumed (Ref. 1). For an evaluation of the possible adverse effects of lead on marine organisms these reductions of several orders of magnitude in lead concentration, both in sea water and marine organisms, are of a great importance. The lowest reported concentrations at which effects were observed are several magnitudes higher than the highest lead concentrations observed in polluted waters. Similarly, the great decrease in the lead body burden of marine organisms which serve as human foods reduces drastically the lead intake by humans through the marine pathways, which becomes negligible in comparison with the intake from non-marine sources. This also reduces, by several orders of magnitude, the possible risks of lead toxicity for marine organisms and for humans from the consumption of marine foods.

LEVELS IN THE MARINE ENVIRONMENT

Patterson and collaborators (Ref. 2, 3) have now provided ample proof that most, if not all, measurements of lead in sea water and marine organisms carried out under not ultraclean laboratory conditions, yield results which are by several orders of magnitude too high. Of the few research workers that have intercalibrated succesfully with the group of Patterson (Ref. 1), Nürnberg and collaborators (Ref. 4) present their data at this meeting.

Prior to the study of Patterson and collaborators lead levels in open-ocean sea water were thought to range from 0.02 to 0.1 μg Pb/l. In coastal waters concentrations up to several micrograms have been frequently reported (for a review see Bernhard and Zattera, Ref. 5). Now the typical open-ocean concentrations of total lead (Pb-T) range from 0.001 to 0.01 μg Pb-T/l (Table 1).

In little polluted coastal waters of the US Pacific coast lead levels range from 0.005 to 0.02 μg Pb-T/l. The highest concentrations outside the sewage plumes are less than 0.05 μg Pb-T/l. In the near shore waters off the north-western Italian coast, between Ventimiglia and Anzio (Rome), higher concentrations (ranging from 0.1 to 0.5 μg Pb-T/l) were observed by Mart et al. (Ref. 4). In the near shore waters of the Belgian-Dutch-German coast the same authors found higher concentrations than in the Italian coastal waters.

Up to now only few reliable data have been published on marine organisms (Table 2). Marine algae have the highest level. Invertebrates such as molluscs and crustaceans have a lower level and in the edible muscle tissue of the more developed teleost fish the level is less than 0.5 μg Pb/kg FW (fresh weight). Lead concentrations in the edible parts of marine organisms were previously believed to be around 1000 μg Pb/kg FW (Ref. 5).

The only data on organic lead levels in marine organisms are those of Sirota and Uthe (Ref. 7), who found up to 90% of the total Pb as tetraalkyl-Pb. Their total Pb levels are, however, several orders of magnitude higher than the levels listed in Table 2.

EFFECTS OF INORGANIC LEAD ON MARINE ORGANISMS

Table 3 lists some effective concentrations in various marine organisms, ranging from unicellular algae to fish.

Unicellular algae and sea urchins seem to be the most sensitive organisms. Growth inhibition has been observed in the most sensitive species of algae, in *Thalassiosira pseudonana* and *Porphyridium marinum*, at 200 μg Pb/l. Sea urchins are sensitive to similar concentrations. On the other hand no effects were observed in molluscs after an exposure to 200 μg Pb/l for 49 and 84 days. This higher sensitivity of marine plants and lower marine organisms to lead could be explained by the results contained in Table 2 showing that these less developed organisms have a higher Pb level than higher organisms, because they are less efficient in discriminating against lead.

Taking these and other data on toxicity into account, the Water Quality Criteria-1972 (Ref. 19) suggests 10 μg Pb/l as the minimal risk concentration. Problems of uncertainty, similar to those encountered in the analysis of lead in sea water and marine organisms should not have such a great effect on the results of toxicity tests, since the solutions tested were usually prepared from standards by dilution. The standards were prepared from weighted lead salts. Contamination of the solutions used in the tests should raise the actual concentration to higher true values. Consequently the experimentor will attribute the observed effect to a lower concentration than actually tested and lower levels of toxicity are wrongly reported.

In testing algae complexing agents, such as EDTA and TRIS, are often used to prepare test media. Some researchers are aware that these substances may complex heavy metals and consequently give wrong results. For example, Berland et al. (Ref. 8) omitted EDTA from their culture media but still used TRIS. TRIS is also a chelator, but a mathematical model of sea water containing lead and TRIS in the concentrations used by these authors indicates that, contrary to copper, which is

TABLE 1 Lead Concentrations (µg/l) in Polluted and Unpolluted Sea Water

Area and sampling depth (m)	n	Range	References
Open waters			
Central NE Pacific (0 - 5000 m)	(14)	0.001 - 0.014 (D)	Schaule and Patterson (Ref. 1)
	(8)	0.0002 - 0.0045(P)	
Coastal waters			
La Jolla, 5 km W Scripps Pier (0.1 m)	(2)	0.016 - 0.036 (T)	Patterson et al. (Ref. 2)
San Juan de Fuca Straits (0.1 m)	(1)	0.024 (T)	
Los Angeles, above JWPCP outfall (0.1-30 m)	(5)	0.1 - 1.3 (T)	
Italian NW coast (surface)	(225)	0.05 - 10.0 (T)	Mart et al. (Ref. 4)
Belgium-German North Sea (surface)	(200)	0.3 - 15.0 (T)	

D = dissolved; P = particulate; T = total

TABLE 2 Lead Concentrations in Marine Organisms

Species	µg/kg FW	References
Valonia	200	Burnett and Patterson (Ref. 1)
Abalone, muscle	4	
Mytilus californianus, muscle	25	
Hinnites multirugosa, muscle	5	
Spiny lobster, muscle	5	
Thunnus alalunga, muscle	0.3	Patterson et al. (Ref. 2)

TABLE 3 Effects of Inorganic Lead on Some Marine Organisms

Species	Effective conc. µg/l	Effect	References
18 unicellular algae	250 - 2000	growth inhibition	Berland et al. (Ref. 8)
Macrocystis pyrifora	50000	50% inhibit. photosyn.	Clendenning and North (Ref. 9)
Cristigera sp.	300	12% growth rate inhibition	Gray (Ref. 10)
Psammechinus miliaris	200	eggs abnormal	McKee and Wolf (Ref. 11)
Eirene viridula	1000	malformation tentacles	Karbe (Ref. 12)
	300	no effects	
Ophryotricha spp.	1000	500 h LC 50	Brown and Ahsanullah (Ref. 13)
Neanthes arenaceodentata			
juveniles	2500	28 d LC 50	Reish et al. (Ref. 14)
adults	3200		
Capitella capitata adults	1000		
Crassostrea virginica	200	no effects after 49 h	Pringle et al. (Ref. 15)
	500	egg develop. normal	Calabrese et al. (Ref. 16)
	2450	48 h LC 50 egg develop.	
Mya arenaria	200	no effects after 84 h	Pringle et al. (Ref. 15)
Uca pugilator	1000	no effect on regeneration	Weis (Ref. 17)
Fundulus heteroclitus	40000	96 h LC 50	Jachim et al. (Ref. 18)

bound to 99% by TRIS, lead is not bound to TRIS (A. Piro, CNEN, pers. comm. 1978). Comparing the effective concentrations in Table 3 with the lead concentrations found one can see that even the highest concentrations observed in polluted areas are 10 times lower than the lowest concentrations which produce an effect on marine organisms. Other heavy metals such as mercury, copper, nickel, etc., become toxic at concentrations which are of the same order of magnitude as the water concentrations observed in the marine environment (Ref. 5). Lead is one of the least toxic heavy metals for marine organisms.

LEAD INTAKE BY HUMANS THROUGH MARINE FOODS

The yearly world average sea fish consumption and per capita is about 8 kg sea fish, 1 kg marine shell fish and 1 kg fresh water fish. Since marine shell fish is much less consumed than sea fish only the consumption of sea fish will be considered here. Sea fish consumption in the Near Eastern and African countries is below the world average (Table 4). Inland countries, such as Niger, have no sea fish consumption. Congo with about 7 kg per year per capita has one of the highest consumptions. The average consumption in Iran and Irak is about 0.25 kg. Nations with the highest per capita consumption are in Europe (Portugal 26 kg, Norway 18.7 kg) and in the Far East (Japan 25.5 kg, the two Koreas 16 and 17 kg). The US has an average of about 4 kg and South American nations consume about 3 kg.

TABLE 4 Consumption* (1972 - 1974 Averages) of Sea Fish (Kg per Year and Capita) by Some Selected Countries (Ref. 20)

Country	Consumption	Country	Consumption
Czechoslovakia	3.0	Irak	0.24
France	8.2	Iran	0.22
Hungary	0.8	Israel	4.3
Germany (FRG)	5.3	Saudi Arabia	2.4
Iceland	23.3	Turkey	2.3
Italy	5.7		
Norway	18.9	Argentina	2.6
Polen	8.4	Brasil	2.6
Portugal	26.2	Chile	5.9
		Peru	7.3
Afghanistan	0	US	4.2
Burma	2.8	Venezuela	4.5
China	1.8		
India	1.5	Angola	5.1
Japan	25.5	Burungi	0
Korea DPR	17.1	Cameroon	3.7
Korea Rep.	16.1	Congo	7.1
Malesia	9.0	Mozambique	1.5
Mongolia	0	Niger	0
Nepal	0	Nigeria	1.1
Thailand	6.1	Rhodesia	1.2
		S. Africa	5.5

* Consumption equals supply. Supply is the amount of food reaching the consumer but is not necessarily the amounts consumed. The amount actually consumed is smaller depending on the loss during food storage and preparation in the household. Supplies of whole fresh and whole frozen have been multiplied with 0.6 to account for unconsumable parts.

However, people who derive most, if not all, of their proteins from marine foods, i.e., the critical groups of heavy fish eaters, such as fishermen, fish processors and vendors and their families, consume much larger amounts of sea fish. Bernhard (Ref. 21) estimates that some individuals in these critical groups consume up to 5.5 kg of edible marine fish per week which equals 285 kg of fish per year. Assuming that the lead concentrations given in Table 2 are for marine organisms in general the intake through marine pathways can be calculated and compared with the intake from non-marine intakes. Burnett and Patterson (Ref. 1) observed a decrease in the Pb level with an increasing position in the food chain. Marine fishes have a much lower Pb level than marine organisms lower in the food chain. Eight times more marine fish than marine shell fish is eaten. Thus, 1 μg Pb/kg can be assumed to be a representative level for calculating the contribution of the marine pathway to humans. In Table 5 this contribution is estimated for the world average, Portugal as the nation with the highest per capita consumption (supply) and individuals of critical groups. For comparison, the typical US total intake for 1975 and an estimation of an international expert group are given. In the general population most of the lead intake comes from food (WHO, 1977). Lead in air is of a minor importance. The WHO estimates (1400 - 2800 μg) are an order of magnitude higher than that (210 μg) of Elias et al. (Ref. 22). This difference is due to the uncertainties in the lead analysis caused by ambient contamination of samples, etc.

TABLE 5 Lead Intake (μg Pb per Week and Capita) through Marine and Non-Marine Pathways Assuming that Sea Fish Contain 1 μg Pb/kg FW (Ref. 20, 22, and 23)

	Consumption g per week and capita	Lead Intake μg per week and capita
Sea Fish		
World average	200	0.2
Portugal *	500	0.5
US *	80	0.08
Individual of critical groups	5500	5.5
Non-Marine		
US typical total (1975)		210
WHO estimate, total		1400 - 2800

* Consumption equals supply as explained in the footnote under Table 4.

According to the lower estimates of Elias et al. (Ref. 22) of 210 μg Pb per week per capita, the Pb intake of the general population with the highest average consumption (Portugal) through the marine pathway accounts for only 0.2 % of the total. The intake with sea fish by the critical groups is also insignificant (2.5%) compared to the Pb intake derived from non-marine sources, even if a large percentage of it is of a more toxic organic form. However, also marine foods can be highly contaminated by lead. These high lead levels, as Burnett and Patterson (Ref. 1) have shown, are due to the Pb contamination of marine foods by industrial lead during the processing of sea foods.

Investigations of the biochemical cycle of lead may significantly contribute to a better understanding of forces and parameters governing the distribution of heavy metals in our environment. However, so far there has been no evidence that lead transmitted through the marine pathway makes a significant contribution to the lead levels in humans.

REFERENCES

(1) M. Burnett and C. Patterson, Analysis of natural and industrial lead in marine ecosystem, Proceedings of the International Group of Experts on "Lead - Pollution, Fate and Occurrence in Marine Environment", Rovinj, October 1977, M. Branica and Z. Konrad (eds.), Pergamon Press, 1979,15-30.

(2) C. C. Patterson and D. M. Settle, The reduction of orders of magnitude errors in lead analysis of biological materials and natural waters by evaluating and controlling the extent and sources of industrial lead contamination introduced during sample collection and analysis, in Accuracy in Trace Analysis: Sampling, Sample Handling and Analysis, P. D. LaFleur, Ed., NBS Special Publ. 422. US Dept. Comm. 1, 321 (1976).

(3) M. Burnett, A. Ng., D. Settle, and C. Patterson, Impact of man on coastal marine ecosystem, Proceedings of the International Group of Experts on "Lead - Pollution, Fate and Occurrence in the Marine Environment",Rovinj, October 1977, M. Branica and Z. Konrad (eds.), Pergamon Press, 1979,7-13.

(4) L. Mart, N. W. Nürnberg, and P. Valenta, Comparative baseline studies on Pb-levels in European coastal waters, Proceedings of the International Group of Experts on "Lead - Pollution, Fate and Occurrence in the Marine Environment", Rovinj, October 1977, M. Branica and Z. Konrad (eds.), Pergamon Press, 1979, 155-179.

(5) M. Bernhard and A. Zattera, Major pollutants in the marine environment, in E. Pearson and E. Frangipane, Eds., Marine Pollution and Marine Waste Disposal, Pergamon Press, 195 (1975).

(6) B. Schaule and C. C. Patterson, The occurence of lead in the Northeast Pacific, and the effects of anthropogenic inputs, in Proceedings of the International Group of Experts on "Lead - Pollution, Fate and Occurrence in the Marine Environment", Rovinj, October 1977, M. Branica and Z. Konrad (eds.), Pergamon Press, 1979, 31-43.

(7) G. R. Sirota and J. F. Uthe, Determination of tetraalkyllead compounds in biological materials, Anal. Chem., 823 (1977).

(8) B. R. Berland, D. J. Bonin, V. I. Kapkev, S. Y. Maestrini and D. P. Arlhac, Action toxique de quatre mètaux lourds sur la croissance d`algeus unicellulaires, C. R. Acad. Paris (Biol. Marine) Serie D, 282, 633 (1976).

(9) K. A. Klendenning and W. J. North, Effects of wastes on the giant kelps, Macrocystis pyrifera in, E. A. Pearson, Ed., 1st Int. Conf. Waste Disposal into the Marine Environment, Pergamon Press, 1960.

(10) J. S. Gray, Synergetic effects of three heavy metals on growth rate of a marine ciliate protozoan, in F. J. Bernberg and W. B. Vernberg, Eds., Pollution and Physiology of Marine Organisms, Acad. Press 465, (1974).

(11) J. E. McKee and H. W. Wolf, Eds., Water quality criteria, 2nd Edition Calif. State Water Quality Control Bd. Sacramento (1963).

(12) L. Karbe, Marine Hydroiden als Testorganismen zur Prüfung der Toxizität von Wasserstoffen. Die Wirkung von Schwermetallen auf Kolonien von Eirene viridula, Mar. Biol. 12, 316 (1972).

(13) B. Brown and M. Ahsanullah, Effect of heavy metals on mortality and growth, Mar. Poll. Bull. 2, 182 (1971).

(14) D. J. Reish, J. M. Martin, F. M. Piltz, and J. Q. Wood, The effect of heavy metals on laboratory populations of two polychaetes with comparisons to the water quality conditions and standards in southern California marine waters, Water Res. 10, 299 (1976).

(15) B. H. Pringle, D. E. Hissong, E. L. Katz, and S. T. Mulawica, Trace metal accumulation by estuarine mollusks, J. Sanitary Eng. Div. Amer. Soc. Civil Eng. 94, 455 (1968).

(16) A. Calabrese, R. S. Colier, D. A. Nelson and J. R. MacInnes, The toxicity of heavy metals to embryos of the American oyster (Crassostrea virginica), Marine Biol. 18, 162 (1973).

(17) J. S. Weis, Effects of mercury, cadmium, and lead salts on regeneration and ecdysis in the fiddler crab, Uca pugilator, Fish. Bull. (US Dept. Comm.) 74, 464.

(18) E. Jachim, J. M. Hamlin and S. Sonis, Effects of metal poisoning in five liver enzymes in the killifish (Fundulus heteroclitus), J. Fish. Res. Bd. Can. 27, 383 (1970).

(19) WQC-1973, Water quality criteria 1972. US Nat. Acd. Sciences, Superint. Doc. US Govern. Print. Office Wash., 249 (1973).

(20) FAO (1977) Provisional Food Balance Sheets, 1972-1974 Average, FAO, Rome.

(21) M. Bernhard, Heavy metals and chlorinated hydrocarbons in the Mediterranean, Ocean Management 3, 353 (1978).

(22) R. Elias, Y. Hirac, and C. Patterson, Impact of present levels of aerosol Pb concentrations on both natural ecosystems and humans, in Intern. Conf. on Heavy Metals in the Environment 257, Toronto (Oct. 1975).

(23) WHO, Environmental Health Criteria, Lead WHO, Geneva (1977).

Author Index

LEAD IN THE MARINE ENVIRONMENT